S*pringer* **M***onographs* in **M***athematics*

For further volumes:
http://www.springer.com/series/3733

Preface

Two subjects are treated in this book. The main subject is the theory of *Bernoulli numbers*, which are a series of rational numbers that appear in various contexts of mathematics, and the other subject is the related theory of *zeta functions*, which are very important in number theory. We hope that these are enjoyable subjects both for amateur mathematics lovers and for professional researchers. There are easy parts as well as difficult parts in this book, but we believe that, according to the taste of the readers, they can enjoy at least some parts of this attractive mathematics. Since the logical relations between the chapters are rather loose and not a straight course from the beginning to the end, it would still be worthwhile for readers who pick only some chapters which fit their tastes and background knowledge. As far as we know, books whose main subject is Bernoulli numbers are rare. Some parts of this book consist of rather standard number theory, but our expositions on these subjects are not always so standard, and some parts are completely new and have not been written in any reference before.

Now, there are many numbers in the world which have names, but what is a Bernoulli number? In high-school we learn the formula for the sum of the integers from 1 to n. Or maybe the formula of the sum of squares $1^2 + 2^2 + \cdots + n^2$ is also written in the standard textbooks for high schools. But if we make the powers bigger, for example, if we take

$$1^6 + 2^6 + \cdots + n^6,$$

then what kind of formula can we have? It seems that many people hit upon this kind of question and it often happens that ambitious high-school students try and find the solution. In fact, the formula was already known for any power at the beginning of the eighteenth century. (See Chap. 1, p. 2.) Bernoulli introduced some special numbers to express this formula. These are the Bernoulli numbers. (For a more precise history, see Chap. 1.) In this book, this formula of the sum of powers appears in various places. In fact, in this book, four alternative proofs of this formula will be given in different places (cf. Sects. 1.2, 4.3, 5.2, 8.3).

Well, is it so interesting to give this formula of the sum of powers? Indeed it is. But if the meaning of Bernoulli numbers were only this, we would not have written this book. The real meaning of Bernoulli numbers lies in a different place. Let's look at the following formulas:

$$1+\frac{1}{2^2}+\frac{1}{3^2}+\frac{1}{4^2}+\frac{1}{5^2}+\frac{1}{6^2}+\cdots=\frac{\pi^2}{6},$$

$$1+\frac{1}{2^4}+\frac{1}{3^4}+\frac{1}{4^4}+\frac{1}{5^4}+\frac{1}{6^4}+\cdots=\frac{\pi^4}{90},$$

$$1-\frac{1}{3}+\frac{1}{5}-\frac{1}{7}+\frac{1}{9}-\frac{1}{11}+\frac{1}{13}\cdots=\frac{\pi}{4}.$$

(You can find the proofs in Corollary 4.12, p. 61, and Theorem 9.6, p. 148.) At first glance, these kinds of formulas could give you a mysterious impression, since a regular sum of simple rational numbers suddenly changes into a transcendental number like π. The left-hand sides of these formulas are values at some special points of so-called zeta functions, or L-functions, which are another important subject of the book. An example of zeta functions is given by

$$\zeta(s)=1+\frac{1}{2^s}+\frac{1}{3^s}+\frac{1}{4^s}+\frac{1}{5^s}+\frac{1}{6^s}+\cdots.$$

Here the sum is regarded as a function with respect to the complex variable s. This function is called the Riemann zeta function. The left-hand sides of the first two formulas above are the values of $\zeta(s)$ at $s=2$ and $s=4$, namely $\zeta(2)$ and $\zeta(4)$. The right-hand sides are related to Bernoulli numbers and generalized Bernoulli numbers (see below). The true reason that Bernoulli numbers are indispensable for number theory lies in this type of relation with special values (values at special points) of zeta functions. A huge number of functions called zeta functions appear in number theory, and they always contain very important information. We can even say that there appears a different zeta function for each important number theoretical object. The zeta functions which are related to Bernoulli numbers are just a part of them. But still they have something to do with many number theoretical phenomena. For example, they are connected with a kind of volume in non-Euclidean geometry, with dimension formulas of spaces of modular forms, and with the number of equivalence classes of quadratic forms (class number formula), and they appear as constant terms of Eisenstein series, which are important in the theory of automorphic forms. Also they are related to the number theory of cyclotomic fields, that is, the systems of numbers obtained by adjoining the roots of unity to the rational numbers. If we wanted to explain all of this, we would need several more books, so it is not possible to do it here, but we have tried to give some hints about those relations.

Now, for the reader's convenience, we sketch the content of each chapter. Comments below written in parentheses will not be explained in this book in detail.

In Chap. 1, after explaining the history of how Bernoulli numbers were introduced and giving the definition of Bernoulli numbers using recurrence relations, we prove the formula for the sum of powers of consecutive natural numbers. In Chap. 2, we give a formula to write down Bernoulli numbers in a simple way. There we use *Stirling numbers*, whose definition is comparatively simple. In Chap. 3, we will prove the Clausen–von Staudt theorem which gives information on the denominators of Bernoulli numbers, and also Kummer's congruence which gives a congruence relation between Bernoulli numbers. (This latter theorem is needed when we define p-adic zeta functions.) In Chap. 4, we define generalized Bernoulli numbers associated to Dirichlet characters. (Through this extension, the connection with quadratic fields and cyclotomic fields will appear.) Also the Bernoulli polynomials (a system of polynomials similar to the generating function of Bernoulli numbers) are defined, and as an application, we give a second proof of the formula of the sum of powers. The formula expressing the values of the Riemann zeta function at even positive integers by Bernoulli numbers is also given here. In Chap. 5, we give the Euler–Maclaurin summation formula, which is very useful to evaluate sum of values of functions at integers. In the coefficients of this formula, Bernoulli numbers or Bernoulli polynomials appear. As an application, the analytic continuation of the Riemann zeta function to the whole plane is proved and the values of the Riemann zeta function at non-positive integers are beautifully expressed by Bernoulli numbers. In Chap. 6, the relation between ideals of a quadratic field and quadratic forms is given and a relation between the generalized Bernoulli numbers and the class numbers of positive definite quadratic forms is explained. In Chap. 7, a congruence relation between the class number of imaginary quadratic fields and the Bernoulli numbers will be given. In Chap. 8, Gauss sums are introduced and then formulas to describe various sums of the roots of unity with characters by Gauss sums and generalized Bernoulli numbers are given. Here we get a third proof of the formula for sums of powers. In Chap. 9, L-functions will be introduced by modifying the usual definition of the Riemann zeta function by adjoining characters; then their functional equation is shown by using a contour integral expression. We will also give a formula to describe values at non-positive integers of L-functions in terms of generalized Bernoulli numbers. This formula gives a "raison d'être" for the generalized Bernoulli numbers. In Chap. 10, among the zeta functions of prehomogeneous vector spaces, we pick an example which apparently does not seem so simple but can be shown to be a very easy function. (This is an example connected with the appearance of Bernoulli numbers in the dimension formula for modular forms.) In Chap. 11, we give an alternative proof of Kummer's congruence by considering p-adic measures. (This has a deep connection with p-adic L-functions.) In Chap. 12, we review the fact that Bernoulli numbers appear in the Taylor expansion of the (co)tangent function, and ask what would emerge if we replace the (co)tangent with elliptic functions (a well-worn device to generalize phenomena on trigonometric functions), and explain the theory of Hurwitz numbers. In Chap. 13, we show the analytic continuation of the Barnes multiple zeta function, which is a generalization of the Hurwitz zeta function, and explain relations between their special values and Bernoulli polynomials. We also show the functional equation of the double zeta functions in the same way as in Chap. 9. In Chap. 14, we define poly-Bernoulli numbers, which are obtained by replacing the usual generating function of Bernoulli numbers by polylogarithm functions. In an Appendix by Don Zagier, some "curious and exotic" identities for Bernoulli numbers are given.

As explained above, Bernoulli numbers are related with a great many things. But Bernoulli numbers do not have so many of their own problems to be solved, except for a few very difficult problems. So, maybe there exist few mathematicians who study only Bernoulli numbers. Since the situation is like that, there are not many books similar to this one, and it was not so easy to have a perspective on the wide range of relations for Bernoulli numbers. Although the present authors are specialists of number theory, we dare not say we are specialists of Bernoulli numbers. This book consists partly of quotations from the papers by authors who occasionally encountered Bernoulli numbers from their different interests, and partly of various memoranda (namely these are observations which are not necessarily very original results worthy to be written as careful scientific papers, but rather lightly checked, interestingly interpreted, or calculated for fun and worth being taken note of so as not to be forgotten). And this kind of connection is indeed the usual common connection of mathematicians to the Bernoulli numbers. In this sense, we believe that this book is useful for any reader who is interested in mathematics.

The original edition of this book was published in Japanese by Makino Publisher in Japan in 2001. We would like to give our sincere thanks to Professor Don B. Zagier for recommending us enthusiastically to write this English version based on that Japanese book, for recommending this book to Springer-Verlag to be published, and also for writing the nice Appendix. We would like to thank Springer-Verlag for publishing this book, in particular Dr. Joachim Heinze of Springer and Ms. Chino Hasebe and Mr. Masayuki Nakamura of Springer Tokyo for their valuable help for the arrangement, and Dr. Alexander Weisse for his help with TeX. We would also like to thank anonymous referees of the English version for a lot of valuable comments. We would like to give our deep thanks to Mr. Suenobu Makino, the owner of Makino Publishers, who originally suggested us the theme of Bernoulli numbers and recommended to write the book in Japanese. Without his recommendation, the Japanese version would not have appeared. Tsuneo Arakawa, one of the co-authors for the Japanese version, passed away on October 3, 2003 at the age of 54 before this English project started. Since some parts of this book are based on his Japanese manuscript, we would like to keep his name as one of the authors. The second- and the third-named authors express their profound regret and sadness over the early death of Tsuneo Arakawa.

November 2013 The Authors

Contents

Chapter 1
Bernoulli Numbers

1.1 Definitions: Introduction from History

In his posthumous book *Ars Conjectandi* [16] published in 1713 (the law of large numbers in probability theory is stated in this book), Jakob Bernoulli[1] introduced the Bernoulli numbers in connection to the study of the sums of powers of consecutive integers $1^k + 2^k + \cdots + n^k$. After listing the formulas for the sums of powers

$$\sum_{i=1}^{n} i = \frac{n(n+1)}{2}, \quad \sum_{i=1}^{n} i^2 = \frac{n(n+1)(2n+1)}{6}, \quad \sum_{i=1}^{n} i^3 = \left(\frac{n(n+1)}{2}\right)^2, \ldots$$

up to $k = 10$ (Bernoulli expresses the right-hand side without factoring), he gives a general formula involving the numbers which are known today as Bernoulli numbers. Bernoulli then explains how these numbers are determined inductively, and emphasizes how his formula ((1.1) below) is useful for computing the sum of powers. He claims that he did not take "a half of a quarter of an hour" to compute the sum of tenth powers of 1 to 1,000, which he computed correctly as 91409924241424243424241924242500.

Using modern notation, his formula is written as

$$\sum_{i=1}^{n} i^k = \sum_{j=0}^{k} \binom{k}{j} B_j \frac{n^{k+1-j}}{k+1-j} \left(= \frac{1}{k+1} \sum_{j=0}^{k} \binom{k+1}{j} B_j n^{k+1-j} \right), \quad (1.1)$$

[1]Born on December 27, 1654 in Basel, Switzerland—died on August 16, 1705 in Basel, Switzerland. Jakob is the eldest among the mathematicians in the famous Bernoulli family. It is said that Jakob, his younger brother Johann (born on July 27, 1667 in Basel, Switzerland—died on January 1, 1748 in Basel, Switzerland), and his second son, Daniel (born on February 8, 1700 in Groningen, Netherlands—died on March 17, 1782 in Basel, Switzerland) are the most distinguished among them.

T. Arakawa et al., *Bernoulli Numbers and Zeta Functions*, Springer Monographs in Mathematics, DOI 10.1007/978-4-431-54919-2_1, © Springer Japan 2014

where $\binom{k}{j}$ is the binomial coefficient

$$\binom{k}{j} = \frac{k(k-1)(k-2)\cdots(k-j+1)}{j!},$$

and B_j is the number determined by the recurrence formula

$$\sum_{j=0}^{k} \binom{k+1}{j} B_j = k+1, \quad k = 0, 1, 2, \ldots.$$

It is this B_j that is subsequently called a Bernoulli number,[2] and is the main theme of this book.

Bernoulli does not give symbols for B_0 and B_1, and he writes $A, B, C, D, \ldots$ for B_{2n}. As we will see later, $B_{2n+1} = 0$ except for B_1. In his book, the right-hand side of (1.1) is written as

$$\frac{1}{c+1}n^{c+1} + \frac{1}{2}n^c + \frac{c}{2}An^{c-1} + \frac{c.c-1.c-2}{2.3.4}Bn^{c-3} +$$
$$\frac{c.c-1.c-2.c-3.c-4}{2.3.4.5.6}Cn^{c-5} + \frac{c.c-1.c-2.c-3.c-4.c-5.c-6}{2.3.4.5.6.7.8}Dn^{c-7}$$
$$+\cdots,$$

where $c = k$.

We would like to mention here that, in the book called *Katsuyo Sanpo* ("essentials of the art of calculation") by the outstanding Japanese mathematician Takakazu Seki,[3] published also posthumously, in 1712 (and thus 1 year before Bernoulli!), the formula for the sums of powers and the inductive definition of the Bernoulli numbers are given. His formula and definition are completely the same as Bernoulli's.

[2]It was apparently de Moivre (Abraham, born on May 26, 1667 in Vitry-le-Francois, Champagne, France—died on November 27, 1754 in London, England) who first called this number a Bernoulli number in the book *Miscellanea analytica de seriebus et quadraturis* (London, 1730). De Moivre is famous for de Moivre's formula in trigonometry.

[3]Born in 1642(?) in Kohzuke(?), Japan—died on October 24, 1708 in Edo, Japan. He is usually considered Japan's greatest mathematician of the Edo period (1600–1857, when the country was closed to essentially all foreign contact). Hardly anything is known about his mathematical education, and he seems to have been largely self-taught. He served under the shoguns Tsunashige Tokugawa and Ienobu Tokugawa, occupying the post of Controller of the Treasury Office, and wrote several treatises in higher mathematics. Apart from discovering Bernoulli numbers simultaneously with or before Bernoulli, he discovered determinants and the rules for calculating them simultaneously with or before Leibniz, solved extraordinarily difficult problems of elimination theory for systems of polynomial equations in many variables, and began a study of calculation procedures for the arc of a circle that was continued and completed by his disciple Katahiro Takebe with the discovery of infinite series expansions for various trigonometric functions.

Seki refers to $B_0, B_1, B_2, \ldots$ as *Shusuu* (which means "numbers to be taken") of the first order, second order etc., labeling odd indexed B_{2n+1} as well. It is not widely known that Seki independently found the Bernoulli numbers, but the collected works of Seki [84] have been published and the volume contains an English translation of each article. A reproduction of Seki's table giving the formula for the sums of powers in terms of binomial coefficients and "Seki–Bernoulli numbers", together with a translation into modern notation, is given as Figs. 1.1 and 1.2.

Neither Seki nor Bernoulli explains in much detail how to deduce the formula (1.1). Prior to their work, the formula for the sums of powers was discussed in *Academia Algebrae* by Faulhaber.[4] He obtained the result saying that, when k is odd, the quantity $\sum_{i=1}^{n} i^k$ is a polynomial in $\frac{n(n+1)}{2} = \sum_{i=1}^{n} i$, (the first example being $\sum_{i=1}^{n} i^3 = \left(\sum_{i=1}^{n} i\right)^2$), whereas when k is even, $\sum_{i=1}^{n} i^k$ is divisible by $\frac{n(n+1)(2n+1)}{6} = \sum_{i=1}^{n} i^2$ as polynomials in n, and the quotient is again a polynomial in $\sum_{i=1}^{n} i$. But apparently he did not reach the Bernoulli numbers. These facts discovered by Faulhaber were rediscovered by Jacobi[5] [52], who gave a rigorous proof. We shall provide a proof of the formula (1.1) in the next section. As for the results of Faulhaber, we only formulate them and do not give proofs, leaving them to the literature. We mention here that a very extensive bibliography on Bernoulli numbers, compiled by Karl Dilcher, is available online [28].

Let us now give the definition of the Bernoulli numbers again. We follow Seki and Bernoulli, and define them using a recurrence formula. The Bernoulli numbers may also be defined by using a generating function. We will see in the next section that these definitions are equivalent.

Definition 1.1 (Bernoulli numbers). Define B_n $(n = 0, 1, 2, \ldots)$ inductively by the formula

$$\sum_{i=0}^{n} \binom{n+1}{i} B_i = n + 1 \quad (n = 0, 1, 2, \ldots). \tag{1.2}$$

Let us compute B_n for small n. For $n = 0$ we have

$$B_0 = 1.$$

Putting $n = 1$ in the formula (1.2) we have

$$B_0 + 2B_1 = 2,$$

from which we obtain

$$B_1 = \frac{1}{2}(2 - B_0) = \frac{1}{2}.$$

---

[4]Johann Faulhaber (born on May 5, 1580 in Ulm, Germany—died in 1635 in Ulm, Germany).

[5]Carl Gustav Jacob Jacobi (born on December 10, 1804 in Potsdam, Prussia (now Germany)—died on February 18, 1851 in Berlin, Germany).

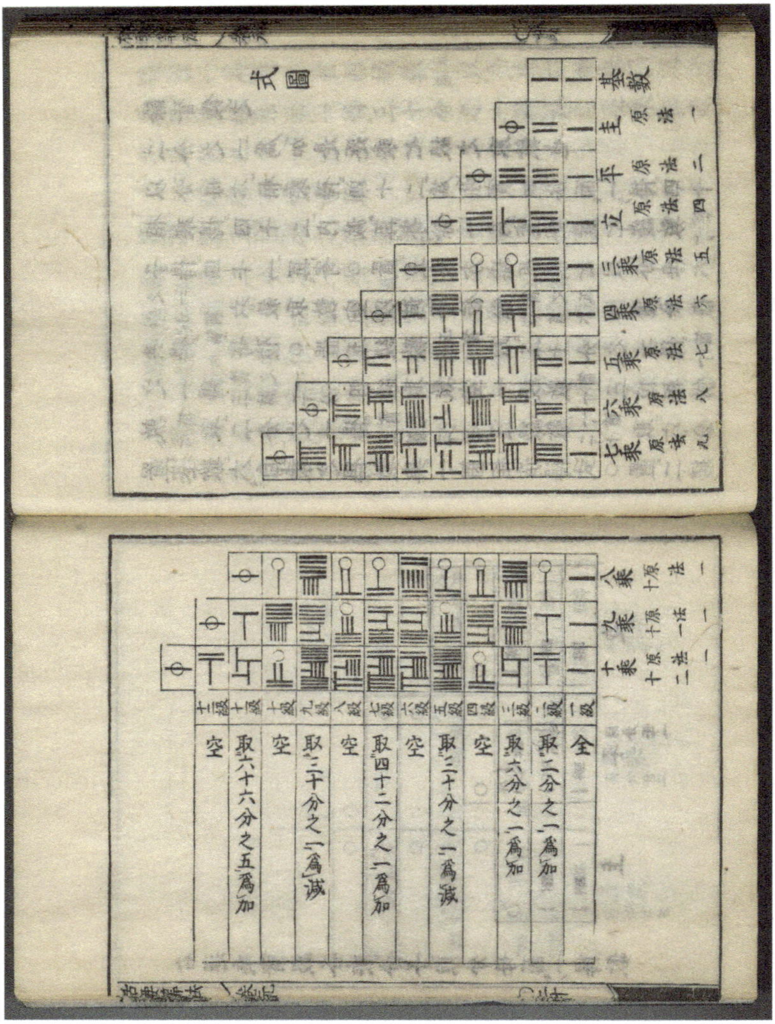

Fig. 1.1 Katsuyou Sanpou, Seki's tabular presentation of the power-sum formula, written not in Chinese characters but using the notation of *Sangi* or counting rods. The table (which is reproduced here with a 90° rotation) contains each of the ingredients of the formula in parentheses in Eq. (1.1): The number designated by "power" in the translation overleaf is the k of this formula, the binomial coefficients are tabulated in the "Pascal's triangle" in the right part of the table (= left part in the translation), the Bernoulli numbers B_j are given on the left (the right in the English translation), the missing coefficient $\binom{k+1}{k+1}$ in (1.1) corresponds to the crossed out "1" at the beginning of each column of the Pascal triangle, and the "denominator" is the number $k+1$ by which the whole expression has to be divided at the end of the calculation

Fig. 1.2 Katsuyou Sanpou (English translation, reflected with respect to the original)

											~~1~~		
										~~1~~	12	level 12	void
									~~1~~	11	66	level 11	take 5/66 and add
								~~1~~	10	55	220	level 10	void
							~~1~~	9	45	165	495	level 9	take 1/30 and subtract
						~~1~~	8	36	120	330	792	level 8	void
					~~1~~	7	28	84	210	462	992[e]	level 7	take 1/42 and add
				~~1~~	6	21	56	126	252	462	792	level 6	void
			~~1~~	5	15	35	70	126	210	330	495	level 5	take 1/30 and subtract
		~~1~~	4	10	20	35	56	84	120	165	220	level 4	void
	~~1~~[a]	3	6	10	15	21	28	36	45	55	66	level 3	take 1/6 and add
1[b]	2	3	4	5	6	7	8	9	10	11	12	level 2	take 1/2 and add
1	1	1	1	1	1	1	1	1	1	1	1	level 1	all
0	1	2	3	4	5	6	7	8	9	10	11		power
	1[c]	2[d]	4	5	6	7	8	9	10	11	12		denominator

a: The crossed-out numbers are to be omitted.
b: This number should also be crossed out.
c: This number should be 2.
d: This number should be 3.
e: This number should be 924.

Putting $n = 2$ in (1.2) we have

$$B_0 + 3B_1 + 3B_2 = 3,$$

from which we obtain

$$B_2 = \frac{1}{3}(3 - B_0 - 3B_1) = \frac{1}{6}.$$

Similarly, putting $n = 3$ in (1.2), we have

$$B_0 + 4B_1 + 6B_2 + 4B_3 = 4,$$

which gives

$$B_3 = \frac{1}{4}(4 - B_0 - 4B_1 - 6B_2) = 0.$$

Table 1.1 Bernoulli numbers

n	0	1	2	3	4	5	6	7	8	9	10	11	12	13	14	15	16	17	18
B_n	1	$\frac{1}{2}$	$\frac{1}{6}$	0	$-\frac{1}{30}$	0	$\frac{1}{42}$	0	$-\frac{1}{30}$	0	$\frac{5}{66}$	0	$-\frac{691}{2730}$	0	$\frac{7}{6}$	0	$-\frac{3617}{510}$	0	$\frac{43867}{798}$

Putting $n = 4$ in (1.2), we have

$$B_0 + 5B_1 + 10B_2 + 10B_3 + 5B_4 = 5,$$

which gives

$$B_4 = \frac{1}{5}(5 - B_0 - 5B_1 - 10B_2 - 10B_3) = -\frac{1}{30},$$

and so on. Table 1.1 is the list of Bernoulli numbers B_n up to $n = 18$. It follows from the definition that B_n is a rational number. Mathematical software packages such as Mathematica and Maple include the algorithm for computing Bernoulli numbers, and it is not hard to obtain several hundred B_n's in a moment.

Remark 1.2. There is another convention on Bernoulli numbers: $B_1 = -\frac{1}{2}$ and everything else is the same. We adopt the first definition because this is the original definition of Seki and Bernoulli for one thing, and it is better suited to the special values of the Riemann zeta function for another, that is, the formula in Theorem 5.4 is valid also for $m = 1$. Although the difference is minuscule, we call readers' attention to it.

Since $B_n = 0$ for every odd n greater than 1 (we prove this later), it is not difficult to translate formulas from one definition to the other by replacing B_n by $(-1)^n B_n$.

1.2 Sums of Consecutive Powers of Integers and Theorem of Faulhaber

In this section we first prove the Seki–Bernoulli formula (1.1) for the sum

$$S_k(n) := \sum_{i=1}^{n} i^k \quad (k, n \text{ integers}, k \geq 0, n \geq 1).$$

Then we will state Faulhaber's theorem and give a recurrence formula for Bernoulli numbers at the end of this section.

Before going into the proof, let us give several explicit formulas for $S_k(n)$:

$$S_0(n) = n,$$

$$S_1(n) = \frac{n^2}{2} + \frac{n}{2}$$

$$\begin{aligned}
&= \frac{1}{2}n(n+1), \\
S_2(n) &= \frac{n^3}{3} + \frac{n^2}{2} + \frac{n}{6} \\
&= \frac{1}{6}n(n+1)(2n+1), \\
S_3(n) &= \frac{n^4}{4} + \frac{n^3}{2} + \frac{n^2}{4} \\
&= \frac{1}{4}n^2(n+1)^2, \\
S_4(n) &= \frac{n^5}{5} + \frac{n^4}{2} + \frac{n^3}{3} - \frac{n}{30} \\
&= \frac{1}{30}n(n+1)(2n+1)(3n^2+3n-1), \\
S_5(n) &= \frac{n^6}{6} + \frac{n^5}{2} + \frac{5n^4}{12} - \frac{n^2}{12} \\
&= \frac{1}{12}n^2(n+1)^2(2n^2+2n-1), \\
S_6(n) &= \frac{n^7}{7} + \frac{n^6}{2} + \frac{n^5}{2} - \frac{n^3}{6} + \frac{n}{42} \\
&= \frac{1}{42}n(n+1)(2n+1)(3n^4+6n^3-3n+1).
\end{aligned}$$

For the proof, we first see immediately that

$$S_0(n) = n.$$

Let $k \geq 1$. From the binomial theorem we obtain

$$(m+1)^{k+1} - m^{k+1} = \sum_{j=0}^{k} \binom{k+1}{j} m^j.$$

Putting $m = 1, 2, \ldots, n$, and summing over all m, we obtain

$$(n+1)^{k+1} - 1 = \sum_{j=0}^{k} \binom{k+1}{j} S_j(n).$$

From this we have

$$S_k(n) = \frac{1}{k+1}\left\{(n+1)^{k+1} - 1 - \sum_{j=0}^{k-1}\binom{k+1}{j}S_j(n)\right\}.$$

Putting $k = 1, 2, \ldots$ in this formula, we obtain

$$S_1(n) = \frac{n^2}{2} + \frac{n}{2},\ S_2(n) = \frac{n^3}{3} + \frac{n^2}{2} + \frac{n}{6}, \ldots.$$

By induction we see that

$S_k(n)$ is a polynomial of degree $k+1$ in n, whose leading term is $\frac{1}{k+1}n^{k+1}$.

Our objective is to write down this polynomial explicitly.

Consider the polynomial $S_k(x)(= \frac{1}{k+1}x^{k+1} + \cdots)$ obtained by replacing the variable n by x. In general, two polynomials $f(x), g(x)$ are identical if $f(n) = g(n)$ for all positive integers n. By definition we have

$$S_k(n+1) - S_k(n) = (n+1)^k \quad (n = 1, 2, 3, \ldots).$$

Thus, we obtain

$$S_k(x+1) - S_k(x) = (x+1)^k.$$

Putting $x = 0$, and using $S_k(1) = 1$, we have

$$S_k(0) = 0.$$

This yields the constant term of $S_k(x)$. The other coefficients may be obtained from the derivatives $S_k^{(j)}(0)$, $1 \leq j \leq k$. Taking the derivative of the relation $S_k(x+1) - S_k(x) = (x+1)^k$, we obtain

$$S_k'(x+1) - S_k'(x) = k(x+1)^{k-1}. \tag{1.3}$$

Putting $x = 0, 1, 2, \ldots, n-1$, and adding them all, we have

$$S_k'(n) - S_k'(0) = kS_{k-1}(n).$$

This holds for any positive integer n. If we let $S_k'(0) = b_k$ ($b_0 = 1$), then we have the relation

$$S_k'(x) = kS_{k-1}(x) + b_k.$$

Taking the derivatives of both sides, we have

$$S_k''(x) = kS_{k-1}'(x). \tag{1.4}$$

Putting $x = 0$, we have

$$S_k''(0) = kb_{k-1}.$$

Taking the derivative of (1.4) once again, and using (1.4), we have

$$S_k'''(x) = kS_{k-1}''(x) = k(k-1)S_{k-2}'(x).$$

Putting $x = 0$, we have

$$S_k'''(0) = k(k-1)b_{k-2}.$$

Similarly, taking the derivatives successively, we obtain

$$S_k^{(j)}(0) = k(k-1)\cdots(k-j+2)b_{k-j+1} \quad (2 \le j \le k+1).$$

Finally, we have

$$\begin{aligned} S_k(x) &= \sum_{j=0}^{k+1} \frac{S_k^{(j)}(0)}{j!} x^j \\ &= \sum_{j=1}^{k+1} \frac{1}{k+1} \binom{k+1}{j} b_{k-j+1} x^j \quad (S_k^{(0)}(0) = 0) \\ &= \frac{1}{k+1} \sum_{j=0}^{k} \binom{k+1}{j} b_j x^{k+1-j}. \end{aligned}$$

Since $S_k(1) = 1$, we obtain, by putting $x = 1$ in the above formula, the recurrence formula

$$k+1 = \sum_{j=0}^{k} \binom{k+1}{j} b_j,$$

which is nothing but the recurrence for the Bernoulli numbers. We therefore conclude $b_j = B_j$. In view of the identity

$$\frac{1}{k+1}\binom{k+1}{j} = \frac{1}{k+1-j}\binom{k}{j},$$

we obtain (1.1). □

Remark 1.3. If we set $S'_k(x-1) = B_k(x)$ ($B_k(x) = x^k + \cdots$), it follows from (1.3) and (1.4) that

$$B_k(x+1) - B_k(x) = kx^{k-1}, \quad B'_k(x) = kB_{k-1}(x).$$

This $B_k(x)$ is called the kth Bernoulli polynomial, which leads to various applications. We will give its definition and some properties in Sect. 4.3 (p. 55). The formula for sums of consecutive powers may be obtained easily from the Bernoulli polynomial. We will show it in Sect. 4.3.

Let us prove the fact that $B_k = 0$ for odd k greater than 1 using the polynomial $S_k(x)$. We will prove this fact later more easily using the generating function, but we prove it here because we need to use it before that.

Proposition 1.4. *If n is an odd integer greater than or equal to 3, then $B_n = 0$. As a consequence, $(-1)^n B_n = B_n$ for all positive integers n except for $n = 1$.*

Proof. Suppose $k \geq 1$. Putting $x = -1$ in the formula $S_k(x+1) - S_k(x) = (x+1)^k$ and using the fact $S_k(0) = 0$, we obtain $S_k(-1) = 0$. Thus, by putting $x = -1$ in the formula

$$(k+1)S_k(x) = \sum_{j=0}^{k} \binom{k+1}{j} B_j x^{k+1-j},$$

we obtain

$$\sum_{j=0}^{k} \binom{k+1}{j} (-1)^j B_j = 0.$$

If we subtract this from the recurrence formula in the definition of Bernoulli numbers

$$\sum_{j=0}^{k} \binom{k+1}{j} B_j = k+1,$$

then only odd indexed terms remain and we have

$$2 \sum_{j=0}^{[\frac{k-1}{2}]} \binom{k+1}{2j+1} B_{2j+1} = k+1,$$

where $[x]$ stands for the greatest integer less than or equal to x. Since $B_1 = 1/2$, the term for $j = 0$ in the left-hand side cancels with the right-hand side, and we obtain

$$\sum_{j=1}^{[\frac{k-1}{2}]} \binom{k+1}{2j+1} B_{2j+1} = 0 \quad (k \geq 3).$$

Putting $k = 3, 5, 7, \ldots$, we obtain $B_k = 0$ inductively for all odd $k \geq 3$. □

Next, we state Faulhaber's theorem.

Theorem 1.5. *Let $k \geq 1$ and set*

$$u = n(n+1)\ (= 2S_1(n)), \quad v = n(n+1)(2n+1)\ (= 6S_2(n)).$$

Then we have

$$S_{2k+1}(n) = \frac{u^2}{2k+2} \sum_{i=0}^{k-1} A_i^{(k)} u^{k-1-i}$$

and

$$S_{2k}(n) = \frac{v}{(2k+1)(2k+2)} \sum_{i=0}^{k-1} (k+1-i) A_i^{(k)} u^{k-1-i},$$

where $A_i^{(k)}$ is a number determined by the formula

$$A_0^{(k)} = 1,\ A_i^{(k)} = -\frac{1}{k+1-i} \sum_{j=0}^{i-1} \binom{k+1-j}{k+j-2i} A_j^{(k)} \quad (1 \leq i \leq k-1).$$

(For $k + j - 2i < 0$, we put $\binom{k+1-j}{2i+1-2j} = 0$.)

For the proof we refer to [30] and [60]. We encourage readers to attempt to prove it by themselves.

Before ending this section we prove an amusing recurrence formula for Bernoulli numbers using $S_k(x)$ [55].

In the defining recurrence formula for Bernoulli numbers (1.2), if we move the right-hand side to the left, and use the fact that $(-1)^i B_i = B_i$ for $i \neq 1$ and $B_1 = 1/2$, then we obtain

$$\sum_{i=0}^{n} (-1)^i \binom{n+1}{i} B_i = 0 \tag{1.5}$$

for $n \geq 1$. Now, we define $\widetilde{B_n} = (n+1)B_n$. Then, we have

Theorem 1.6. *For $n \geq 1$, we have*

$$\sum_{i=0}^{n}(-1)^i \binom{n+1}{i} \widetilde{B_{n+i}} = 0.$$

Remark 1.7. Though this formula is similar to (1.5), it requires only half the number of terms to compute $\widetilde{B_{2n}}$ (and thus B_{2n}).

Proof. From the binomial theorem we have

$$x^{n+1}(1-x)^{n+1} = \sum_{i=0}^{n+1}(-1)^i \binom{n+1}{i} x^{n+i+1}.$$

Taking the derivatives of both sides with respect to x, we have

$$(n+1)\bigl(x^n(1-x)^{n+1} - x^{n+1}(1-x)^n\bigr) = \sum_{i=0}^{n+1}(-1)^i \binom{n+1}{i}(n+i+1)x^{n+i}.$$

From this we have

$$\begin{aligned}(n+1)x^n(1-x)^n(1-2x) &= (-1)^{n+1}(2n+2)x^{2n+1} \\ &\quad + \sum_{i=0}^{n}(-1)^i \binom{n+1}{i}(n+i+1)x^{n+i}. \qquad (1.6)\end{aligned}$$

Putting $x = (1+y)/2$ in this formula, we have

$$\begin{aligned}\text{l.h.s.} &= -(n+1)\left(\frac{1+y}{2}\right)^n \left(\frac{1-y}{2}\right)^n y \\ &= -2^{-2n}(n+1)y(1-y^2)^n \\ &= -2^{-2n}(n+1)\sum_{i=0}^{n}(-1)^i \binom{n}{i} y^{2i+1}.\end{aligned}$$

We put $x = 1, 2, \ldots, m$ in (1.6), add them up, and use the equations

$$\sum_{x=1}^{m} y^{2i+1} = \sum_{x=1}^{m}(2x-1)^{2i+1} = S_{2i+1}(2m) - 2^{2i+1}S_{2i+1}(m)$$

to obtain

$$- 2^{-2n}(n+1)\sum_{i=0}^{n}(-1)^i\binom{n}{i}\left(S_{2i+1}(2m) - 2^{2i+1}S_{2i+1}(m)\right)$$
$$= (-1)^{n+1}(2n+2)S_{2n+1}(m) + \sum_{i=0}^{n}(-1)^i\binom{n+1}{i}(n+i+1)S_{n+i}(m).$$

Since this holds for all positive integers m, we can use the same argument as before to conclude that the formula obtained by replacing m by x,

$$- 2^{-2n}(n+1)\sum_{i=0}^{n}(-1)^i\binom{n}{i}\left(S_{2i+1}(2x) - 2^{2i+1}S_{2i+1}(x)\right)$$
$$= (-1)^{n+1}(2n+2)S_{2n+1}(x) + \sum_{i=0}^{n}(-1)^i\binom{n+1}{i}(n+i+1)S_{n+i}(x),$$

holds as a formula for polynomials. We take the derivative of both sides of this with respect to x and set $x = 0$. The desired formula follows from the facts that $S_k'(0) = B_k$ (previous section), that the term for $i = 0$ in the left-hand side vanishes due to the fact $2S_1'(0) - 2S_1'(0) = 0$, and that $B_k = 0$ for odd $k \geq 3$ (Proposition 1.4). □

1.3 Formal Power Series

We have defined Bernoulli numbers by a recurrence formula. However, it is also common to define Bernoulli numbers using the generating function

$$\frac{te^t}{e^t - 1} = \sum_{n=0}^{\infty} B_n \frac{t^n}{n!}. \tag{1.7}$$

For the other definition of B_n we mentioned in Remark 1.2 in Sect. 1.1, we need to replace the left-hand side by $\frac{t}{e^t-1} = \frac{te^t}{e^t-1} - t$. In fact, the generating function is very useful for the study of various properties of Bernoulli numbers.

In this section we will explain fundamental facts about formal power series for those who are not familiar with such objects. In the next section we will prove the above formula as Theorem 1.12. Those who are not familiar with abstract algebra are advised to take a glance at the next section and see how computations go before reading this section.

Let R be a commutative integral domain with unit (written 1). (Although we write in this general fashion, readers may think of R as the rational number field $\mathbf{Q}$ for the time being. In this book R is either $\mathbf{Q}$, or its finite extension, except for Chap. 11, where we need p-adic numbers. A commutative integral domain is a commutative ring with the property that $ab = 0$ implies $a = 0$ or $b = 0$.) A formal sum

$$\sum_{n=0}^{\infty} a_n t^n = a_0 + a_1 t + a_2 t^2 + a_3 t^3 + \cdots$$

with an indeterminate t and coefficients in R is called a formal power series with coefficients in R, and the set of all such formal power series is denoted by $R[[t]]$. Two formal power series are defined to be equal if and only if all coefficients of t^n coincide. Typical examples which will appear in this book are (taking $R = \mathbf{Q}$)

$$e^t = \sum_{n=0}^{\infty} \frac{t^n}{n!} = 1 + \frac{t}{1!} + \frac{t^2}{2!} + \frac{t^3}{3!} + \cdots$$

and

$$\log(1+t) = \sum_{n=1}^{\infty} (-1)^{n-1} \frac{t^n}{n} = t - \frac{t^2}{2} + \frac{t^3}{3} - \cdots .$$

These are Taylor expansions of complex functions e^x and $\log(1+x)$ around $x = 0$, regarded as formal sums. We use the notation $e^t, \log(1+t)$, but these are nothing but the formal power series of the right-hand sides, and we will not consider them as functions in t.

The sum and the product of two formal power series are defined to be

$$\sum_{n=0}^{\infty} a_n t^n + \sum_{n=0}^{\infty} b_n t^n = \sum_{n=0}^{\infty} (a_n + b_n) t^n$$

and

$$\left(\sum_{n=0}^{\infty} a_n t^n\right) \cdot \left(\sum_{n=0}^{\infty} b_n t^n\right) = \sum_{n=0}^{\infty} c_n t^n, \quad c_n = \sum_{i=0}^{n} a_i b_{n-i} .$$

The product is defined using the distributive law formally, and the first few terms of it are

$$\begin{aligned} &(a_0 + a_1 t + a_2 t^2 + a_3 t^3 + \cdots) \cdot (b_0 + b_1 t + b_2 t^2 + b_3 t^3 + \cdots) \\ &= a_0 b_0 + (a_0 b_1 + a_1 b_0) t + (a_0 b_2 + a_1 b_1 + a_2 b_0) t^2 \\ &+ (a_0 b_3 + a_1 b_2 + a_2 b_1 + a_3 b_0) t^3 + \cdots . \end{aligned}$$

It is readily verified that $R[[t]]$ is a commutative ring with unit. The zero element and the unit are the formal power series

$$0 + 0 \cdot t + 0 \cdot t^2 + 0 \cdot t^3 + \cdots$$

and

$$1 + 0 \cdot t + 0 \cdot t^2 + 0 \cdot t^3 + \cdots,$$

respectively. They are denoted simply by 0 and 1, respectively. The set of formal power series all of whose coefficients are 0 except for the constant term is naturally identified with R. Furthermore, regarding a polynomial $P(t)$ with coefficients in R as a formal power series whose coefficients a_n are 0 when n is greater than the degree of $P(t)$, we can consider $R[t]$ as a subring of $R[[t]]$. The definitions of the sum and the product of $R[[t]]$ are natural generalizations of those of polynomials.

Since we assume that R is an integral domain, $R[[t]]$ is also an integral domain. Namely, we have

Proposition 1.8. *The ring $R[[t]]$ does not have a zero divisor (the product of two non-zero elements is again non-zero). In particular, if $A \neq 0$ and $AB = AC$ $(A, B, C \in R[[t]])$, then $B = C$.*

Proof. Let $A = a_0 + a_1 t + a_2 t^2 + a_3 t^3 + \cdots$ and $B = b_0 + b_1 t + b_2 t^2 + b_3 t^3 + \cdots$ be non-zero elements, and let a_k, b_l be the first non-zero coefficients. Then the coefficient c_{k+l} of t^{k+l} in $A \cdot B = c_0 + c_1 t + c_2 t^2 + \cdots$ is

$$c_{k+l} = a_0 b_{k+l} + a_1 b_{k+l-1} + \cdots + a_k b_l + \cdots + a_{k+l-1} b_1 + a_{k+l} b_0 = a_k b_l.$$

Since R is an integral domain, $c_{k+l} = a_k b_l$ is not 0. Thus $A \cdot B$ is not 0.

If $AB = AC$, then $A(B - C) = 0$, which implies $A = 0$ or $B - C = 0$. Thus, if $A \neq 0$, then we have $B = C$. □

Proposition 1.9. *A formal power series $\sum_{n=0}^{\infty} a_n t^n$ is invertible in $R[[t]]$ if and only if the constant term a_0 is invertible.*

Proof. If $\sum_{n=0}^{\infty} a_n t^n$ is invertible, i.e., if there exists a formal power series $\sum_{n=0}^{\infty} b_n t^n$ such that $(\sum_{n=0}^{\infty} a_n t^n) \cdot (\sum_{n=0}^{\infty} b_n t^n) = 1$, then the constant term $a_0 b_0$ of the product must be 1, and thus a_0 is invertible. Conversely, suppose that a_0 is invertible. We define $b_n \in R$ $(n \geq 0)$ by

$$b_0 = a_0^{-1} \quad \text{and} \quad b_n = -a_0^{-1} \cdot \sum_{i=1}^{n} a_i b_{n-i} \ (n \geq 1),$$

and we compute the product $(\sum_{n=0}^{\infty} a_n t^n) \cdot (\sum_{n=0}^{\infty} b_n t^n)$. Then, we have $a_0 b_0 = 1$ and $\sum_{i=0}^{n} a_i b_{n-i} = 0$ for $n \geq 1$. This shows that $\sum_{n=0}^{\infty} b_n t^n$ is the multiplicative inverse of $\sum_{n=0}^{\infty} a_n t^n$. □

We denote by $A(0)$ the constant term of a formal power series $A(t)$. As we mentioned before, a formal power series is not a function and we do not generally replace t by any special value. However, replacing t by 0 is always valid, and we use this notation exceptionally for this case.

Next, we show that we can define an infinite sum of elements of $R[[t]]$ under certain conditions. Let $\{A_k(t)\}_{k=1}^{\infty}$, $A_k(t) = \sum_{i=0}^{\infty} a_i^{(k)} t^i$ be an infinite family of formal power series. Suppose that these series satisfy the condition

$$\text{for each } i, \text{ there exist only finitely many } k \text{ such that } a_i^{(k)} \neq 0.$$

In other words, if we write $A_k(t) = a_{n_k}^{(k)} t^{n_k} + (\text{higher-order terms})$ $(a_{n_k}^{(k)} \neq 0)$, we have $n_k \to \infty$ as $k \to \infty$. Then we define $A_1 + A_2 + A_3 + \cdots$ to be

$$A_1 + A_2 + A_3 + \cdots = \sum_{i=0}^{\infty} a_i t^i,$$

where

$$a_i = \sum_{k=1}^{\infty} a_i^{(k)}.$$

In short, we take the sum of the coefficients in a common degree, and the above condition ensures that each sum is a finite sum.

As a special case, we can define the "substitution" of a formal power series with vanishing constant term into another formal power series. Namely, if $A(t) = \sum_{i=0}^{\infty} a_i t^i$, and $B(t) = \sum_{i=1}^{\infty} b_i t^i$ $(B(0) = b_0 = 0)$, then we can give a meaning to the expression

$$A(B(t)) = \sum_{i=0}^{\infty} a_i (B(t))^i,$$

because, from $b_0 = 0$, the polynomial $a_i(B(t))^i$ begins at least in the degree i term, and the family $\left\{a_i(B(t))^i\right\}_{i=0}^{\infty}$ satisfies the above condition. As an easy example, we have

$$\begin{aligned} e^{-t} &= 1 + \frac{(-t)}{1!} + \frac{(-t)^2}{2!} + \frac{(-t)^3}{3!} + \cdots \\ &= 1 - \frac{t}{1!} + \frac{t^2}{2!} - \frac{t^3}{3!} + \cdots \end{aligned}$$

and

$$\begin{aligned} \log(1-t) = \log(1+(-t)) &= -t - \frac{(-t)^2}{2} + \frac{(-t)^3}{3} - \cdots \\ &= -t - \frac{t^2}{2} - \frac{t^3}{3} - \cdots . \end{aligned}$$

The identities

$$e^{\log(1+t)} = 1 + t$$

and

$$\log(1 + (e^t - 1)) = t$$

can be considered as the identities obtained by substituting one formal power series to the other. (We will give proofs of these identities after Remark 2.7 on p. 33.)

We can also compute the reciprocal of a formal power series with constant term 1 by "substitution". First we remark that the reciprocal of $1 - t$ is $1 + t + t^2 + \cdots$:

$$\frac{1}{1-t} = 1 + t + t^2 + \cdots.$$

We can see this by computing the product $(1-t)(1+t+t^2+\cdots)$. Now, let $1 + B(t)$ be a formal power series with constant term 1. Then $B(t)$ is a formal power series without constant term ($B(0) = 0$), and thus the reciprocal of $1 + B(t)$ can be computed by

$$\frac{1}{1+B(t)} = \frac{1}{1-(-B(t))} = 1 + (-B(t)) + (-B(t))^2 + \cdots.$$

As an example, we compute the first few Bernoulli numbers using (1.7). Since

$$\begin{aligned} e^t - 1 &= t + \frac{t^2}{2!} + \frac{t^3}{3!} + \frac{t^4}{4!} + \frac{t^5}{5!} + \cdots \\ &= t\left(1 + \frac{t}{2} + \frac{t^2}{6} + \frac{t^3}{24} + \frac{t^4}{120} + \cdots\right), \end{aligned}$$

we have

$$\begin{aligned} \frac{1}{\left(1 + \frac{t}{2} + \frac{t^2}{6} + \frac{t^3}{24} + \frac{t^4}{120} + \cdots\right)} &= 1 - \left(\frac{t}{2} + \frac{t^2}{6} + \frac{t^3}{24} + \frac{t^4}{120} + \cdots\right) \\ &\quad + \left(\frac{t}{2} + \frac{t^2}{6} + \frac{t^3}{24} + \frac{t^4}{120} + \cdots\right)^2 \\ &\quad - \left(\frac{t}{2} + \frac{t^2}{6} + \frac{t^3}{24} + \frac{t^4}{120} + \cdots\right)^3 \\ &\quad + \left(\frac{t}{2} + \frac{t^2}{6} + \frac{t^3}{24} + \frac{t^4}{120} + \cdots\right)^4 + \cdots \\ &= 1 - \frac{t}{2} + \frac{t^2}{12} + 0 \cdot t^3 - \frac{t^4}{720} + \cdots. \end{aligned}$$

From this we have

$$\frac{te^t}{e^t-1} = e^t \cdot \frac{1}{\left(1+\frac{t}{2}+\frac{t^2}{6}+\frac{t^3}{24}+\frac{t^4}{120}+\cdots\right)}$$
$$= \left(1+t+\frac{t^2}{2}+\frac{t^3}{6}+\frac{t^4}{24}+\cdots\right)\left(1-\frac{t}{2}+\frac{t^2}{12}+0\cdot t^3-\frac{t^4}{720}+\cdots\right)$$
$$= 1+\frac{t}{2}+\frac{t^2}{12}+0\cdot t^3-\frac{t^4}{720}+\cdots.$$

Thus, we have $B_0 = 1$, $B_1 = \frac{1}{2}$, $B_2 = \frac{1}{6}$, $B_3 = 0$, $B_4 = -\frac{1}{30}, \ldots$.

In this "substitution" operation, the following proposition gives the "inverse".

Proposition 1.10. *For a formal power series $A(t) = a_0 + a_1 t + a_2 t^2 + \cdots$, there exists a formal power series $B(t)$ such that*

$$B(0) = 0, \ A(B(t)) = t$$

if and only if

$$a_0 = 0 \text{ and } a_1 \text{ is invertible.}$$

In this case $B(t)$ is unique, and we have $B(A(t)) = t$. In other words $A(t)$ and $B(t)$ are inverse to each other with respect to composition.

Proof. If $B(t) = b_1 t + b_2 t^2 + \cdots$ exists and satisfies $A(B(t)) = t$, then by comparing the constant terms and the degree 1 terms, we have $a_0 = 0$ and $a_1 b_1 = 1$. This shows that the condition is necessary.

Conversely, suppose that $A(t)$ satisfies $a_0 = 0$ and a_1 is invertible. We would like to determine the coefficients of $B(t) = b_1 t + b_2 t^2 + b_3 t^3 + \cdots$ satisfying $A(B(t)) = t$. We first see from the coefficient of t that $a_1 b_1 = 1$. Since a_1 is invertible, we put $b_1 = a_1^{-1}$. For $n \geq 2$, the coefficient of t^n in $A(B(t))$ equals the coefficient of t^n in

$$a_1(B(t)) + a_2(B(t))^2 + a_3(B(t))^3 + \cdots + a_n(B(t))^n$$

since there is no term t^n after this due to the fact that $B(0) = 0$. This can be written as

$$a_1 b_n + (\text{polynomial in } a_2, a_3, \ldots, a_n, b_1, b_2, \ldots, b_{n-1}).$$

If $b_1, b_2, \ldots, b_{n-1}$ are already determined, then we can determine b_n uniquely from the fact that the above formula equals 0 and a_1 is invertible. This proves the existence as well as the uniqueness of $B(t)$.

The $B(t)$ obtained in this way satisfies $B(0) = 0$, and b_1 is invertible. We thus see that there is a $C(t)$ ($C(0) = 0$) such that

$$B(C(t)) = t.$$

Then we substitute $C(t)$ for t in the formula $t = A(B(t))$, and use the fact that $B(C(t)) = t$ to get

$$C(t) = A(B(C(t))) = A(t).$$

Thus, we have

$$B(A(t)) = t.$$

□

The formulas we mentioned earlier,

$$e^{\log(1+t)} - 1 = t \text{ and } \log(1 + (e^t - 1)) = t,$$

can be interpreted as saying $\log(1 + t)$ and $e^t - 1$ are inverse to each other with respect to composition.

The derivative of formal power series $\sum_{n=0}^{\infty} a_n t^n$, written $\frac{d}{dt}\left(\sum_{n=0}^{\infty} a_n t^n\right)$ (or $\left(\sum_{n=0}^{\infty} a_n t^n\right)'$), is defined formally by term-by-term differentiation:

$$\frac{d}{dt}\left(\sum_{n=0}^{\infty} a_n t^n\right) = \sum_{n=1}^{\infty} n a_n t^{n-1} = a_1 + 2a_2 t + 3a_3 t^2 + 4a_4 t^3 + \cdots.$$

For example, $(e^t)' = e^t$, and $(\log(1 + t))' = 1 - t + t^2 - t^3 + \cdots = \frac{1}{1+t}$. This definition satisfies the usual rule of derivatives regarding sum and product. We give a proof of the product rule $(f(t)g(t))' = f'(t)g(t) + f(t)g'(t)$. Let $f(t) = \sum_{n=0}^{\infty} a_n t^n$, and $g(t) = \sum_{n=0}^{\infty} b_n t^n$. Then

$$\begin{aligned}
(f(t)g(t))' &= \left(\sum_{n=0}^{\infty}\left(\sum_{i=0}^{n} a_i b_{n-i}\right) t^n\right)' = \sum_{n=1}^{\infty}\left(n \sum_{i=0}^{n} a_i b_{n-i}\right) t^{n-1} \\
&= \sum_{n=1}^{\infty}\left(\sum_{i=0}^{n} i a_i b_{n-i} + \sum_{i=0}^{n}(n-i) a_i b_{n-i}\right) t^{n-1} \\
&= \sum_{n=1}^{\infty}\left(\sum_{i=1}^{n} i a_i b_{n-i}\right) t^{n-1} + \sum_{n=1}^{\infty}\left(\sum_{i=0}^{n-1} a_i (n-i) b_{n-i}\right) t^{n-1} \\
&= f'(t)g(t) + f(t)g'(t).
\end{aligned}$$

Also, when $R \supset \mathbf{Q}$, the integral $\displaystyle\int_0^t \sum_{n=0}^{\infty} a_n t^n \, dt$ is defined formally by term-by-term integration:

$$\int_0^t \sum_{n=0}^{\infty} a_n t^n \, dt = \sum_{n=0}^{\infty} a_n \frac{t^{n+1}}{n+1} = a_0 t + a_1 \frac{t^2}{2} + a_2 \frac{t^3}{3} + a_3 \frac{t^4}{4} + \cdots.$$

In order to treat the generating function of Bernoulli numbers, it is convenient to generalize formal power series to Laurent[6] series (with a finite number of terms with negative powers).

Definition 1.11. The set of formal Laurent series with coefficients in R,

$$\sum_{i=-N}^{\infty} a_i t^i \quad \text{(for some integer } N\text{)},$$

is denoted by $R((t))$.

The sum and the product in $R((t))$ are defined in the same way as in $R[[t]]$ and with these operations $R((t))$ is a commutative integral domain that contains $R[[t]]$ as a subdomain. Also, $\sum_{i=-N}^{\infty} a_i t^i$ is invertible in $R((t))$ if and only if the first non-zero coefficient a_{-N} is invertible. These can be proved in the same way as in the proof of Propositions 1.8 and 1.9. In particular, if R is a field, then so is $R((t))$. Furthermore, the formal derivative and integral (when $R \supset \mathbf{Q}$) are also defined, and the quotient rule $(f(t)/g(t))' = (f'(t)g(t) - f(t)g'(t))/g(t)^2$ holds. This follows from the product rule. (Put $f(t)/g(t) = h(t)$, and take the derivative of $f(t) = g(t)h(t)$.)

1.4 The Generating Function of Bernoulli Numbers

Theorem 1.12. *Let B_n $(n = 0, 1, 2, \ldots)$ be the Bernoulli numbers. We have the following formula in $\mathbf{Q}((t))$:*

$$\frac{te^t}{e^t - 1} = \sum_{n=0}^{\infty} B_n \frac{t^n}{n!}.$$

As a matter of fact, the right-hand side belongs to $\mathbf{Q}[[t]]$, but we consider the quotient of the formal power series te^t by $e^t - 1$ in $\mathbf{Q}((t))$.

[6] Pierre Alphonse Laurent (born on July 18, 1813 in Paris, France—died on September 2, 1854 in Paris, France).

Proof. It is sufficient to show $(\sum_{n=0}^{\infty} B_n \frac{t^n}{n!})(e^t - 1) = te^t$. The formula in question is obtained by dividing both sides by $e^t - 1$. From the definition of the product of formal power series we have

$$\left(\sum_{n=0}^{\infty} B_n \frac{t^n}{n!}\right)(e^t - 1) = \left(\sum_{n=0}^{\infty} \frac{B_n}{n!} t^n\right)\left(\sum_{n=1}^{\infty} \frac{t^n}{n!}\right)$$

$$= \sum_{n=1}^{\infty}\left(\sum_{i=0}^{n-1} \frac{B_i}{i!} \frac{1}{(n-i)!}\right) t^r$$

$$= \sum_{n=1}^{\infty}\left(\sum_{i=0}^{n-1} \binom{n}{i} B_i\right)\frac{t^n}{n!}.$$

From the recurrence formula (1.2) we have $\sum_{i=0}^{n-1} \binom{n}{i} B_i = n$ for all $n \geq 1$. Thus we have

$$\sum_{n=1}^{\infty}\left(\sum_{i=0}^{n-1} \binom{n}{i} B_i\right)\frac{t^n}{n!} = \sum_{n=1}^{\infty} \frac{t^n}{(n-1)!} = te^t,$$

which concludes the proof. □

Remark 1.13. Conversely, if we define B_n by the formula in the above theorem, then we have

$$\left(\sum_{n=0}^{\infty} B_n \frac{t^n}{n!}\right)(e^t - 1) = te^t.$$

Expanding the left-hand side as in the above proof, and comparing it with the right-hand side, we obtain the recurrence formula (1.2). This shows that Definition 1.1 and the definition using the generating function in Theorem 1.12 are equivalent.

Using the generating function, we can give a simpler proof of Proposition 1.4.

Proposition 1.14 (Proposition 1.4 revisited). *If n is an odd integer greater than or equal to 3, then $B_n = 0$.*

Proof. It suffices to show that the formal power series $\frac{te^t}{e^t-1} - \frac{t}{2}$ does not have any odd-degree terms. Since we have

$$\frac{te^t}{e^t - 1} - \frac{t}{2} = \frac{t(e^t - 1 + 1)}{e^t - 1} - \frac{t}{2} = \frac{t}{e^t - 1} + \frac{t}{2}$$

and

$$\frac{(-t)e^{-t}}{e^{-t} - 1} - \frac{(-t)}{2} = \frac{-t}{1 - e^t} + \frac{t}{2} = \frac{t}{e^t - 1} + \frac{t}{2},$$

$\frac{te^t}{e^t-1} - \frac{t}{2}$ is invariant under the substitution $t \to -t$. This shows that the coefficients of odd-degree terms are all 0. □

Looking at B_n for even n in Table 1.1, we see that they are non-zero and, from B_2 on, the signs alternate. This may be seen from the formula for the value of the Riemann zeta function $\zeta(s) = \sum_{n=1}^{\infty} \frac{1}{n^s}$ at positive even integers (replace k by $2k$ on p. 61) (4.4):

$$\zeta(2k) = \frac{(-1)^{k-1}}{2} \frac{(2\pi)^{2k}}{(2k)!} B_{2k} \quad (k \geq 1).$$

But here, we give a direct proof using the following recurrence formula for the Bernoulli numbers due to Euler[7] [33].

Proposition 1.15.

$$(2n+1)B_{2n} = -\sum_{m=1}^{n-1} \binom{2n}{2m} B_{2m} B_{2(n-m)} \quad (n \geq 2).$$

Proof. By the previous proposition, if we subtract the term of degree 1 from the generating function of the Bernoulli numbers, we obtain the generating function of even indexed B_{2n}:

$$\frac{te^t}{e^t-1} - \frac{t}{2} = \sum_{n=0}^{\infty} B_{2n} \frac{t^{2n}}{(2n)!}.$$

Let $f(t)$ be the left-hand side. Taking the derivative of $f(t)$ as was explained on p. 19, we have

$$f(t) - tf'(t) = f(t)^2 - \frac{t^2}{4}.$$

Substituting $\sum_{n=0}^{\infty} B_{2n} \frac{t^{2n}}{(2n)!}$ for $f(t)$, we have

$$\sum_{n=0}^{\infty} (1-2n) B_{2n} \frac{t^{2n}}{(2n)!} = \sum_{n=0}^{\infty} \left(\sum_{m=0}^{n} \binom{2n}{2m} B_{2m} B_{2(n-m)} \right) \frac{t^{2n}}{(2n)!} - \frac{t^2}{4}.$$

[7]Leonhard Euler (Born on April 15, 1707 in Basel, Switzerland—died on September 18, 1783 in St. Petersburg, Russia).

Comparing the coefficients of both sides, we have

$$(1-2n)B_{2n} = \sum_{m=0}^{n} \binom{2n}{2m} B_{2m} B_{2(n-m)}$$

for $n \geq 2$. Since the terms for $m = 0$ and $m = n$ yield $2B_{2n}$, we subtract them from both sides. Multiplying by -1, we obtain the formula in the proposition. □

Corollary 1.16. *For $n \geq 1$, we have $(-1)^{n-1}B_{2n} > 0$.*

Proof. We prove this by induction. For $n = 1$, $B_2 = \frac{1}{6} > 0$. Suppose that the proposition is true for all integers less than n. By multiplying the formula in the proposition by $(-1)^{n-1}$, we get

$$(2n+1)\cdot(-1)^{n-1}B_{2n} = \sum_{m=1}^{n-1} \binom{2n}{2m} (-1)^{m-1}B_{2m}(-1)^{n-m-1}B_{2(n-m)}.$$

The right-hand side is positive by assumption. Thus, $(-1)^{n-1}B_{2n} > 0$, which settles the proof. □

To conclude this chapter, we give expansions of $\tan x$ and $\cot x$ as alternative generating functions for Bernoulli numbers.

Proposition 1.17. *The Taylor expansion of* $\tan x$ *and the Laurent expansion of* $\cot x$ *around $x = 0$ are*

$$\tan x = \sum_{n=1}^{\infty} (-1)^{n-1}(2^{2n}-1)2^{2n}B_{2n}\frac{x^{2n-1}}{(2n)!},$$

$$\cot x = \frac{1}{x} + \sum_{n=1}^{\infty} (-1)^n 2^{2n} B_{2n}\frac{x^{2n-1}}{(2n)!}.$$

The right-hand sides converge for $|x| < \frac{\pi}{2}$ and $0 < |x| < \pi$, respectively.

Proof. The formal power series $f(t)$ in Proposition 1.15 can be written as $f(t) = \frac{t}{2}\coth(\frac{t}{2})$, where $\coth t = \frac{e^t+e^{-t}}{e^t-e^{-t}}$ is the hyperbolic cotangent. Replacing t by x, we have

$$\frac{x}{2}\coth\left(\frac{x}{2}\right) = \sum_{n=0}^{\infty} B_{2n}\frac{x^{2n}}{(2n)!},$$

which is regarded as the Laurent expansion of $\coth(\frac{x}{2})$. Replacing x by $2ix$ ($i = \sqrt{-1}$), dividing both sides by x, and using the fact $i\coth(ix) = \cot x$, we obtain the Laurent expansion of $\cot x$. From the duplication formula for $\cot x$ which reads

$$\cot 2x = \frac{1}{2}(\cot x - \tan x),$$

we have

$$\tan x = \cot x - 2\cot(2x).$$

This shows that the Taylor expansion of $\tan x$ can be obtained from the expansion of $\cot x$. As for the radii of convergence, we see it from the fact that the poles nearest to 0 are located at $\pm\pi/2$ and $\pm\pi$, respectively. □

Remark 1.18. The coefficient of $\frac{x^{2n-1}}{(2n-1)!}$ in the Taylor expansion of $\tan x$,

$$T_n := (-1)^{n-1}(2^{2n} - 1)2^{2n}\frac{B_{2n}}{2n},$$

is sometimes called the tangent number. It is a positive integer. The fact that it is positive is seen from Corollary 1.16. We will prove that it is an integer in Remark 7.6 on p. 98. Knuth and Buckholtz [61] give another proof of this fact and a method to compute Bernoulli numbers using T_n. See also Exercise 3.6 in Chap. 3.

Exercise 1.19. Compute the Bernoulli numbers B_n up to $n = 12$ by using Theorem 1.6.

Exercise 1.20. Guess any pattern of prime numbers appearing in denominators of Bernoulli numbers. (This is one of the topics covered in Chap. 3.)

Exercise 1.21. Compute the first several terms of the formal power series $e^{\log(1+t)}$ and $\log(1 + (e^t - 1))$ using the definition of composition.

Exercise 1.22. Prove the quotient rule

$$(f(t)/g(t))' = (f'(t)g(t) - f(t)g'(t))/g(t)^2$$

of formal Laurant series.

Exercise 1.23. Compute first several tangent numbers T_n, and check that they are positive integers.

Chapter 2
Stirling Numbers and Bernoulli Numbers

In this chapter we give a formula that describes Bernoulli numbers in terms of Stirling numbers. This formula will be used to prove a theorem of Clausen and von Staudt in the next chapter. As an application of this formula, we also introduce an interesting algorithm to compute Bernoulli numbers.

We first summarize the facts on Stirling numbers. Those Stirling numbers we need in this chapter are of the second kind, but we also introduce Stirling numbers of the first kind, which will be needed later.[1]

2.1 Stirling Numbers

We first define the Stirling numbers of the second kind, restricting ourselves to the case where a combinatorial meaning can be given. We adopt Knuth's notation [59].

Definition 2.1 (Stirling numbers of the second kind (Stirling's subset numbers)).

For positive integers n and m, define

$$\left\{ {n \atop m} \right\} := \text{the number of ways to divide a set of } n \text{ elements into } m \text{ nonempty sets.}$$

(Knuth proposes to read this as "n subset m".)

[1]According to Knuth [59], Stirling (James, born in May, 1692 in Garden, Scotland—died on December 5, 1770 in Edinburgh, Scotland) first introduced the second kind. The names "first kind" and "second kind" are due to Nielsen (Niels, born on December 2, 1865 in Orslev, Denmark—died on September 16, 1931 in Copenhagen, Denmark), who first used these names in his book on the Gamma function [73, §26].

T. Arakawa et al., *Bernoulli Numbers and Zeta Functions*, Springer Monographs in Mathematics, DOI 10.1007/978-4-431-54919-2_2, © Springer Japan 2014

For example, there are seven ways to divide the set $\{1, 2, 3, 4\}$ into two non-empty sets:

$$\{1\} \cup \{2,3,4\},\ \{2\} \cup \{1,3,4\},\ \{3\} \cup \{1,2,4\},\ \{4\} \cup \{1,2,3\},$$
$$\{1,2\} \cup \{3,4\},\ \{1,3\} \cup \{2,4\},\ \{1,4\} \cup \{2,3\}.$$

Thus, $\left\{ {4 \atop 2} \right\} = 7$.

From the definition, $\left\{ {n \atop m} \right\} = 0$ if $m > n$. Also, from the definition we have the recurrence formula:

$$\left\{ {n+1 \atop m} \right\} = \left\{ {n \atop m-1} \right\} + m \left\{ {n \atop m} \right\}. \tag{2.1}$$

This formula can be proved in the same way as in the combinatorial proof of the relation of binomial coefficients: $\binom{n+1}{m} = \binom{n}{m-1} + \binom{n}{m}$. Suppose we divide a set of $n+1$ elements into m sets. We look at one particular element. If this element forms a set by itself, there are $\left\{ {n \atop m-1} \right\}$ ways to divide the remaining n elements into $m-1$ sets. If this element belongs to a set with other elements, there are $\left\{ {n \atop m} \right\}$ ways to divide other n elements into m sets, and there are m ways to put this particular element into one of these sets, and thus there are $m\left\{ {n \atop m} \right\}$ ways altogether.

We now proceed to define $\left\{ {n \atop m} \right\}$ by the recurrence formula (2.1) for any integers m and n.

Definition 2.2 (Stirling numbers of the second kind (general case)). For any integers n and m, define $\left\{ {n \atop m} \right\}$ by the recurrence formula (2.1) with the initial conditions $\left\{ {0 \atop 0} \right\} = 1$, and $\left\{ {n \atop 0} \right\} = \left\{ {0 \atop m} \right\} = 0\ (n, m \neq 0)$.

From (2.1), if we know two out of three values $\left\{ {n+1 \atop m} \right\}$, $\left\{ {n \atop m-1} \right\}$, and $\left\{ {n \atop m} \right\}$, we can determine the third except for the case $m = 0$, where we cannot divide $m\left\{ {n \atop m} \right\}$ by m. In this case the value $\left\{ {n \atop 0} \right\} = 0$ is already given in the definition. We give the values of $\left\{ {n \atop m} \right\}$ for $-7 \leq m, n \leq 7$ in Table 2.1 on p. 27. It is easy to see how all values of $\left\{ {n \atop m} \right\}$ are determined from the initial conditions, considering how the positions $(n+1, m)$, $(n, m-1)$, and (n, m) are located in the table.

The new definition of $\left\{ {n \atop m} \right\}$ coincides with the previous combinatorial definition when $n, m > 0$. To see this, it suffices to verify $\left\{ {n \atop 1} \right\} = 1\ (n \geq 1)$ and $\left\{ {1 \atop m} \right\} = 0\ (m \geq 2)$ under the new definition. Since these values are obvious under the old definition, and the recurrence formulas are the same for both definitions, the claim follows. We put $m = 1$ in (2.1) to obtain

$$\left\{ {n+1 \atop 1} \right\} = \left\{ {n \atop 0} \right\} + \left\{ {n \atop 1} \right\}.$$

Table 2.1 $\left\{ {n \atop m} \right\}$

$m\backslash n$	−7	−6	−5	−4	−3	−2	−1	0	1	2	3	4	5	6	7
7	0	0	0	0	0	0	0	0	0	0	0	0	0	0	1
6	0	0	0	0	0	0	0	0	0	0	0	0	0	1	21
5	0	0	0	0	0	0	0	0	0	0	0	0	1	15	140
4	0	0	0	0	0	0	0	0	0	0	0	1	10	65	350
3	0	0	0	0	0	0	0	0	0	0	1	6	25	90	301
2	0	0	0	0	0	0	0	0	0	1	3	7	15	31	63
1	0	0	0	0	0	0	0	0	1	1	1	1	1	1	1
0	0	0	0	0	0	0	0	1	0	0	0	0	0	0	0
−1	0	0	0	0	0	0	1	0	0	0	0	0	0	0	0
−2	0	0	0	0	0	1	1	0	0	0	0	0	0	0	0
−3	0	0	0	0	1	3	2	0	0	0	0	0	0	0	0
−4	0	0	0	1	6	11	6	0	0	0	0	0	0	0	0
−5	0	0	1	10	35	50	24	0	0	0	0	0	0	0	0
−6	0	1	15	85	225	274	120	0	0	0	0	0	0	0	0
−7	1	21	175	735	1624	1764	720	0	0	0	0	0	0	0	0

Putting $n = 0$ in this formula, we obtain $\left\{ {1 \atop 1} \right\} = 1$. Also, since $\left\{ {n \atop 0} \right\} = 0$, we have $\left\{ {n+1 \atop 1} \right\} = \left\{ {n \atop 1} \right\}$, and thus $\left\{ {n \atop 1} \right\} = 1$ $(n \geq 1)$. Similarly, $\left\{ {1 \atop m} \right\} = 0$ $(m \geq 2)$ can be verified by putting $n = 0$ in (2.1).

Remark 2.3. Logically, we only need Definition 2.2. However, in order to make the meaning clearer, we began with Definition 2.1.

We now define the Stirling numbers of the first kind.

Definition 2.4 (Stirling numbers of the first kind (Stirling's cycle numbers)). For positive integers n and m, define

$$\left[{n \atop m} \right] := \text{number of permutations of } n \text{ letters (elements of the symmetric group of degree } n\text{) that consist of } m \text{ disjoint cycles.}$$

(We read "n cycle m".)

An element of the symmetric group of degree n (the group of all permutations (bijections) of the set $\{1, 2, \ldots, n\}$) can be decomposed uniquely into the product of disjoint cycles. (See any textbook of group theory, for example [42, Chap. I, Th. 6.3].) Among the permutations of n letters, we count the number of permutations that decompose into m disjoint cycles (including the cycles of length 1), and that is $\left[{n \atop m} \right]$. For example, the permutations of four letters that decompose into two disjoint cycles are

$$(1)(2\,3\,4),\ (1)(2\,4\,3),\ (2)(1\,3\,4),\ (2)(1\,4\,3),$$
$$(3)(1\,2\,4),\ (3)(1\,4\,2),\ (4)(1\,2\,3),\ (4)(1\,3\,2),$$
$$(1\,2)(3\,4),\ (1\,3)(2\,4),\ (1\,4)(2\,3).$$

Thus, $\left[{4 \atop 2}\right] = 11$.

It follows immediately from the definition that $\left[{n \atop n}\right] = 1$ (only the identity), $\left[{n \atop 1}\right] = (n-1)!$ (number of n-cycles).

The Stirling numbers of the first kind satisfy the recurrence formula

$$\left[{n+1 \atop m}\right] = \left[{n \atop m-1}\right] + n\left[{n \atop m}\right]. \tag{2.2}$$

We can see this in the same way as before. A permutation of $n+1$ letters may fix the letter $n+1$. In this case the letter $n+1$ forms a cycle by itself, and there are $\left[{n \atop m-1}\right]$ permutations of the remaining n letters that decompose into $m-1$ disjoint cycles. A permutation that moves the letter $n+1$ can be obtained by inserting the letter $n+1$ in one of the m disjoint cycles in the decomposition of permutations of n letters. There are j ways to form a cycle of length $j+1$ out of a cycle of length j by inserting another letter. Therefore, there are n ways to insert the letter $n+1$, and thus there are in total $n\left[{n \atop m}\right]$ permutations that move the letter $n+1$. This shows (2.2).

Just as we did for the Stirling numbers of the second kind, we define $\left[{n \atop m}\right]$ for any m and n as follows.

Definition 2.5 (Stirling numbers of the first kind (general case)). For any integers n and m, define $\left[{n \atop m}\right]$ by the recurrence formula (2.2) with the initial conditions $\left[{0 \atop 0}\right] = 1$, and $\left[{n \atop 0}\right] = \left[{0 \atop m}\right] = 0\ (n, m \neq 0)$.

Table 2.2 shows the values of $\left[{n \atop m}\right]$ in the same range as Table 2.1. Readers will certainly notice the remarkable relations between these two tables. As Knuth asserts, there is only *one* kind of Stirling number, and the recurrence formulas (2.1) and (2.2) are essentially the same formula (see the next proposition).

We list some formulas which Stirling numbers satisfy. There are more formulas, but we only list the ones we need later in this book. For others, see [38] for example.

Proposition 2.6.

(1)
$$\left[{n \atop m}\right] = \left\{{-m \atop -n}\right\}.$$

(2)
$$x^n = \sum_{m=0}^{n} \left\{{n \atop m}\right\} x^{\underline{m}} \quad (n \geq 0),$$

Table 2.2 $\begin{bmatrix} n \\ m \end{bmatrix}$

$m\backslash n$	−7	−6	−5	−4	−3	−2	−1	0	1	2	3	4	5	6	7
7	0	0	0	0	0	0	0	0	0	0	0	0	0	0	1
6	0	0	0	0	0	0	0	0	0	0	0	0	0	1	21
5	0	0	0	0	0	0	0	0	0	0	0	0	1	15	175
4	0	0	0	0	0	0	0	0	0	0	0	1	10	85	735
3	0	0	0	0	0	0	0	0	0	0	1	6	35	225	1624
2	0	0	0	0	0	0	0	0	0	1	3	11	50	274	1764
1	0	0	0	0	0	0	0	0	1	1	2	6	24	120	720
0	0	0	0	0	0	0	0	1	0	0	0	0	0	0	0
−1	0	0	0	0	0	0	1	0	0	0	0	0	0	0	0
−2	0	0	0	0	0	1	1	0	0	0	0	0	0	0	0
−3	0	0	0	0	1	3	1	0	0	0	0	0	0	0	0
−4	0	0	0	1	6	7	1	0	0	0	0	0	0	0	0
−5	0	0	1	10	25	15	1	0	0	0	0	0	0	0	0
−6	0	1	15	65	90	31	1	0	0	0	0	0	0	0	0
−7	1	21	140	350	301	63	1	0	0	0	0	0	0	0	0

where $x^{\underline{m}}$ *stands for*

$$x^{\underline{m}} = x(x-1)(x-2)\cdots(x-m+1)\ \ (m > 0),\ x^{\underline{0}} = 1.$$

(3)
$$x^{\underline{n}} = (-1)^n \sum_{m=0}^{n} (-1)^m \begin{bmatrix} n \\ m \end{bmatrix} x^m.$$

(4)
$$\left(x\frac{d}{dx}\right)^n = \sum_{m=1}^{n} \left\{ {n \atop m} \right\} x^m \left(\frac{d}{dx}\right)^m \quad (n \geq 1).$$

(5.1) *For* $m, n \geq 0$,
$$\sum_{l \geq 0} (-1)^l \left\{ {n \atop l} \right\} \begin{bmatrix} l \\ m \end{bmatrix} = (-1)^m \delta_{m,n}.$$

(5.2) *For* $m, n \geq 0$,
$$\sum_{l \geq 0} (-1)^l \begin{bmatrix} n \\ l \end{bmatrix} \left\{ {l \atop m} \right\} = (-1)^m \delta_{m,n}.$$

Here, $\delta_{m,n}$ *stands for Kronecker's delta; i.e.,* $\delta_{m,n} = 1$ *for* $m = n$ *and* $\delta_{m,n} = 0$ *for* $m \neq n$*. The sums on the left-hand sides of* (5.1) *and* (5.2) *are finite sums.*

(6) *For* $m, n \geq 0$, $$\left\{ {n \atop m} \right\} = \frac{(-1)^m}{m!} \sum_{l=0}^{m} (-1)^l \binom{m}{l} l^n.$$

(7) $$\frac{(e^t - 1)^m}{m!} = \sum_{n=m}^{\infty} \left\{ {n \atop m} \right\} \frac{t^n}{n!} \quad (m \geq 0).$$

(8) $$\frac{t^m}{(1-t)(1-2t)\cdots(1-mt)} = \sum_{n=m}^{\infty} \left\{ {n \atop m} \right\} t^n \quad (m \geq 1).$$

(9) $$\frac{(-\log(1-t))^m}{m!} = \sum_{n=m}^{\infty} \left[{n \atop m} \right] \frac{t^n}{n!} \quad (m \geq 0).$$

Proof. (1) Write $a_{n,m} = \left\{ {-m \atop -n} \right\}$. It suffices to verify that $a_{n,m}$ satisfies the same recurrence formula with the same initial conditions as those for $\left[{n \atop m} \right]$. It is easy for the initial conditions. In the recurrence formula (2.1), replacing m by $-n$ and n by $-m$, we have

$$\left\{ {-m+1 \atop -n} \right\} = \left\{ {-m \atop -n-1} \right\} - n \left\{ {-m \atop -n} \right\}.$$

Adding $n\left\{ {-m \atop -n} \right\}$ to both sides, we have

$$\left\{ {-m+1 \atop -n} \right\} + n \left\{ {-m \atop -n} \right\} = \left\{ {-m \atop -n-1} \right\}.$$

This implies $a_{n,m-1} + n a_{n,m} = a_{n+1,m}$, which is nothing but the recurrence formula (2.2).

(2) We use the same strategy. Namely, write $x^n = \sum_{m=0}^{n} a_{n,m} x^{\underline{m}}$, and verify that $a_{n,m}$ satisfies the same recurrence formula as $\left\{ {n \atop m} \right\}$ within the range $m, n \geq 0$ and the same initial conditions. It is easy for the initial conditions (put $a_{0,m} = 0$ $(m > 0)$). From $x^{\underline{m+1}} = x^{\underline{m}}(x - m)$ and $x^{\underline{m}} \cdot x = x^{\underline{m+1}} + m x^{\underline{m}}$, we have

$$\begin{aligned} \sum_{m=0}^{n+1} a_{n+1,m} x^{\underline{m}} = x^{n+1} &= x^n \cdot x \\ &= \sum_{m=0}^{n} a_{n,m} x^{\underline{m}} \cdot x \end{aligned}$$

$$= \sum_{m=0}^{n} a_{n,m} (x^{\underline{m+1}} + m x^{\underline{m}})$$

$$= \sum_{m=1}^{n+1} a_{n,m-1} x^{\underline{m}} + \sum_{m=0}^{n} m a_{n,m} x^{\underline{m}}.$$

Comparing the coefficients of $x^{\underline{m}}$ of both sides, we obtain the recurrence formula (2.1). (Here, we put $a_{n,-1} = a_{n,n+1} = 0$.)

(3) is similarly proved and we omit it.

(4) We prove it by induction. For $n = 1$, both sides equal $x\frac{d}{dx}$. Suppose the formula is true for n. Then

$$\left(x\frac{d}{dx}\right)^{n+1} = \left(x\frac{d}{dx}\right)\sum_{m=1}^{n} \left\{ {n \atop m} \right\} x^m \left(\frac{d}{dx}\right)^m$$

$$= x\sum_{m=1}^{n} \left\{ {n \atop m} \right\} \left(m x^{m-1} \left(\frac{d}{dx}\right)^m + x^m \left(\frac{d}{dx}\right)^{m+1} \right)$$

$$= \sum_{m=1}^{n} m \left\{ {n \atop m} \right\} x^m \left(\frac{d}{dx}\right)^m + \sum_{m=1}^{n} \left\{ {n \atop m} \right\} x^{m+1} \left(\frac{d}{dx}\right)^{m+1}$$

$$= \sum_{m=1}^{n+1} \left(m \left\{ {n \atop m} \right\} + \left\{ {n \atop m-1} \right\} \right) x^m \left(\frac{d}{dx}\right)^m$$

$$= \sum_{m=1}^{n+1} \left\{ {n+1 \atop m} \right\} x^m \left(\frac{d}{dx}\right)^m .$$

(To prove the next-to-last equality, we use $\left\{ {n \atop n+1} \right\} = 0$ and $\left\{ {n \atop 0} \right\} = 0$.) This shows that the formula is true for $n + 1$.

(5) We prove it using (2) and (3). For (5.1), we put the formula

$$x^{\underline{l}} = (-1)^l \sum_{m=0}^{l} (-1)^m \left[{l \atop m} \right] x^m$$

obtained by replacing n by l in (3) into the formula

$$x^n = \sum_{l=0}^{n} \left\{ {n \atop l} \right\} x^{\underline{l}},$$

which is obtained by replacing m by l in (2). Then we have

$$x^n = \sum_{l=0}^{n} \left\{ {n \atop l} \right\} (-1)^l \sum_{m=0}^{l} (-1)^m \left[{l \atop m} \right] x^m$$
$$= \sum_{m=0}^{n} (-1)^m \sum_{l=m}^{n} (-1)^l \left\{ {n \atop l} \right\} \left[{l \atop m} \right] x^m.$$

Comparing the coefficients on both sides, we obtain (5.1). The formula (5.2) can be obtained by substituting (2) into (3).

(6) We prove it by verifying that the right-hand side satisfies the recurrence formula for $\left\{ {n \atop m} \right\}$. Denote the right-hand side by $a_{n,m}$ once again. It is easy to see that $a_{0,0} = 1$ and $a_{n,0} = 0$ for $n \geq 1$. (Define $0^0 = 1$.) If $n = 0$, and $m \geq 1$, then $\sum_{l=0}^{m} (-1)^l \binom{m}{l} = (1-1)^m = 0$. Thus, we have $a_{0,m} = 0$. As for the recurrence formula, we verify it as follows:

$$m a_{n,m} + a_{n,m-1}$$
$$= \frac{(-1)^m}{(m-1)!} \sum_{l=0}^{m} (-1)^l \binom{m}{l} l^n + \frac{(-1)^{m-1}}{(m-1)!} \sum_{l=0}^{m-1} (-1)^l \binom{m-1}{l} l^n$$
$$= \frac{(-1)^m}{(m-1)!} \sum_{l=0}^{m} (-1)^l \left\{ \binom{m}{l} - \binom{m-1}{l} \right\} l^n$$
$$= \frac{(-1)^m}{(m-1)!} \sum_{l=0}^{m} (-1)^l \frac{l}{m} \binom{m}{l} l^n$$
$$= \frac{(-1)^m}{m!} \sum_{l=0}^{m} (-1)^l \binom{m}{l} l^{n+1}$$
$$= a_{n+1,m}.$$

One can prove (6) more naturally using the difference calculus. See for example [53, §58].

(7) First, note that the right-hand side of (7) can be expressed as the sum from $n = 0$. (If $n < m$, then $\left\{ {n \atop m} \right\} = 0$.) Substitute (6) into the right-hand side and simplify:

$$\text{r.h.s} = \sum_{n=0}^{\infty} \left\{ \frac{(-1)^m}{m!} \sum_{l=0}^{m} (-1)^l \binom{m}{l} l^n \right\} \frac{t^n}{n!}$$
$$= \frac{(-1)^m}{m!} \sum_{l=0}^{m} (-1)^l \binom{m}{l} \left(\sum_{n=0}^{\infty} \frac{(lt)^n}{n!} \right)$$

$$
\begin{aligned}
&= \frac{(-1)^m}{m!} \sum_{l=0}^{m} (-1)^l \binom{m}{l} e^{lt} \\
&= \frac{(-1)^m}{m!} (1 - e^t)^m \\
&= \frac{(e^t - 1)^m}{m!} \\
&= \text{l.h.s.}
\end{aligned}
$$

It can also be proved by writing $(e^t - 1)^m/m! = \sum_{n=m}^{\infty} a_{n,m} \frac{t^n}{n!}$, and verifying that $a_{n,m}$ satisfies the same recurrence formula as $\left\{ {n \atop m} \right\}$.

(8) Denote the right-hand side by f_m. From the recurrence formula (2.1) for $\left\{ {n \atop m} \right\}$, we have (noting that $\left\{ {n \atop m} \right\} = 0$ for $n < m$)

$$
\begin{aligned}
f_m &= \sum_{n=m}^{\infty} \left\{ {n \atop m} \right\} t^n \\
&= \sum_{n=m}^{\infty} \left(\left\{ {n-1 \atop m-1} \right\} + m \left\{ {n-1 \atop m} \right\} \right) t^n \\
&= t \sum_{n=m-1}^{\infty} \left\{ {n \atop m-1} \right\} t^n + mt \sum_{n=m}^{\infty} \left\{ {n \atop m} \right\} t^n \\
&= t f_{m-1} + mt f_m.
\end{aligned}
$$

Thus, $f_m = \frac{t}{1-mt} f_{m-1}$. From $\left\{ {n \atop 1} \right\} = 1$ $(n \geq 1)$, we have $f_1 = \frac{t}{1-t}$, which shows that f_m is equal to the left-hand side of (8).

(9) Writing $(-\log(1-t))^m/m! = \sum_{n=m}^{\infty} a_{n,m} \frac{t^n}{n!}$, and taking the derivatives of both sides, we have

$$
\frac{(-\log(1-t))^{m-1}}{(m-1)!} \cdot \frac{1}{1-t} = \sum_{n=m}^{\infty} a_{n,m} \frac{t^{n-1}}{(n-1)!}.
$$

From this we have

$$
\sum_{n=m-1}^{\infty} a_{n,m-1} \frac{t^n}{n!} = (1-t) \sum_{n=m}^{\infty} a_{n,m} \frac{t^{n-1}}{(n-1)!}.
$$

This shows that $a_{n,m}$ satisfies the same recurrence formula as $\left[{n \atop m} \right]$. It is easy to verify the initial conditions. □

Remark 2.7. Sometimes the Stirling numbers are defined by the formulas (2) and (3) in this proposition.

Using (5.1) and (7) in this proposition, we give a proof of the formulas that were left unproved in the previous chapter (p. 17). We give a proof of

$$\log(1 + (e^t - 1)) = t.$$

Once this is proved, the converse

$$e^{\log(1+t)} - 1 = t$$

follows from Proposition 1.10. (Of course, we can prove it similarly using (9) in this proposition.)

$$\begin{aligned}
\log(1 + (e^t - 1)) &= \sum_{m=1}^{\infty} (-1)^{m-1} \frac{(e^t - 1)^m}{m} \\
&= \sum_{m=1}^{\infty} (-1)^{m-1} (m-1)! \sum_{n=0}^{\infty} \left\{ {n \atop m} \right\} \frac{t^n}{n!} \\
&= \sum_{n=0}^{\infty} \left(\sum_{m=1}^{\infty} (-1)^{m-1} \left[{m \atop 1} \right] \left\{ {n \atop m} \right\} \right) \frac{t^n}{n!} \quad \left((m-1)! = \left[{m \atop 1} \right] \right) \\
&= \sum_{n=0}^{\infty} (-1)^{n-1} \delta_{1,n} \frac{t^n}{n!} \\
&= t.
\end{aligned}$$

2.2 Formulas for the Bernoulli Numbers Involving the Stirling Numbers

A formula expressing the Bernoulli numbers as a finite sum involving the Stirling numbers of the second kind has been known at least since Kronecker[2] [62]. (See also an account of von Staudt in Sect. 3.3.) By Proposition 2.6 (6), the Stirling numbers can be expressed as a finite sum involving only the binomial coefficients. Therefore, the Bernoulli numbers can be expressed as an elementary (i.e., involving only binomial coefficients and simple polynomials) double sum. Gould [36] studied various historical sources for analogous formulas, and he concluded that, at least until around 1970, the very fact that there are closed finite formulas (without involving infinite sums and integrals) for the Bernoulli numbers was not widely known. He also conjectures that there is no "elementary" (in the above sense)

[2]Leopold Kronecker (born on December 7, 1823 in Liegnitz, Prussia (now Legnica, Poland)—died on December 29, 1891 in Berlin, Germany).

formula for the Bernoulli numbers expressed as a finite *single* sum. As far as we know, such a formula has not been discovered.

We prove the following formula.

Theorem 2.8.

$$B_n = (-1)^n \sum_{m=0}^{n} \frac{(-1)^m m! \left\{ {n \atop m} \right\}}{m+1} \qquad (n \geq 0).$$

Proof. Rewriting the generating function of B_n as

$$\frac{te^t}{e^t - 1} = \frac{t}{1 - e^{-t}} = \frac{-\log(1 - (1 - e^{-t}))}{1 - e^{-t}},$$

and substituting $1 - e^{-t} (= t + \cdots)$ for t in $-\log(1 - t) = \sum_{m=1}^{\infty} \frac{t^m}{m}$, we have

$$\begin{aligned}
\sum_{n=0}^{\infty} B_n \frac{t^n}{n!} &= \frac{-\log(1 - (1 - e^{-t}))}{1 - e^{-t}} \\
&= \sum_{m=1}^{\infty} \frac{(1 - e^{-t})^{m-1}}{m} \\
&= \sum_{m=0}^{\infty} \frac{(-1)^m (e^{-t} - 1)^m}{m+1} \\
&= \sum_{m=0}^{\infty} \frac{(-1)^m m!}{m+1} \sum_{n=m}^{\infty} \left\{ {n \atop m} \right\} \frac{(-t)^n}{n!} \qquad \text{(Proposition 2.6 (7))} \\
&= \sum_{n=0}^{\infty} (-1)^n \left(\sum_{m=0}^{n} \frac{(-1)^m m! \left\{ {n \atop m} \right\}}{m+1} \right) \frac{t^n}{n!}.
\end{aligned}$$

Comparing the coefficients of $\frac{t^n}{n!}$ on both sides, we obtain the formula. □

Remark 2.9. By Proposition 2.6 (6), we can also write

$$B_n = (-1)^n \sum_{m=0}^{n} \frac{1}{m+1} \sum_{l=0}^{m} (-1)^l \binom{m}{l} l^n.$$

Proposition 2.10 (A variation of Theorem 2.8).

$$B_n = \sum_{m=0}^{n} \frac{(-1)^m m! \left\{ {n+1 \atop m+1} \right\}}{m+1} \qquad (n \geq 0).$$

Proof. We can prove this by computing the generating function on the right-hand side. Here, we use Theorem 2.8. From the recurrence formula (2.1) for the Stirling numbers we have $\left\{ {n+1 \atop m+1} \right\} = \left\{ {n \atop m} \right\} + (m+1)\left\{ {n \atop m+1} \right\}$, and thus

$$\text{r.h.s.} = \sum_{m=0}^{n} \frac{(-1)^m m! \left(\left\{ {n \atop m} \right\} + (m+1)\left\{ {n \atop m+1} \right\} \right)}{m+1}.$$

On the other hand, since $\left[{m+1 \atop 1} \right] = m!$, it follows from Proposition 2.6 (5.1) that

$$\sum_{m=0}^{n} \frac{(-1)^m m! (m+1) \left\{ {n \atop m+1} \right\}}{m+1} = \sum_{m=0}^{n} (-1)^m \left[{m+1 \atop 1} \right] \left\{ {n \atop m+1} \right\} = \delta_{1,n}.$$

Hence, the right-hand side of the proposition equals

$$\sum_{m=0}^{n} \frac{(-1)^m m! \left\{ {n \atop m} \right\}}{m+1} + \delta_{1,n} = (-1)^n B_n + \delta_{1,n}.$$

(Theorem 2.8.) For $n = 1$, this formula equals $-B_1 + 1 = \frac{1}{2} = B_1$. For $n \neq 1$, it equals $(-1)^n B_n$, which coincides with B_n due to Proposition 1.4 on p. 10. □

This proposition can be interpreted as the following algorithm, which first appeared in the study of multiple zeta values at non-positive integers by S. Akiyama and Y. Tanigawa [2]. First, define the 0th row by $1, \frac{1}{2}, \frac{1}{3}, \frac{1}{4}, \frac{1}{5}, \ldots$. (We count the number starting from 0 instead of 1 for convenience.) Then, we define the first row by $1 \cdot (1 - \frac{1}{2}), 2 \cdot (\frac{1}{2} - \frac{1}{3}), 3 \cdot (\frac{1}{3} - \frac{1}{4}), \ldots$. In general, if we denote by $a_{n,m}$ the mth entry of the nth row, define the mth entry of the $(n+1)$st row $a_{n+1,m}$ by $(m+1) \cdot (a_{n,m} - a_{n,m+1})$. In other words, we take the difference of the adjacent entries and multiply by (position of the entry +1), and place them in the next row. Then, the leftmost entries of the rows, $1, \frac{1}{2}, \frac{1}{6}, 0, -\frac{1}{30}, \ldots$, are nothing but the Bernoulli numbers.

Taking Proposition 2.10 into account, it suffices to show the following proposition to prove this fact.

Proposition 2.11. *Given* $a_{0,m}$ $(m = 0, 1, 2, \ldots)$, *define* $a_{n,m}$ $(n \geq 1)$ *by*

$$a_{n,m} = (m+1) \cdot (a_{n-1,m} - a_{n-1,m+1}) \quad (n \geq 1, m \geq 0).$$

Then,

$$a_{n,0} = \sum_{m=0}^{n} (-1)^m m! \left\{ {n+1 \atop m+1} \right\} a_{0,m}.$$

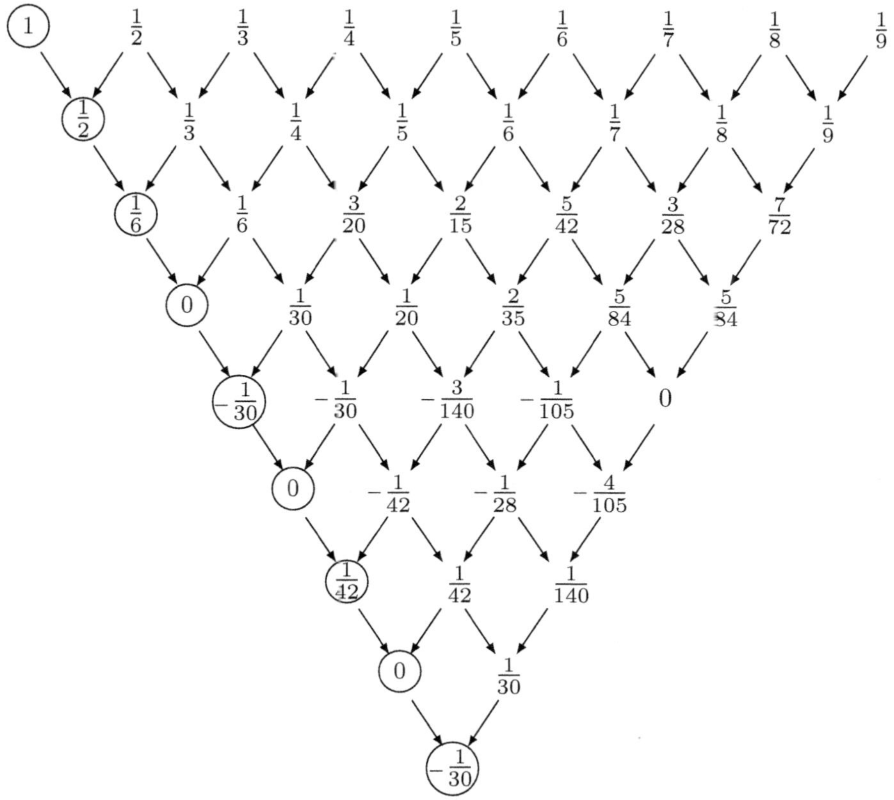

Proof. We use the generating function

$$g_n(t) = \sum_{m=0}^{\infty} a_{n,m} t^m.$$

By the recurrence formula for $a_{n,m}$, we have for $n \geq 1$

$$\begin{aligned} g_n(t) &= \sum_{m=0}^{\infty} (m+1)(a_{n-1,m} - a_{n-1,m+1}) t^m \\ &= \frac{d}{dt}\Big(\sum_{m=0}^{\infty} a_{n-1,m} t^{m+1}\Big) - \frac{d}{dt}\Big(\sum_{m=0}^{\infty} a_{n-1,m+1} t^{m+1}\Big) \\ &= \frac{d}{dt}(t g_{n-1}(t)) - \frac{d}{dt}(g_{n-1}(t) - a_{n-1,0}) \end{aligned}$$

$$= g_{n-1}(t) + (t-1)\frac{d}{dt}(g_{n-1}(t))$$
$$= \frac{d}{dt}((t-1)g_{n-1}(t)).$$

Thus, if we put $(t-1)g_n(t) = h_n(t)$, we have

$$h_n(t) = (t-1)\frac{d}{dt}(h_{n-1}(t)) \quad (n \geq 1).$$

From this we obtain

$$h_n(t) = \left((t-1)\frac{d}{dt}\right)^n (h_0(t)).$$

It follows from Proposition 2.6 (4) (with x replaced by $t-1$) that

$$h_n(t) = \sum_{m=0}^{n} \left\{ {n \atop m} \right\} (t-1)^m \left(\frac{d}{dt}\right)^m h_0(t).$$

Putting $t = 0$, we have

$$\begin{aligned}
-a_{n,0} &= \sum_{m=0}^{n} \left\{ {n \atop m} \right\} (-1)^m m!(a_{0,m-1} - a_{0,m}) \\
&= \sum_{m=0}^{n-1} \left\{ {n \atop m+1} \right\} (-1)^{m+1}(m+1)!a_{0,m} - \sum_{m=0}^{n} \left\{ {n \atop m} \right\} (-1)^m m!a_{0,m} \\
&= -\sum_{m=0}^{n} (-1)^m m!a_{0,m} \left((m+1)\left\{ {n \atop m+1} \right\} + \left\{ {n \atop m} \right\} \right) \\
&= -\sum_{m=0}^{n} (-1)^m m! \left\{ {n+1 \atop m+1} \right\} a_{0,m}.
\end{aligned}$$

This proves the formula in the proposition. □

Remark 2.12. We note that, if we use instead the algorithm

$$a_{n+1,m} = (m+1)\cdot(a_{n,m+1} - a_{n,m}),$$

we obtain Bernoulli numbers with the other convention ($B_1 = \frac{1}{2}$).

Exercise 2.13. Prove the following.

(1) $\left\{ {n \atop m} \right\} = 0$ when $mn < 0$.

(2) $\left\{ {-1 \atop -m} \right\} = (m-1)!$ for $m = 1, 2, 3, \ldots$.

(3) $\left\{ {n+1 \atop n} \right\} = \left\{ {-n \atop -n-1} \right\} = \frac{n(n+1)}{2}$ for $n = 0, 1, 2, \ldots$.

Exercise 2.14. Formulate the statements corresponding to the previous exercise for the Stirling numbers of the first kind, and prove them.

Exercise 2.15. Prove the formula $e^{\log(1+t)} - 1 = t$ for the composition of formal power series by using Proposition 2.6 (9).

Exercise 2.16. Prove the following formula for the sum of powers:

$$\sum_{i=1}^{n} i^k = \sum_{j=0}^{n} j! \left\{ {k \atop j} \right\} \binom{n+1}{j+1}.$$

Exercise 2.17. Compute the first several B_n using the formula in Theorem 2.8.

Exercise 2.18. In the Akiyama–Tanigawa triangle (on p. 37), what are the numbers next to the leftmost entries?

Chapter 3
Theorem of Clausen and von Staudt, and Kummer's Congruence

3.1 Theorem of Clausen and von Staudt

The denominators of the Bernoulli numbers can be completely determined. This is due to Clausen[1] [26] and von Staudt[2] [96]. More precisely, the "fractional part" of B_n is given by the following theorem. This result gives a foundation for studying p-adic properties of the Bernoulli numbers. It also plays a fundamental role in the theory of p-adic modular forms through the Eisenstein[3] series [82]

Theorem 3.1. *For $n = 1$ and for any even integer $n \geq 2$, B_n can be written as*

$$B_n = -\sum_{\substack{p:\text{prime} \\ p-1 \mid n}} \frac{1}{p} + C_n \quad (C_n \text{ is an integer}).$$

Here, the sum runs over all the prime numbers p such that $p - 1$ divides n. (For integers a and b, the symbol $a \mid b$ means that a divides b, and $a \nmid b$ means that a does not divide b.)

Proof. The statement is clearly true for $n = 1$. Let n be an even integer greater than or equal to 2. Since the numerator of each term $(-1)^m m! \left\{ {n \atop m} \right\} / (m + 1)$ in the right-hand side of the formula in Theorem 2.8 is an integer, the only prime numbers

[1] Thomas Clausen (born on January 16, 1801 in Snogebaek, Denmark—died on May 23, 1885 in Dorpat, Russia (now Tartu, Estonia)).

[2] Karl Georg Christian von Staudt (born on January 24, 1798 in Imperial Free City of Rothenburg (now Rothenburg ob der Tauber, Germany)—died on June 1, 1867 in Erlangen, Bavaria (now Germany).

[3] Ferdinand Gotthold Max Eisenstein (born on April 16, 1823 in Berlin, Germany—died on October 11, 1852 in Berlin, Germany).

T. Arakawa et al., *Bernoulli Numbers and Zeta Functions*, Springer Monographs in Mathematics, DOI 10.1007/978-4-431-54919-2_3,

that contribute to the denominator of B_n are divisors of $m+1$. It follows from this that a prime number p such that $p > n+1$ cannot be a divisor of the denominator of B_n.

First suppose that $m+1$ is a composite number, and let $m+1 = ab$, $1 < a, b < m$. If $a \neq b$, then ab divides $m!$, and thus $(-1)^m m!\left\{{n \atop m}\right\}/(m+1)$ is an integer. If $a = b$ and $2a \le m$, then a and $2a$ divide $m!$, and thus $a^2 = m+1$ divides $m!$. This implies that $(-1)^m m!\left\{{n \atop m}\right\}/(m+1)$ is an integer. If $a = b$ and $2a > m$, then $2a \ge m+1$, and thus $m+1 = a^2 \ge 2a \ge m+1$. This implies $a^2 = 2a$, or $a = 2$. In this case $m = 3$, and this term appears when $n \ge 3$. In this case (n even), from Proposition 2.6 (6) on p. 30 we have

$$\begin{aligned}(-1)^m m!\left\{{n \atop m}\right\} &= \sum_{l=0}^{3}(-1)^l\binom{3}{l}l^n \\ &= 0 - 3 + 3\cdot 2^n - 3^n \\ &\equiv 1 - (-1)^n \equiv 0 \mod 4.\end{aligned}$$

(For integers a, b and a natural number[4] n, if $n|(a-b)$, then we say that a is congruent to b modulo n, and we write $a \equiv b \mod n$. If $n \nmid (a-b)$, then we write $a \not\equiv b \mod n$.) This shows that $(-1)^m m!\left\{{n \atop m}\right\}/(m+1)$ is an integer. We have thus shown that if $m+1$ is a composite number, $(-1)^m m!\left\{{n \atop m}\right\}/(m+1)$ is an integer.

Next, suppose that $m+1$ equals a prime number p. Since $p-1 \le n$, it follows from Proposition 2.6 (6) that

$$\begin{aligned}(-1)^m m!\left\{{n \atop m}\right\} &= \sum_{l=0}^{p-1}(-1)^l\binom{p-1}{l}l^n \\ &\equiv \sum_{l=0}^{p-1} l^n \mod p \quad \left(\text{since } \tbinom{p-1}{l} \equiv (-1)^l \mod p\right) \\ &\equiv \begin{cases} -1 \mod p & \text{if } p-1 \text{ divides } n, \\ 0 \mod p & \text{if } p-1 \text{ does not divide } n. \end{cases} \qquad (3.1)\end{aligned}$$

The last congruence can be seen as follows. If $p-1$ divides n, then the fact that $\sum_{l=0}^{p-1} l^n \equiv p-1 \equiv -1 \mod p$ follows from Fermat's[5] Little Theorem; i.e., $l^{p-1} \equiv 1 \mod p$ if l and p are relatively prime. If $p-1$ does not divide n, then choose c such that $c^n \not\equiv 1 \mod p$ (for example, a primitive root mod p). When l runs from 1 through $p-1$, $cl \mod p$ also runs from the classes of 1 through $p-1$, and thus we have $\sum_{l=0}^{p-1}(cl)^n \equiv \sum_{l=0}^{p-1} l^n$. Therefore, we have

[4] Throughout the book, the term "natural number" means a positive integer.

[5] Pierre de Fermat (born on August 17, 1601 in Beaumont-de-Lomagne, France—died on January 12, 1665 in Castres, France).

$(c^n - 1)\sum_{l=0}^{p-1} l^n \equiv 0 \mod p$, and $\sum_{l=0}^{p-1} l^n \equiv 0 \mod p$. We thus see that if $m + 1 = p$, $(-1)^m m!\left\{{n \atop m}\right\}/(m+1)$ is an integer if $p - 1$ does not divide n. If, on the other hand, $p - 1$ divides n, then $(-1)^m m!\left\{{n \atop m}\right\}/(m+1)$ is not an integer but the above congruence shows that it becomes an integer if we add $\frac{1}{p}$ to it. This completes the proof. □

The theorem of Clausen and von Staudt completely describes the denominators of Bernoulli numbers. Then how about the numerators? Not much is known about this. Here we briefly explain the notion of (ir)regular primes, which are related to the numerators of Bernoulli numbers.

A prime p is said to be *regular* if p does not divide any of the numerators of the Bernoulli numbers $B_2, B_4, \ldots, B_{p-3}$. Otherwise p is called *irregular*. According to Kummer, p is regular if and only if the class number of the pth cyclotomic field $\mathbf{Q}(\zeta_p)$ is not divisible by p (cf. Washington [100, §5.3]). It is well known that this work of Kummer originated from his attempt to solve Fermat's last theorem. For this topic, see [31] or [79].

For instance, there are only three irregular primes less than 100: $37, 59, 67$. Kummer determined (by hand!) all irregular primes less than or equal to 163, and moreover determined the indices of B_n whose numerators are divisible by a given irregular prime. The next few irregular primes after 67 are $101, 103, 131, 149, 157$, and $p = 157$ is the first prime which divides two of the numerators of $B_2, B_4, \ldots, B_{p-3}$ (see the table in [100] for irregular primes p less than 4,001 and indices of Bernoulli numbers divisible by p).

3.2 Kummer's Congruence

The goal of this section is to prove the following theorem, which plays a crucial role in the arithmetic of cyclotomic fields and the theory of p-adic L-functions. For a prime number p, we denote by $\mathbf{Z}_{(p)}$ the ring of rational numbers whose denominator is prime to p. Using such a ring, it is easy to describe how many times p divides the numerator of a rational number, or decide whether or not two rational numbers are congruent modulo a power of p. The set of invertible elements, $\mathbf{Z}_{(p)}^{\times}$, of the ring $\mathbf{Z}_{(p)}$ is the set of rational numbers whose numerator and denominator are both relatively prime to p. For two elements x and y, we say that the congruence relation $x \equiv y \mod p^a$ holds if the numerator of $x - y$ is divisible by p^a.

Theorem 3.2. *Let p be an odd prime.*

(1) *Suppose n is a positive even integer not divisible by $p - 1$. Then we have*

$$\frac{B_n}{n} \in \mathbf{Z}_{(p)}.$$

(2) *Let $a \geq 1$ be a natural number and m, n be two even integers that satisfy $a + 1 \leq m \leq n$. Suppose that m and n are not divisible by $p - 1$, and that $n \equiv m \mod (p-1)p^{a-1}$. Then we have*

$$\frac{B_n}{n} \equiv \frac{B_m}{m} \mod p^a.$$

If we remove the condition $m \geq a + 1$, then the above congruence relation does not hold in general (except for the case $a = 1$, where $m \geq a + 1$ gives no restriction) and a certain correction is necessary (see Theorem 11.6 on p. 198 for a more precise statement). This corrected version is what is known today as Kummer's congruence. However, the version that Kummer[6] himself gave was different (except when $a = 1$) from these theorems we are presenting (see [64]). The formulation of Kummer's congruence seems to have been changed since a natural interpretation of the congruence by p-adic integrals was discovered. Kummer's own proof is elementary, and we can derive the above congruence relation using his method. The complete form of the congruence will be proved using a p-adic integral in Chap. 11.

We first prove the following lemma.

Lemma 3.3. *Let p be an odd prime number, and let $\varphi(t) \in \mathbf{Z}_{(p)}[[t]]$, $r, s \in \mathbf{Z}_{(p)}$. Develop the formal power series obtained by substituting $e^{rt} - e^{st} = (r-s)t + \frac{(r^2-s^2)}{2}t^2 + \cdots$ for t in $\varphi(t)$:*

$$\varphi(e^{rt} - e^{st}) = \sum_{n=0}^{\infty} A_n \frac{t^n}{n!}.$$

Then we have $A_n \in \mathbf{Z}_{(p)}$, and for any m, a $(m \geq a \geq 1)$, the congruence

$$A_{m+(p-1)p^{a-1}} \equiv A_m \mod p^a$$

holds.

Proof. Write $\varphi(t) = \sum_{k=0}^{\infty} a_k t^k$. Then, we have

$$\begin{aligned}
\varphi(e^{rt} - e^{st}) &= \sum_{k=0}^{\infty} a_k (e^{rt} - e^{st})^k \\
&= \sum_{k=0}^{\infty} a_k \sum_{h=0}^{k} (-1)^h \binom{k}{h} e^{rt(k-h)} e^{sth} \\
&= \sum_{k=0}^{\infty} \sum_{h=0}^{k} (-1)^h a_k \binom{k}{h} e^{(r(k-h)+sh)t}.
\end{aligned}$$

---

[6]Ernst Eduard Kummer (born on January 29, 1810 in Sorau, Brandenburg, Prussia (now Germany)—died on May 14, 1893 in Berlin, Germany).

We obtain A_m by taking the derivative m times with respect to t, and putting $t = 0$:

$$A_m = \sum_{k=0}^{\infty}\sum_{h=0}^{k}(-1)^h a_k \binom{k}{h}(r(k-h)+sh)^m.$$

Since $(e^{rt} - e^{st})^k$ begins with the term t^k, the sum with respect to h in the above formula equals 0 if $k > m$. Thus, the above sum with respect to k is a finite sum.

Replacing m by $m + (p-1)p^{a-1}$, and subtracting A_m, we have

$$\begin{aligned} &A_{m+(p-1)p^{a-1}} - A_m \\ &= \sum_{k=0}^{\infty}\sum_{h=0}^{k}(-1)^h a_k \binom{k}{h}(r(k-h)+sh)^m((r(k-h)+sh)^{(p-1)p^{a-1}} - 1). \end{aligned}$$

If $r(k-h)+sh$ is divisible by p, then since $m \geq a$, $(r(k-h)+sh)^m$ is divisible by p^a. If $r(k-h)+sh$ is not divisible by p, then $(r(k-h)+sh)^{(p-1)p^{a-1}} - 1$ is divisible by p^a since the order of $(\mathbf{Z}/p^a\mathbf{Z})^\times$ is $(p-1)p^{a-1}$. In either case the right-hand side is divisible by p^a, and thus

$$A_{m+(p-1)p^{a-1}} \equiv A_m \mod p^a$$

holds. □

Proof (of Theorem 3.2). For a natural number $c \neq 1$ relatively prime to p, consider

$$\varphi(t) = \frac{1}{t} - \frac{c}{(1+t)^c - 1}.$$

Since $c \in \mathbf{Z}_{(p)}^\times$, the expression

$$\frac{(1+t)^c - 1}{c} = t + \frac{c-1}{2}t^2 + \cdots$$

belongs to $\mathbf{Z}_{(p)}[[t]]$, and it is invertible in $\mathbf{Z}_{(p)}((t))$ (at the end of Sect. 1.3 on p. 20). From

$$\frac{c}{(1+t)^c - 1} = \frac{1}{t} - \frac{c-1}{2} + \frac{c^2-1}{12}t + \cdots,$$

we have $\varphi(t) \in \mathbf{Z}_{(p)}[[t]]$. Since

$$\frac{t}{e^t - 1} = \frac{te^t}{e^t - 1} - t = 1 - \frac{t}{2} + \sum_{n=2}^{\infty} B_n \frac{t^n}{n!},$$

we have

$$\begin{aligned}\varphi(e^t-1) &= \frac{1}{e^t-1}-\frac{c}{e^{ct}-1}=\frac{1}{t}\left(\frac{t}{e^t-1}-\frac{ct}{e^{ct}-1}\right)\\ &= -\frac{1-c}{2}+\sum_{n=2}^{\infty}\left((1-c^n)\frac{B_n}{n}\cdot\frac{t^{n-1}}{(n-1)!}\right).\end{aligned}$$

Thus it follows from the lemma ($r=1, s=0$) that $(1-c^n)\frac{B_n}{n}\in\mathbf{Z}_{(p)}$, and if $n\equiv m \mod (p-1)p^{a-1}$, then

$$(1-c^n)\frac{B_n}{n}\equiv(1-c^m)\frac{B_m}{m} \mod p^a.$$

Here, we choose c to be a primitive root modulo p ($c \mod p$ is a generator of $(\mathbf{Z}/p\mathbf{Z})^\times$). If n is not divisible by $p-1$, then m is not divisible by $p-1$, and we have $1-c^n$, $1-c^m\in(\mathbf{Z}/p^a\mathbf{Z})^\times$ with $1-c^n\equiv 1-c^m \mod p^a$. This gives $\frac{B_n}{n}, \frac{B_m}{m}\in\mathbf{Z}_{(p)}$ and the congruence in the theorem. □

3.3 Short Biographies of Clausen, von Staudt and Kummer

Thomas Clausen, an astronomer and a mathematician, was born on January 16, 1801 in Snogbæk, Denmark.[7] He was educated by a local pastor Georg Holst before he got a position as an assistant at the Altona observatory, where Schumacher[8] was the head. He then moved to the Joseph von Utzschneider Optical Institute in Munich as the successor to Fraunhofer,[9] who is famous for "Fraunhofer lines" in physics and optics. In 1842, he was appointed observer at the Dorpat observatory in Russia, and there he remained until his retirement in 1872. The Copenhagen Academy awarded Clausen a prize for his work on the determination of the orbit of the 1770 comet (published in *Astronomische Nachrichten* in 1842). Bessel[10] highly praised this

[7]The description in this section is based on Biermann [17, 18], Noether [74], Hensel [40], and Lampe [65].

[8]Heinrich Christian Schumacher (born on September 3, 1780 in Bramstedt, Germany—died in 1850). His first degree was in law, and after that he studied astronomy under Gauss and became an astronomer. He launched the journal *Astronomische Nachrichten*, which is famous for the contributions of Abel and Jacobi on the theory of elliptic functions, as the managing editor. Incidentally, Abel once met Clausen during his stay in Hamburg (cf. [19]).

[9]Joseph von Fraunhofer (born on March 6, 1787 in Straubing, Germany—died on June 7, 1826 in Munich, Germany).

[10]Friedrich Wilhelm Bessel (born on July 22, 1784 in Minden, Westphalia (now Germany)—died on March 17, 1846 in Königsberg, Prussia (now Kaliningrad, Russia)), an astronomer, who is famous for the Bessel function.

work. In 1854 he received through Gauss[11] a corresponding membership from the Göttingen Academy. In 1856 he received the same class of membership from the St. Petersburg Academy. Clausen published approximately 150 papers on a multitude of subjects. Still today one sees his name in such terms as Clausen's identity[12] in the theory of hypergeometric series [23], and Clausen's function (a variant of the dilogarithm function) [24]. Also, he computed π up to 250 decimal places,[13] and found the prime factorization of the sixth Fermat number $2^{2^6}+1$.[14] The actual factorization is $2^{2^6}+1 = 274177 \cdot 67280421310721$ and, as Clausen commented, the second factor of this is the largest prime number known to that day (letter to Gauss on January 1, 1855). He wrote to his friend and astronomer C. A. F. Peters in Königsberg that he had found a new method of factoring a large number, and as an example he wrote $(10^{17}-1)/9 = 2071723 \cdot 5363222357$, which was not known before and had taken some interest in connection to the period length of the periodic decimal of $1/p$. However, he never published his "new method" and we have no idea what kind of method he had found.

Clausen never married, and died on May 23, 1885 in Dorpat, now Tartu in Estonia.

Karl Georg Christian von Staudt was born on January 24, 1798 in Rothenburg-ob-der-Tauber, Germany, as a son of a wealthy family. From 1818 to 1822, von Staudt studied mathematics at Göttingen under Gauss. He was also interested in astronomy (at that time Gauss was also the head of the observatory), and, as early as in 1820, he computed the positions of Mars and the asteroid Pallas, and in 1821 he computed the orbit of a comet. The accuracy of the computation was highly praised (described as an "outstanding proficiency") by his master Gauss. He had been holding positions as a high-school teacher at Würzburg and Nürnberg until he was appointed professor at Erlangen in 1835, where he remained throughout his life.

His main contribution in mathematics is in *Geometrie der Lage* (1847, *Geometry of Position*), which is the title of his book and is now called projective geometry. As for Bernoulli numbers, he discovered not only the theorem introduced in Sect. 3.1 [96, found independently of Clausen], but also, according to Noether[15] [74], several other results including the formula in Theorem 2.8 (p. 35) and even Kummer's

---

[11]Johann Carl Friedrich Gauss (born on April 30, 1777 in Brunswick, Duchy of Brunswick (now Germany)—died on February 23, 1855 in Göttingen, Hanover (now Germany)).

[12]$_2F_1(\alpha,\beta,\alpha+\beta+\frac{1}{2};x)^2 = {}_3F_2\left(\begin{matrix}2\alpha,2\beta,\alpha+\beta\\ \alpha+\beta+\frac{1}{2},2\alpha+2\beta\end{matrix};x\right)$

[13]It was correct up to 248 decimal places. This was the world record from 1847 to 1853. For a history of computation of π, see for example [13].

[14] Fermat conjectured that all the numbers of the form $2^{2^n}+1$ (which are called Fermat numbers) are prime. This is true for $n \le 4$, but Euler showed in 1732 that this is not the case for $n = 5$ by giving the factorization. As of November 2013, it is known that every Fermat number $2^{2^n}+1$ with $5 \le n \le 32$ is not prime. (Complete factorization is known up to $n = 11$.)

[15]Max Noether (born on September 24, 1844 in Mannheim, Germany—died on December 13, 1921 in Erlangen, Germany).

congruence (modulo prime number case). He presented these results in two papers[16] submitted to Erlangen University on the occasion of becoming a member of the senate and the faculty. These results, however, caught no attention because, as Noether put it, von Staudt was so modest and did not care about the circulation of the papers (cf. [88] for the contents of these two papers). He also published a paper [97] on a proof of the "fundamental theorem on algebra" of Gauss (Clausen also wrote a paper on the same subject [25]). He wrote up a paper one week before his death, while suffering from asthma. He died on June 1, 1867. Noether quotes from the funeral address the passage "his endurance, prepared for the lifelong effort, not for months or years, to pursue a distinct goal", which he thinks describes well the character of von Staudt.

Ernst Eduard Kummer was born on January 29, 1810 in Sorau, Germany (now Zary, Poland). Kummer entered the Gymnasium in Sorau in 1819 and the University of Halle in 1828. He soon gave up his original study, Protestant theology, under the influence of the mathematics professor Scherk[17] and applied himself to mathematics. In 1831 he received a prize for his essay on the question posed by Scherk: "De cosinuum et sinuum potestatibus secundum cosinus et sinus arcuum multiplicium evolvendis".[18] Kummer taught from 1832 until 1842 at the Gymnasium in Liegnitz (now Legnia, Poland), mainly mathematics and physics. His students during this period included Kronecker and Joachimsthal,[19] both of whom became interested in mathematics through Kummer's encouragement and stimulation. In particular, the influence of Kummer upon Kronecker was enormous and they became lifelong friends. One can see the fragments of their friendship in their correspondence collected in Kummer's *Werke*.

The subjects of Kummer's mathematical study are rather clearly divided according to the periods. The first is in function theory, the second in number theory and the third in geometry. His period of Gymnasium teaching coincided with his creative period in function theory. The most important was the famous paper on the hypergeometric series [63]. Kummer sent this paper to Jacobi, which led to his scientific connection with Jacobi and with Dirichlet.[20] In 1839, through Dirichlet's proposal, Kummer became a member of the Berlin Academy of Sciences, and in 1840, he married Ottilie Mendelssohn, a cousin of Dirichlet's wife.[21] (She died in 1848 and later Kummer married Bertha Cauer.) On the recommendation of Dirichlet and Jacobi, Kummer was appointed full professor at the University of Breslau (now

---

[16]Under the same title "De numeris Bernoullianis", Erlangen, 1845.

[17]Heinrich Ferdinand Scherk (born on October 27, 1798 in Poznan, Poland—died on October 4, 1885 in Bremen, Germany), he also wrote a paper on Bernoulli numbers [81].

[18]"On expansions of powers of cosine and sine by cosine and sine with their arguments multiplied."

[19]Ferdinand Joachimsthal (born on March 9, 1818 in Goldberg, Prussian Silesia (now Zlotoryja, Poland)—died on April 5, 1861 in Breslau, Germany (now Wroclaw, Poland)).

[20]Johann Peter Gustav Lejeune Dirichlet (born on February 13, 1805 in Düren, French Empire (now Germany)—died on May 5, 1859 in Göttingen, Hanover (now Germany)).

[21]Dirichlet's wife Rebecca Mendelssohn is a younger sister of the composer Felix Mendelssohn.

Wroclaw, Poland) in 1842. His second period of research, dominated especially by number theory and lasting approximately 20 years, began about this time.

In 1855, Gauss died and Dirichlet left Berlin to succeed him at Göttingen. In the same year Kummer was appointed professor at Berlin to succeed Dirichlet.

In his third period, starting from around 1860, he studied geometry, his most famous work being the study of the quartic surface known today as the Kummer surface. (He also published papers on the arithmetic of cyclotomic fields until the mid 1870s.)

Kummer's popularity as a professor, based not only on the clarity of his lectures but on his charm and sense of humor as well, attracted a great number of students. Kummer was first *Gutachter* for 39 dissertations at Berlin. Of his doctoral students, seventeen later became university teachers, several of them famous mathematicians: Fuchs,[22] du Bois-Reymond,[23] Gordan,[24] Bachmann,[25] Schwarz[26] (Kummer's son in law), Cantor,[27] Schönflies.[28] Kummer was also second *Gutachter* for thirty dissertations at Berlin.

He retired in 1883 and was succeeded by Fuchs. Kummer died on May 14 in 1893, 10 years after his retirement.

Exercise 3.4. Compute the integer C_n in Theorem 3.1 for several n. What is the first n such that $C_n \neq 1$? (Note that $\lim_{n\to\infty} |C_n| = \infty$ because C_n is very close to B_n.)

Exercise 3.5. Suppose n is even and $n \geq 4$. Prove the following.

(1) $2B_n \in \mathbf{Z}_{(2)}$.
(2) The congruence $2B_n \equiv 1 \bmod 4$ holds. Hint: Use the formula in Theorem 2.8 and look at each term on the right modulo 4.

Exercise 3.6. Use Theorems 3.1 and 3.2 (1) to prove that the tangent number T_n at the end of Chap. 1 is an integer.

---

[22] Lazarus Immanuel Fuchs (born on May 5, 1833 in Moschin, Prussia (now Poznan, Poland)—died on April 26, 1902 in Berlin, Germany).

[23] Paul David Gustav du Bois-Reymond (born on December 2, 1831 in Berlin, Germany—died on April 7, 1889 in Freiburg, Germany).

[24] Paul Albert Gordan (born on April 27, 1837 in Breslau, Germany (now Wroclaw, Poland)—died on December 21, 1912 in Erlangen, Germany).

[25] Paul Gustav Heinrich Bachmann (born on June 22, 1837 in Berlin, Germany—died on March 31, 1920 in Weimar, Germany).

[26] Karl Herman Amandus Schwarz (born on January 25, 1843 in Hermsdorf, Silesia (now Poland)—died on November 30, 1921 in Berlin, Germany).

[27] Georg Ferdinand Ludwig Philipp Cantor (born on March 3, 1845 in St. Petersburg, Russia—died on January 6, 1918 in Halle, Germany).

[28] Arthur Moritz Schönflies (born on April 17, 1853 in Landsberg an der Warthe, Germany (now Gorzów, Poland)—died on May 27, 1928 in Frankfurt am Main, Germany).

Chapter 4
Generalized Bernoulli Numbers

In this chapter we introduce generalized Bernoulli numbers and Bernoulli polynomials. Generalized Bernoulli numbers are Bernoulli numbers twisted by a Dirichlet character, which we define at the beginning of the first section. Bernoulli polynomials are generalizations of Bernoulli numbers with an indeterminate. These two generalizations are related, and they will appear in various places in the following chapters.

4.1 Dirichlet Characters

Let us define a Dirichlet character as a map from the set of integers $\mathbf{Z}$ to the set of complex numbers $\mathbf{C}$.

Definition 4.1. Let f be a natural number. A function $\chi : \mathbf{Z} \longrightarrow \mathbf{C}$ is called a Dirichlet character modulo f if it satisfies the following three conditions:

(i) χ is multiplicative; i.e. $\chi(ab) = \chi(a)\chi(b)$ for any $a,\ b \in \mathbf{Z}$.
(ii) The value of $\chi(a)$ only depends on $a \bmod f$; i.e. $\chi(a + bf) = \chi(a)$ for any $a, b \in \mathbf{Z}$.
(iii) If $(a, f) = 1$, then $\chi(a) \neq 0$, and if $(a, f) \neq 1$, then $\chi(a) = 0$. Here, (a, f) stands for the greatest common divisor of a and f.

In particular, the function χ_0 defined by

$$\chi_0(a) = \begin{cases} 1 & (a, f) = 1, \\ 0 & (a, f) \neq 1 \end{cases}$$

satisfies the three conditions. We call this the trivial character (modulo f).

T. Arakawa et al., *Bernoulli Numbers and Zeta Functions*, Springer Monographs in Mathematics, DOI 10.1007/978-4-431-54919-2_4, © Springer Japan 2014

For a natural number f, let $\varphi(f)$ be the number of integers from 1 to f which are relatively prime to f. This is called the Euler (totient) function. If a is an integer relatively prime to f, then we have

$$a^{\varphi(f)} \equiv 1 \bmod f.$$

This is known as Euler's theorem in elementary number theory. For a proof see for example Ireland and Rosen [50].

A Dirichlet character χ modulo f satisfies the following properties.

Lemma 4.2. (1) *If* $a \equiv 1 \bmod f$, *then* $\chi(a) = 1$.
(2) *If* $(a, f) = 1$, *then* $\chi(a)^{\varphi(f)} = 1$.
(3)

$$\sum_{a \bmod f} \chi(a) = \begin{cases} \varphi(f) & \text{if } \chi = \chi_0 \text{ (trivial character)}, \\ 0 & \text{if } \chi \neq \chi_0. \end{cases}$$

The summation is taken over all representatives a modulo f.

Proof. (1) Putting $b = 1$ in condition (i), we have $\chi(a) = \chi(a)\chi(1)$. If $a \equiv 1 \bmod f$, then we have $\chi(a) \neq 0$ from (iii). Thus, we have $\chi(1) = 1$, which in turn implies $\chi(a) = 1$ from (ii).

(2) Suppose $(a, f) = 1$. From Euler's theorem we have $a^{\varphi(f)} \equiv 1 \bmod f$. Thus,

$$\chi(a)^{\varphi(f)} = \chi(a^{\varphi(f)}) = \chi(1) = 1.$$

(3) Since $\varphi(f)$ is the number of $a \bmod f$ such that $(a, f) = 1$, the assertion holds if χ is the trivial character. If χ is not the trivial character, then there exists b such that $\chi(b) \neq 0, 1$. If we put $S = \sum_{a \bmod f} \chi(a)$, then we have $\chi(b)S = \sum_{a \bmod f} \chi(b)\chi(a) = \sum_{a \bmod f} \chi(ba)$. When a runs through all the representatives $\bmod f$, so does ba since $(b, f) = 1$. Thus, the sum on the right equals S. This implies $(1-\chi(b))S = 0$ and hence $S = 0$, as $1-\chi(b) \neq 0$. □

We note that a Dirichlet character defines naturally by definition a character of the multiplicative group $(\mathbf{Z}/f\mathbf{Z})^\times$.

Now suppose $f' \mid f$, and let χ' be a Dirichlet character modulo f'. If χ satisfies

$$\chi(a) = \chi'(a) \quad \text{if} \quad (a, f) = 1,$$

then we say that χ is defined $\bmod f'$. For a Dirichlet character χ modulo f, the smallest f' such that χ is defined $\bmod f'$ is called the conductor of χ, and is written as f_χ. By definition, f_χ is a divisor of f.

Definition 4.3 (primitive character). A Dirichlet character χ modulo f is called primitive if $f = f_\chi$.

Let χ be a Dirichlet character modulo f. Since $\chi(-1)^2 = \chi((-1)^2) = \chi(1) = 1$, we see that $\chi(-1) = \pm 1$.

A Dirichlet character with $\chi(-1) = 1$ is called an even character. A character with $\chi(-1) = -1$ is called an odd character. Given a Dirichlet character χ, we define the complex conjugate character $\overline{\chi}$ by

$$\overline{\chi}(a) = \overline{\chi(a)},$$

where $\overline{\chi(a)}$ stands for the complex conjugate of $\chi(a)$. We have $f_\chi = f_{\overline{\chi}}$.

Example 4.4. Define $\chi_j : \mathbf{Z} \longrightarrow \mathbf{C}$ $(j = 0, 1, 2, 3)$ as follows.
If $(a, 2) \neq 1$ (i.e. $2|a$), then $\chi_j(a) = 0$ $(j = 0, 1, 2, 3)$. When a is odd, for each a mod 8, $\chi(a)$ is given in the following table:

$\chi\backslash a$	1 mod 8	3 mod 8	5 mod 8	7 mod 8
$\chi_0(a)$	1	1	1	1
$\chi_1(a)$	1	1	-1	-1
$\chi_2(a)$	1	-1	-1	1
$\chi_3(a)$	1	-1	1	-1

Then, χ_j $(j = 0, 1, 2, 3)$ are all the Dirichlet characters modulo 8. The character χ_0 is the trivial character with $f_{\chi_0} = 1$ and χ_1, χ_2 are primitive characters modulo 8, but $\chi_3 (= \chi_1\chi_2)$ is not primitive, with $f_{\chi_3} = 4$.

4.2 Generalized Bernoulli Numbers

Given a Dirichlet character χ modulo f, the generalized Bernoulli numbers $B_{n,\chi}$ are defined by using the generating function

$$\sum_{a=1}^{f} \frac{\chi(a)te^{at}}{e^{ft} - 1} = \sum_{n=0}^{\infty} B_{n,\chi} \frac{t^n}{n!}.$$

In the following we assume $f > 1$ and χ is not the trivial character unless otherwise stated.

Since we have the expansion

$$\frac{te^{at}}{e^{ft} - 1} = \frac{1}{f} + \left(\frac{a}{f} - \frac{1}{2}\right) t + \left(\frac{a^2}{f} - a + \frac{f}{6}\right) \frac{t^2}{2} + \left(\frac{a^3}{f} - \frac{3a^2}{2} + \frac{af}{2}\right) \frac{t^3}{6} + \cdots,$$

it follows from Lemma 4.2 (3) that

$$B_{0,\chi} = 0,$$

$$B_{1,\chi} = \frac{1}{f}\sum_{a=1}^{f}\chi(a)a,$$

$$B_{2,\chi} = \frac{1}{f}\sum_{a=1}^{f}\chi(a)a^2 - \sum_{a=1}^{f}\chi(a)a,$$

$$B_{3,\chi} = \frac{1}{f}\sum_{a=1}^{f}\chi(a)a^3 - \frac{3}{2}\sum_{a=1}^{f}\chi(a)a^2 + \frac{f}{2}\sum_{a=1}^{f}\chi(a)a.$$

A similar formula for $B_{n,\chi}$ is given as

$$B_{n,\chi} = f^{n-1}\sum_{a=1}^{f}\chi(a)B_n\left(a/f\right), \tag{4.1}$$

where $B_n(x)$ is the Bernoulli polynomial, which will be defined in the next section. (See Remark 4.10 (2) in the next section.)

We have seen that all the Bernoulli numbers with odd indices greater than 1 are 0. For generalized Bernoulli numbers we have the following.

Proposition 4.5. *Let χ be a non-trivial character. Then, for any n satisfying $(-1)^{n-1} = \chi(-1)$, we have $B_{n,\chi} = 0$. In other words, if χ is an even character, then $B_{n,\chi}$ with odd indices n are 0; if χ is an odd character, then $B_{n,\chi}$ with even indices n are 0.*

Proof. Since χ is non-trivial, we can rewrite the generating function as follows:

$$\begin{aligned}\sum_{a=1}^{f-1}\frac{\chi(a)te^{at}}{e^{ft}-1} &= \sum_{a=1}^{f-1}\frac{\chi(f-a)te^{(f-a)t}}{e^{ft}-1}\\ &= \chi(-1)\sum_{a=1}^{f-1}\frac{\chi(a)te^{-at}}{1-e^{-ft}}\\ &= \chi(-1)\sum_{a=1}^{f-1}\frac{\chi(a)(-t)e^{a(-t)}}{e^{f(-t)}-1}.\end{aligned}$$

It follows immediately from this that the generating function is an even function if $\chi(-1) = 1$, and an odd function if $\chi(-1) = -1$. □

Remark 4.6. For any n satisfying $(-1)^n = \chi(-1)$, it is known that $B_{n,\chi}$ is non-zero. This can be seen as follows. From Theorem 9.6 on p. 148, $B_{n,\chi}$ is non-zero if and only if $L(n, \overline{\chi})$ is non-zero. We can prove $L(n, \overline{\chi}) \neq 0$ using the Euler product formula for the L-function with $n > 1$. The fact that $L(1, \overline{\chi}) \neq 0$, i.e., $B_{1,\chi} \neq 0$ for all χ, is essentially equivalent to Dirichlet's theorem on prime numbers in arithmetic progressions. Dirichlet proved his theorem using the class number formula. See Dirichlet [29] or Serre [83, Chap. 6].

4.3 Bernoulli Polynomials

Bernoulli polynomials are often defined by means of a generating function. Here, we define them in a different manner.[1] Their generating function will be given in Proposition 4.9 (5).

First, define a $\mathbf{Q}$-linear map $I : \mathbf{Q}[x] \to \mathbf{Q}[x]$ from the polynomial ring $\mathbf{Q}[x]$ to itself by

$$I(f) = \int_x^{x+1} f(y)\,dy \quad (f(x) \in \mathbf{Q}[x])$$

(difference of the primitive function). Since we see immediately that

$$I(x^n) = \frac{1}{n+1}\left((x+1)^{n+1} - x^{n+1}\right) = x^n + (\text{lower terms}),$$

the matrix representing I with respect to the basis $1, x, x^2, x^3, \ldots$ of $\mathbf{Q}[x]$ is an upper triangular matrix with the diagonal components 1. In particular, I is invertible. We thus define the Bernoulli polynomials as follows.

Definition 4.7. Define the Bernoulli polynomials $B_n(x)$ $(n = 0, 1, 2, \ldots)$ by

$$B_n(x) := I^{-1}(x^n).$$

In other words, $B_n(x)$ is the unique polynomial satisfying

$$I(B_n(x)) = \int_x^{x+1} B_n(y)\,dy = x^n.$$

Because of the form of $I(x)$, $B_n(x)$ is a monic polynomial (a polynomial in which the coefficient of the highest-degree term is 1) of degree n with coefficients in $\mathbf{Q}$.

[1] This was suggested by Don Zagier.

Example 4.8.

$$B_0(x) = 1, \quad B_1(x) = x - \frac{1}{2}, \quad B_2(x) = x^2 - x + \frac{1}{6}, \quad B_3(x) = x^3 - \frac{3}{2}x^2 + \frac{1}{2}x,$$

$$B_4(x) = x^4 - 2x^3 + x^2 - \frac{1}{30}, \quad B_5(x) = x^5 - \frac{5}{2}x^4 + \frac{5}{3}x^3 - \frac{1}{6}x.$$

Proposition 4.9. (1) $B_n(1) = B_n$ $(n \geq 0)$. *Also, if* $n \neq 1$ *, then* $B_n(0) = B_n(1) = B_n$.

(2) $B_n(x+1) - B_n(x) = nx^{n-1} \qquad (n \geq 0)$.

(3) $B_n'(x) = nB_{n-1}(x) \qquad (n \geq 1)$.

(4) $B_n(1-x) = (-1)^n B_n(x) \qquad (n \geq 0)$.

(5) (Generating function)

$$\sum_{n=0}^{\infty} B_n(x)\frac{t^n}{n!} = \frac{te^{xt}}{e^t - 1}. \tag{4.2}$$

(6) (Explicit formula with Bernoulli numbers)

$$B_n(x) = \sum_{j=0}^{n}(-1)^j \binom{n}{j} B_j x^{n-j}.$$

(7)

$$\sum_{i=0}^{k-1} B_n\left(x + \frac{i}{k}\right) = k^{1-n} B_n(kx) \ (k \geq 1).$$

Proof. (2) It is obtained by differentiating the defining equation $\int_x^{x+1} B_n(y)\,dy = x^n$ with respect to x.

(3) It follows from (2) that

$$\int_x^{x+1} \frac{B_n'(y)}{n}\,dy = \frac{1}{n}\left(B_n(x+1) - B_n(x)\right) = x^{n-1}.$$

Thus, it follows from the definition that

$$\frac{B_n'(x)}{n} = B_{n-1}(x),$$

or

$$B_n'(x) = nB_{n-1}(x).$$

(4) Since we have

$$\int_x^{x+1} B_n(1-y)\,dy = -\int_{1-x}^{-x} B_n(y)\,dy = \int_{-x}^{1-x} B_n(y)\,dy = (-x)^n = (-1)^n x^n,$$

it follows from the definition that $(-1)^n B_n(1-x) = B_n(x)$.

(5) Applying I to the coefficient of each term of the left-hand side of (4.2), we have $\sum_{n=0} x^n t^n/n! = e^{xt}$ as formal power series in t. On the other hand, regarding the right-hand side of (4.2) as a formal power series, and applying I term by term, we have $I(te^{xt}/(e^t-1)) = tI(e^{xt})/(e^t-1)$ because of the linearity. Furthermore, since we have

$$I(e^{xt}) = \int_x^{x+1} e^{yt}\,dy = \frac{e^{(x+1)t} - e^{xt}}{t} = \frac{e^t-1}{t}e^{xt},$$

we see $I(te^{xt}/(e^t-1)) = e^{xt}$. The injectivity of I implies that both sides are equal.

(1) The first half of (1) can be seen by putting $x = 1$ in (5) and by using the generating function of Bernoulli numbers (Theorem 1.12 on p. 20). Putting $x = 0$ in (2), we see that $B_n(1) = B_n(0)$ for $n \neq 1$. The second half of the assertion follows from this and the first half.

(6) We compute the product $t/(e^t-1)\times e^{xt}$ on the right-hand side of the generating function (4.2). Since

$$\frac{t}{e^t-1} = \frac{(-t)e^{-t}}{e^{-t}-1} = \sum_{n=0}^{\infty}(-1)^n B_n \frac{t^n}{n!},$$

we have

$$\begin{aligned}\frac{t}{e^t-1}\cdot e^{xt} &= \left(\sum_{n=0}^{\infty}(-1)^n B_n\frac{t^n}{n!}\right)\left(\sum_{n=0}^{\infty}\frac{(xt)^n}{n!}\right)\\ &= \sum_{n=0}^{\infty}\left(\sum_{j=0}^{n}(-1)^j\binom{n}{j}B_j x^{n-j}\right)\frac{t^n}{n!}.\end{aligned}$$

This proves the formula (6).

(7) We have

$$\begin{aligned}\int_x^{x+1}\sum_{i=0}^{k-1} B_n\left(\frac{y}{k}+\frac{i}{k}\right)dy &= \sum_{i=0}^{k-1}\int_x^{x+1} B_n\left(\frac{y+i}{k}\right)dy\\ &= \sum_{i=0}^{k-1}\int_{\frac{x+i}{k}}^{\frac{x+1+i}{k}} B_n(y)\,k\,dy \quad \left(\tfrac{y+i}{k}\to y\right)\end{aligned}$$

$$= k \int_{\frac{x}{k}}^{\frac{x}{k}+1} B_n(y)\,dy$$
$$= k\left(\frac{x}{k}\right)^n = k^{1-n}x^n.$$

From this we have

$$\sum_{i=0}^{k-1} B_n\left(\frac{x}{k}+\frac{i}{k}\right) = k^{1-n}B_n(x).$$

The formula is obtained by letting $x \to kx$. □

Remark 4.10. (1) As we mentioned in Remark 1.3 on p. 10, it is easy to prove the formula for the sum of powers ((1.1) on p. 1) using the Bernoulli polynomials. Replacing n by $k+1$ in Proposition 4.9 (2), we have

$$x^k = \frac{1}{k+1}\left(B_{k+1}(x+1) - B_{k+1}(x)\right).$$

Putting $x = 1, 2, \ldots, n$, and adding them all up, we have

$$\sum_{i=1}^{n} i^k = \frac{1}{k+1}\left(B_{k+1}(n+1) - B_{k+1}(1)\right).$$

We then use Proposition 4.9 (1), (4), (6) to obtain

$$\begin{aligned} B_{k+1}(n+1) - B_{k+1}(1) &= (-1)^{k+1}B_{k+1}(-n) - B_{k+1} \\ &= (-1)^{k+1}\sum_{j=0}^{k+1}(-1)^j\binom{k+1}{j}B_j(-n)^{k+1-j} - B_{k+1} \\ &= \sum_{j=0}^{k}\binom{k+1}{j}B_j n^{k+1-j}. \end{aligned}$$

Since $\frac{1}{k+1}\binom{k+1}{j} = \binom{k}{j}\frac{1}{k+1-j}$, the formula (1.1) follows.

(2) The formula (4.1) which expresses generalized Bernoulli numbers in terms of Bernoulli polynomials is obtained as follows. In (4.2) replace x by a/f and t by ft. Then we obtain

$$\frac{te^{ta}}{e^{ft}-1} = \frac{1}{f}\sum_{n=0}^{\infty} B_n(a/f)\frac{(ft)^n}{n!}.$$

Thus, we have

$$\sum_{n=0}^{\infty} B_{n,\chi}\frac{t^n}{n!} = \sum_{a=1}^{f} \frac{\chi(a)te^{ta}}{e^{ft}-1} = \sum_{a=1}^{f}\chi(a)\frac{1}{f}\sum_{n=0}^{\infty} B_n(a/f)\frac{(ft)^n}{n!}$$

$$= \sum_{n=0}^{\infty}\sum_{a=1}^{f}\chi(a)B_n(a/f)\frac{f^{n-1}t^n}{n!}.$$

Comparing coefficients, we obtain

$$B_{n,\chi} = f^{n-1}\sum_{a=1}^{f}\chi(a)B_n(a/f).$$

Next we take up a beautiful formula involving Bernoulli polynomials. For any real number x, we denote by $[x]$ the greatest integer less than or equal to x. This is sometimes called the Gauss symbol .

Theorem 4.11. *Let k be a natural number. The formula*

$$B_k(x-[x]) = -\frac{k!}{(2\pi i)^k}\sum_{\substack{n\in\mathbf{Z}\\ n\neq 0}}\frac{e^{2\pi i n x}}{n^k} \tag{4.3}$$

holds for all real numbers x if $k \geq 2$; it holds for all real numbers $x \notin \mathbf{Z}$, if $k = 1$. Here, the sum is taken for all integers different from 0. If $k = 1$, the infinite sum on the right-hand side should be understood as

$$\lim_{N\to\infty}\sum_{\substack{n=-N\\ n\neq 0}}^{N}\frac{e^{2\pi i n x}}{n}.$$

If $k \geq 2$, then the right-hand side of (4.3) converges absolutely and uniformly with respect to x.

Proof. Let $0 < x < 1$. Decompose the function

$$f(\zeta) = \frac{e^{x\zeta}}{e^{\zeta}-1}$$

into partial fractions as a meromorphic function in ζ. This $f(\zeta)$ has poles at $\zeta = 2\pi i n$ ($n \in \mathbf{Z}$) and all of them are of order 1. The residue at $\zeta = 2\pi i n$ is given by

$$\lim_{\zeta\to 2\pi i n}(\zeta - 2\pi i n)f(\zeta) = \lim_{t\to 0} t\cdot\frac{e^{x(t+2\pi i n)}}{e^{t+2\pi i n}-1} = e^{2\pi i n x}.$$

Let N be a sufficiently large natural number, and put $R = 2\pi(N+1/2)$. Let C_N be a square path passing through four corner points $R+iR,\ -R+iR,\ -R-iR,\ R-iR$ in this order in the ζ-plane. If t is a point inside C_N such that $t \neq 2\pi i n$ ($n \in \mathbf{Z}$), then it follows from the residue theorem that

$$\int_{C_N} \frac{f(\zeta)}{\zeta - t}\, d\zeta = 2\pi i \left(f(t) + \sum_{n=-N}^{N} \frac{e^{2\pi i n x}}{2\pi i n - t} \right)$$
$$= 2\pi i \left(f(t) - \frac{1}{t} - \sum_{\substack{n=-N \\ n\neq 0}}^{N} \frac{e^{2\pi i n x}}{t - 2\pi i n} \right).$$

As $N \to \infty$, one can show that

$$\int_{C_N} \frac{f(\zeta)}{\zeta - t}\, d\zeta \to 0.$$

Therefore, if $0 < |t| < 2\pi$, we have

$$f(t) = \frac{1}{t} + \lim_{N\to\infty} \sum_{\substack{n=-N \\ n\neq 0}}^{N} \frac{e^{2\pi i n x}}{t - 2\pi i n}$$
$$= \frac{1}{t} - \lim_{N\to\infty} \sum_{\substack{n=-N \\ n\neq 0}}^{N} \frac{e^{2\pi i n x}}{2\pi i n\left(1 - \frac{t}{2\pi i n}\right)}$$
$$= \frac{1}{t} - \lim_{N\to\infty} \sum_{\substack{n=-N \\ n\neq 0}}^{N} \frac{e^{2\pi i n x}}{2\pi i n} \sum_{k=1}^{\infty} \left(\frac{t}{2\pi i n} \right)^{k-1}.$$

Since the sum

$$\sum_{\substack{n=-\infty \\ n\neq 0}}^{\infty} \sum_{k=2}^{\infty} \frac{e^{2\pi i n x}}{(2\pi i n)^k} t^{k-1}$$

converges absolutely, the above formula becomes

$$f(t) = \frac{1}{t} - \lim_{N\to\infty} \sum_{\substack{n=-N \\ n\neq 0}}^{N} \frac{e^{2\pi i n x}}{2\pi i n} - \sum_{\substack{n=-\infty \\ n\neq 0}}^{\infty} \sum_{k=2}^{\infty} \frac{e^{2\pi i n x}}{(2\pi i n)^k} t^{k-1}.$$

Since the generating function (4.2) of Bernoulli polynomials can be regarded as the Taylor expansion of $te^{xt}/(e^t-1)$ at $t=0$ for each real number x, we have

$$f(t)=\sum_{k=0}^{\infty} B_k(x)\frac{t^{k-1}}{k!}.$$

Thus, comparing the coefficients, we obtain

$$B_1(x)=-\lim_{N\to\infty}\sum_{\substack{n=-N\\ n\neq 0}}^{N}\frac{e^{2\pi inx}}{2\pi in},$$

$$B_k(x)=-k!\sum_{\substack{n=-\infty\\ n\neq 0}}^{\infty}\frac{e^{2\pi inx}}{(2\pi in)^k}\qquad (k\geq 2).$$

Since the sum on the right-hand side converges uniformly over the closed interval $[0,1]$ if $k\geq 2$, and since $B_k(x)$ is continuous in the same interval, the above formula holds in $[0,1]$. Furthermore, the right-hand side is periodic with respect to x with period 1. Thus, (4.3) is obtained by extending the left-hand side. □

From this theorem we can obtain the values of the Riemann zeta function at positive even integers. (The Riemann zeta function will be treated in the next section.)

Corollary 4.12. *Let k be an even integer greater than or equal to 2. Then we have*

$$\sum_{n=1}^{\infty}\frac{1}{n^k}=-\frac{1}{2}\frac{B_k}{k!}(2\pi i)^k. \tag{4.4}$$

For example,

$$\sum_{n=1}^{\infty}\frac{1}{n^2}=\frac{\pi^2}{6},\quad \sum_{n=1}^{\infty}\frac{1}{n^4}=\frac{\pi^4}{90},\quad \sum_{n=1}^{\infty}\frac{1}{n^6}=\frac{\pi^6}{945},\quad \sum_{n=1}^{\infty}\frac{1}{n^8}=\frac{\pi^8}{9450},\quad etc.$$

Proof. It suffices to put $x=0$ in the theorem. The left-hand side is equal to B_k from Proposition 4.9 (1). Since the right-hand side is twice the sum over the positive n since k is even, the formula in question holds. □

Remark 4.13. The formula in the theorem can be rewritten as follows.

(1) If k is even,

$$B_k(x-[x])=2(-1)^{k/2-1}k!\sum_{n=1}^{\infty}\frac{\cos 2\pi nx}{(2\pi n)^k}.$$

(2) If k is odd,

$$B_k(x-[x]) = 2(-1)^{(k+1)/2}k!\sum_{n=1}^{\infty}\frac{\sin 2\pi nx}{(2\pi n)^k}. \tag{4.5}$$

Here, we assume x is not an integer if $k = 1$.

Theorem 4.11 has an interesting application.[2] Let us consider the special values of the following zeta function. Let a, b be integers different from 0, and let p, q, r be positive even integers. Define

$$S(p,q,r;a,b) = \sum_{\substack{m\neq 0,\, n\neq 0\\ am+bn\neq 0}}\frac{1}{m^p n^q (am+bn)^r}.$$

Here, the sum runs over all integers $m,\ n$ satisfying $m \neq 0,\ n \neq 0,\ am+bn \neq 0$.

Proposition 4.14. *Let $p,\ q,\ r$ be positive integers. Then, we have*

$$\pi^{-p-q-r}S(p,q,r;a,b) \in \mathbf{Q}.$$

Proof. Define $l = -(am+bn)$. We can express $S(p,q,r;a,b)$ as an integral as follows (we use the notation $\mathbf{e}(w) = e^{2\pi i w}$):

$$\begin{aligned}
S(p,q,r;a,b) &= \sum_{\substack{m,n,l\in\mathbf{Z}\\ m,n,l\neq 0,\ am+bn+l=0}}\frac{1}{m^p n^q l^r}\\
&= \sum_{\substack{m,n,l\in\mathbf{Z}\\ m,n,l\neq 0}}\int_0^1 \frac{\mathbf{e}((am+bn+l)x)}{m^p n^q l^r}\,dx\\
&= \int_0^1 \sum_{\substack{m,n,l\in\mathbf{Z}\\ m,n,l\neq 0}}\frac{\mathbf{e}((am+bn+l)x)}{m^p n^q l^r}\,dx\\
&= \int_0^1 \sum_{\substack{m\in\mathbf{Z}\\ m\neq 0}}\frac{\mathbf{e}(amx)}{m^p}\sum_{\substack{n\in\mathbf{Z}\\ n\neq 0}}\frac{\mathbf{e}(bnx)}{n^q}\sum_{\substack{l\in\mathbf{Z}\\ l\neq 0}}\frac{\mathbf{e}(lx)}{l^r}\,dx.
\end{aligned}$$

Here, we used the fact that the series $\sum_{m\neq 0}\mathbf{e}(amx)/m^p$ converges uniformly over the closed interval [0, 1], because $p,\ q,\ r \geq 2$. Using the formula (4.3), we have

[2]This is based on a lecture of Don Zagier at Kyushu University in 1999.

$$S(p,q,r;a,b) = \frac{(-1)^{\frac{p+q+r-2}{2}}(2\pi)^{p+q+r}}{p!q!r!}\int_0^1 B_p(ax-[ax])B_q(bx-[bx])B_r(x)\,dx.$$

Thus, the value of the integral is a rational number (recall that the Bernoulli polynomial is a polynomial with rational coefficients), and the assertion of the proposition follows. □

Example 4.15. ($p = q = r = 2, a = b = 1$.) From the above formula,

$$S(2,2,2;1,1) = \frac{(2\pi)^6}{8}\int_0^1 B_2(x)^3\,dx.$$

Integration by parts using $B_1'(x) = 1$, $B_2'(x) = 2B_1(x)$ gives

$$\begin{aligned}\int_0^1 B_2(x)^3\,dx &= \int_0^1 B_1'(x)B_2(x)^3\,dx \\ &= \left[B_1(x)B_2(x)^3\right]_0^1 - \int_0^1 6B_1(x)^2B_2(x)^2\,dx \\ &= B_2^3 - 2\int_0^1 \left(B_1(x)^3\right)' B_2(x)^2\,dx \\ &= B_2^3 - 2\left[B_1(x)^3B_2(x)^2\right]_0^1 + \frac{8}{5}\int_0^1 \left(B_1(x)^5\right)' B_2(x)\,dx \\ &= B_2^3 - \frac{1}{2}B_2^2 + \frac{8}{5}\left[B_1(x)^5B_2(x)\right]_0^1 - \frac{16}{5}\int_0^1 B_1(x)^6\,dx \\ &= B_2^3 - \frac{1}{2}B_2^2 + \frac{1}{10}B_2 - \frac{16}{35}\left[B_1(x)^7\right]_0^1 \\ &= \frac{1}{2^2\cdot 3^3\cdot 5\cdot 7}.\end{aligned}$$

Thus, we have

$$S(2,2,2;1,1) = \frac{2\pi^6}{3^3\cdot 5\cdot 7}.$$

Exercise 4.16. For a non-trivial character χ modulo $f > 1$, write $B_{4,\chi}$ and $B_{5,\chi}$ by $\sum_{a=1}^{f}\chi(a)a^i$ with $i = 1, 2, 3, 4, 5$.

Exercise 4.17. Prove that the Bernoulli polynomial $B_n(x)$ for odd $n > 1$ is always divisible by $x(x - \frac{1}{2})(x - 1)$.

Chapter 5
The Euler–Maclaurin Summation Formula and the Riemann Zeta Function

Aside from the formula for the sum of powers, the most basic topics in which Bernoulli numbers appear is the Euler–Maclaurin[1] summation formula and the values of the Riemann zeta function at integer arguments. In this chapter we survey these topics.

5.1 Euler–Maclaurin Summation Formula

The Euler–Maclaurin summation formula gives a very effective tool for evaluating a sum of values of a function at integers.

Theorem 5.1. *Let a and b be integers satisfying $a \leq b$, and let M be a natural number. Suppose $f(x)$ is an M times continuously differentiable function*[2] *over $[a,b]$. Then, we have*

$$\sum_{n=a}^{b} f(n) = \int_a^b f(x)dx + \frac{1}{2}(f(a)+f(b)) + \sum_{k=1}^{M-1} \frac{B_{k+1}}{(k+1)!}(f^{(k)}(b) - f^{(k)}(a))$$
$$- \frac{(-1)^M}{M!} \int_a^b B_M(x-[x]) f^{(M)}(x)dx.$$

Here, B_{k+1} is the Bernoulli number and $B_M(x)$ is the Bernoulli polynomial, and the sum on the right-hand side is understood to be 0 *if $M = 1$.*

[1] Colin Maclaurin (born in February 1698 in Kilmodan, Cowal, Argyllshire, Scotland—died on June 14, 1746 in Edinburgh, Scotland).

[2] Differentiable M times and the Mth derivative is continuous.

T. Arakawa et al., *Bernoulli Numbers and Zeta Functions*, Springer Monographs in Mathematics, DOI 10.1007/978-4-431-54919-2_5,

Remark 5.2. (1) Regard $f(n)$ as the area of a rectangle whose base is the interval $[n-\frac{1}{2}, n+\frac{1}{2}]$ and whose height is $f(n)$. Then the sum of the first two terms on the right-hand side is the first approximation, so to say. The next term can be seen as the higher approximation in terms of the data at the end points of the interval.

(2) If we use the values of the Riemann zeta function at negative integers $\zeta(-k) = -B_{k+1}/(k+1)$, which will be proved later, then the formula looks like

$$\sum_{n=a}^{b} f(n) = \int_a^b f(x)dx + \frac{1}{2}(f(a)+f(b)) - \sum_{k=1}^{M-1} \frac{\zeta(-k)}{k!}(f^{(k)}(b)-f^{(k)}(a))$$
$$-\frac{(-1)^M}{M!}\int_a^b B_M(x-[x])f^{(M)}(x)dx.$$

Proof. Let $g(x)$ be an M times continuously differentiable function on $[0, 1]$. Since $B_1(x) = x - 1/2$, we have $B_1'(x) = 1$. Thus, using integration by parts, we have

$$\int_0^1 g(x)dx = [B_1(x)g(x)]_0^1 - \int_0^1 B_1(x)g'(x)\,dx$$
$$= \frac{1}{2}(g(1)+g(0)) - \int_0^1 B_1(x)g'(x)\,dx.$$

Iterating integration by parts using $B_k(x) = B_{k+1}'(x)/(k+1)$ (Proposition 4.9 (3) on p. 56), we obtain

$$\begin{aligned}\int_0^1 g(x)dx &= \frac{1}{2}(g(1)+g(0)) - \frac{1}{2}\left[B_2(x)g'(x)\right]_0^1 + \frac{1}{2}\int_0^1 B_2(x)g''(x)\,dx \\ &= \frac{1}{2}(g(1)+g(0)) - \frac{1}{2}\left[B_2(x)g'(x)\right]_0^1 \\ &\quad + \frac{1}{2\cdot 3}\left[B_3(x)g''(x)\right]_0^1 - \frac{1}{2\cdot 3}\int_0^1 B_3(x)g'''(x)\,dx \\ &= \quad \cdots\cdots \\ &= \frac{1}{2}(g(1)+g(0)) + \sum_{k=1}^{M-1}\frac{(-1)^k}{(k+1)!}\left[B_{k+1}(x)g^{(k)}(x)\right]_0^1 \\ &\quad + \frac{(-1)^M}{M!}\int_0^1 B_M(x)g^{(M)}(x)\,dx.\end{aligned}$$

Now, if $k \geq 2$, then $B_k(0) = B_k(1) = B_k$ (Proposition 4.9 (1)) and $(-1)^k B_k = B_k$ (Proposition 1.4 on p. 10). Thus, we have

$$\int_0^1 g(x)dx = \frac{1}{2}(g(1)+g(0)) - \sum_{k=1}^{M-1} \frac{B_{k+1}}{(k+1)!}(g^{(k)}(1) - g^{(k)}(0))$$
$$+ \frac{(-1)^M}{M!}\int_0^1 B_M(x)g^{(M)}(x)\,dx.$$

Therefore,

$$\frac{1}{2}(g(1)+g(0)) = \int_0^1 g(x)dx + \sum_{k=1}^{M-1} \frac{B_{k+1}}{(k+1)!}(g^{(k)}(1) - g^{(k)}(0))$$
$$- \frac{(-1)^M}{M!}\int_0^1 B_M(x)g^{(M)}(x)dx.$$

Now, we take $f(x+n)$ $(a \le n \le b-1)$ for the function $g(x)$. Since the function $B_M(x-[x])$, which is obtained by extending $B_M(x)$ from the interval $[0, 1]$ to the entire real line by periodicity, satisfies $B_M(x-n) = B_M(x-[x])$ for any integer n and $n \le x \le n+1$, we obtain

$$\frac{1}{2}(f(n+1)+f(n)) = \int_n^{n+1} f(x)dx + \sum_{k=1}^{M-1} \frac{B_{k+1}}{(k+1)!}(f^{(k)}(n+1) - f^{(k)}(n))$$
$$- \frac{(-1)^M}{M!}\int_n^{n+1} B_M(x-[x])f^{(M)}(x)dx$$

The formula in the theorem is obtained by adding up the above formula from $n = a$ to $n = b-1$ and then adding $\frac{1}{2}(f(a)+f(b))$ to both sides. □

5.2 The Riemann Zeta Function

The exponential function e^z with the complex variable z is defined by the formula

$$e^z = 1 + z + \frac{z^2}{2!} + \cdots + \frac{z^n}{n!} + \cdots.$$

The power series on the right-hand side converges absolutely for any $z \in \mathbf{C}$, and thus it is a holomorphic function over the entire z-plane. From the formula for the product of two power series, we have the exponential law

$$e^{z+w} = e^z e^w.$$

In particular, for $x, y \in \mathbf{R}$, we have Euler's formula

$$e^{x+iy} = e^x e^{iy} = e^x(\cos y + i \sin y).$$

For any positive real number a, the complex power a^s for $s \in \mathbf{C}$ is defined by

$$a^s = e^{s \log a}.$$

Its absolute value is given by

$$|a^s| = e^{\sigma \log a},$$

where $\sigma = \mathrm{Re}(s)$.

Let s be a complex number. Consider the sum

$$\sum_{n=1}^{N} \frac{1}{n^s} = 1 + \frac{1}{2^s} + \cdots + \frac{1}{N^s}.$$

Here, n^s is a complex power of n just defined above. Let $f : (0, \infty) \to \mathbf{C}$ be the function defined by

$$f(x) = x^{-s}.$$

We apply the Euler–Maclaurin summation formula to this $f(x)$ with $a = 1, b = N$. The k-th derivative of f is given as

$$\begin{aligned} f^{(k)}(x) &= (-s)(-s-1)\cdots(-s-k+1)x^{-s-k} \\ &= (-1)^k s(s+1)\cdots(s+k-1)x^{-s-k}. \end{aligned}$$

Also, if $s \neq 1$, then we have

$$\int_1^N x^{-s}\,dx = \frac{1}{s-1}\left(1 - \frac{1}{N^{s-1}}\right).$$

It follows from Theorem 5.1 that for $s \neq 1$

$$\begin{aligned} \sum_{n=1}^{N} \frac{1}{n^s} =& \frac{1}{s-1}\left(1 - \frac{1}{N^{s-1}}\right) \\ &+ \frac{1}{2}\left(1 + \frac{1}{N^s}\right) + \sum_{k=1}^{M-1} \frac{B_{k+1}}{k+1} \cdot \binom{s+k-1}{k}\left(1 - \frac{1}{N^{s+k}}\right) \\ &- \binom{s+M-1}{M} \int_1^N B_M(x-[x])x^{-s-M}\,dx, \end{aligned} \tag{5.1}$$

where $\binom{x}{i}$ is the binomial coefficient $x(x-1)\cdots(x-i+1)/i!$, and we used the fact $(-1)^k B_{k+1} = -B_{k+1}$. If $s = 1$, then it follows from $\int_1^N x^{-1}\,dx = \log N$ that

$$\sum_{n=1}^{N} \frac{1}{n} = \log N + \frac{1}{2}\left(1 + \frac{1}{N}\right) + \sum_{k=1}^{M-1} \frac{B_{k+1}}{k+1}\left(1 - \frac{1}{N^{1+k}}\right)$$
$$- \int_1^N B_M(x - [x])x^{-1-M}\,dx. \tag{5.2}$$

As an application of these formulas, we obtain the following elementary result.

Proposition 5.3. (1) *The series* $\sum_{n=1}^{\infty} \frac{1}{n^s}$ *converges absolutely if* $\mathrm{Re}(s) > 1$, *and the series* $\sum_{n=1}^{\infty} \frac{1}{n}$ *diverges.*

(2) *Let* $\{a_N\}$ *be the sequence defined by* $a_N = \sum_{n=1}^{N} \frac{1}{n} - \log N$. *Then* $\{a_N\}$ *converges. The limit* $\gamma = \lim_{N\to\infty} a_N$ *is called Euler's constant.*[3] *For any natural number* M, *we have*

$$\gamma = \sum_{k=0}^{M-1} \frac{B_{k+1}}{k+1} - \int_1^\infty \frac{B_M(x - [x])}{x^{M+1}}\,dx.$$

In particular, setting $M = 1$, *we have*

$$\gamma = 1 - \int_1^\infty \frac{x - [x]}{x^2}\,dx. \tag{5.3}$$

Proof. (1) Let $\mathrm{Re}(s) = \sigma$. Since $\sum_{n=1}^{N} \left|\frac{1}{n^s}\right| = \sum_{n=1}^{N} \frac{1}{n^\sigma}$, by putting $s = \sigma$, $M = 1$ in formula (5.1), we obtain

$$\sum_{n=1}^{N} \frac{1}{n^\sigma} = \frac{1}{\sigma - 1}\left(1 - \frac{1}{N^{\sigma-1}}\right) + \frac{1}{2}\left(1 + \frac{1}{N^\sigma}\right) - \sigma \int_1^N B_1(x - [x])x^{-\sigma-1}\,dx.$$

Since $\sigma > 1$, both $1/N^{\sigma-1}$ and $1/N^\sigma$ converge to 0 as $N \to \infty$. As for the integral in the last term, first note that the function $B_M(x - [x])$ is bounded since $0 \le x - [x] \le 1$. Thus, if

$$\int_1^\infty x^{-\sigma-1}\,dx = \lim_{N\to\infty} \int_1^N x^{-\sigma-1}\,dx \tag{5.4}$$

[3] It is unknown whether $\gamma = 0.57721566490153286060651209008 24\cdots$ is an irrational number or not.

converges, then so does

$$\int_1^N B_M(x-[x])x^{-\sigma-1}\,dx$$

as $N \to \infty$. The integral (5.4) converges since if $\sigma > 0$, we have

$$\lim_{N\to\infty}\int_1^N x^{-\sigma-1}\,dx = \lim_{N\to\infty}\frac{1}{\sigma}\left(1-\frac{1}{N^\sigma}\right) = \frac{1}{\sigma}.$$

Summing it all up, we see that $\sum_{n=1}^{\infty}\frac{1}{n^s}$ converges absolutely if $\sigma > 1$. Also, the fact that $\sum_{n=1}^{\infty}\frac{1}{n}$ diverges follows immediately from (5.2).

(2) It follows from the formula (5.2) that

$$a_N = \sum_{n=1}^{N}\frac{1}{n} - \log N = \frac{1}{2}\left(1+\frac{1}{N}\right) + \sum_{k=1}^{M-1}\frac{B_{k+1}}{k+1}\left(1-\frac{1}{N^{1+k}}\right)$$
$$-\int_1^N B_M(x-[x])x^{-1-M}\,dx.$$

Similarly to (1), the last integral converges, and thus $\{a_N\}$ converges. Letting $N \to \infty$, we obtain

$$\gamma = \lim_{N\to\infty} a_N = \frac{1}{2} + \sum_{k=1}^{M-1}\frac{B_{k+1}}{k+1} - \int_1^\infty B_M(x-[x])x^{-1-M}\,dx$$
$$= \sum_{k=0}^{M-1}\frac{B_{k+1}}{k+1} - \int_1^\infty B_M(x-[x])x^{-1-M}\,dx.$$

(Here we used $B_1 = 1/2$.) If $M = 1$, then we have $B_1(x-[x]) = x-[x]-\frac{1}{2}$ and

$$\int_1^\infty \frac{dx}{x^2} = \left[-\frac{1}{x}\right]_1^\infty = 1.$$

Thus, we have

$$\gamma = \frac{1}{2} - \int_1^\infty \frac{x-[x]-\frac{1}{2}}{x^2}\,dx$$
$$= 1 - \int_1^\infty \frac{x-[x]}{x^2}\,dx.$$

□

The famous Riemann[4] zeta function $\zeta(s)$ is defined for all complex numbers s satisfying $\mathrm{Re}(s) > 1$ by

$$\zeta(s) = \sum_{n=1}^{\infty} \frac{1}{n^s}.$$

As we have already seen, the infinite series on the right-hand side converges absolutely for $\mathrm{Re}(s) > 1$. If σ is a positive number greater than 1, then this series converges uniformly in $\mathrm{Re}(s) \geq \sigma$, and thus it is a holomorphic function of s in $\mathrm{Re}(s) > 1$. Let us study properties of $\zeta(s)$, using the summation formula (5.1). Suppose $\mathrm{Re}(s) > 1$, and let $N \to \infty$ in (5.1). Then,

$$\sum_{n=1}^{\infty} \frac{1}{n^s} = \frac{1}{s-1} + \frac{1}{2} + \sum_{k=1}^{M-1} \frac{B_{k+1}}{k+1} \cdot \binom{s+k-1}{k} - \binom{s+M-1}{M} \int_1^{\infty} B_M(x-[x]) x^{-s-M}\, dx. \tag{5.5}$$

Now, take the difference between (5.1) and (5.5). Then, for $\mathrm{Re}(s) > 1$, we have

$$\zeta(s) - \sum_{n=1}^{N} \frac{1}{n^s} = \frac{1}{s-1} \cdot \frac{1}{N^{s-1}} - \frac{1}{2N^s} + \sum_{k=1}^{M-1} \frac{B_{k+1}}{k+1} \cdot \binom{s+k-1}{k} \cdot \frac{1}{N^{s+k}} - \binom{s+M-1}{M} \int_N^{\infty} B_M(x-[x]) x^{-s-M}\, dx. \tag{5.6}$$

Since the integral of the right-hand side converges absolutely in the region $\mathrm{Re}(s) > 1 - M$, the formula (5.6) gives an analytic continuation of $\zeta(s)$ to a meromorphic function in the region $\mathrm{Re}(s) > 1 - M$. In other words, $\zeta(s)$ can be extended to $\mathrm{Re}(s) > 1 - M$ through the formula (5.6). Since M can be any natural number, $\zeta(s)$ has an analytic continuation to the entire s-plane as a meromorphic function. It also follows from (5.6) that its only pole is a simple pole located at $s = 1$, and the residue at $s = 1$ equals 1.

We summarize all this, and find the values of $\zeta(s)$ at integers less than or equal to 0.

Theorem 5.4. (1) *The function $\zeta(s)$ has an analytic continuation to the entire s-plane as a meromorphic function, is holomorphic if $s \neq 1$, and has a pole of order* 1 *at $s = 1$ with residue* 1.

[4]Georg Friedrich Bernhard Riemann (born on September 17, 1826 in Breselenz, Germany—died on July 20, 1866 in Selasca, Italy).

(2) *Let m be a positive integer. The value of* $\zeta(s)$ *at* $s = 1 - m$ *is given by*

$$\zeta(1-m) = -\frac{B_m}{m}.$$

Proof. Since we have already seen (1), we prove (2). If $s = 1 - m$, taking M in (5.6) with $M \geq m$, the binominal coefficient in front of the integral on the right-hand side vanishes. This implies that $\zeta(1-m)$ is expressed as a finite sum. Our goal is to express it in a simpler form. In (5.6), let $s = 1 - m$, and $M = m$. Then we have

$$\zeta(1-m) - \sum_{n=1}^{N} n^{m-1} = -\frac{N^m}{m} - \frac{N^{m-1}}{2} + \sum_{k=1}^{m-1} \frac{B_{k+1}}{(k+1)!} \times (1-m)(2-m)\cdots(k-m)\cdot N^{m-k-1}. \tag{5.7}$$

If $m = 1$, then there is no term of summation, and

$$\zeta(0) = \sum_{n=1}^{N} 1 - N - \frac{1}{2} = -\frac{1}{2} = -B_1.$$

Suppose $m \geq 2$. Now, by using the formulas $(1-m)(2-m)\cdots(k-m) = (-1)^k k!\binom{m-1}{k}$, $\frac{1}{k+1}\binom{m-1}{k} = \frac{1}{m-k-1}\binom{m-1}{k+1}$ if $k < m-1$, and $(-1)^k B_k = B_k$ for $k \geq 2$, we have

$$\begin{aligned}
&\sum_{k=1}^{m-1} \frac{B_{k+1}}{(k+1)!}\cdot(1-m)(2-m)\cdots(k-m)\cdot N^{m-k-1} \\
&= \sum_{k=1}^{m-1}(-1)^k \frac{B_{k+1}}{k+1}\binom{m-1}{k} N^{m-k-1} \\
&= \sum_{k=1}^{m-2}(-1)^k \binom{m-1}{k+1} B_{k+1}\frac{N^{m-k-1}}{m-k-1} + (-1)^{m-1}\frac{B_m}{m} \\
&= \sum_{k=2}^{m-1}(-1)^{k-1} \binom{m-1}{k} B_k\frac{N^{m-k}}{m-k} + (-1)^{m-1}\frac{B_m}{m} \\
&= -\sum_{k=2}^{m-1} \binom{m-1}{k} B_k\frac{N^{m-k}}{m-k} - \frac{B_m}{m}.
\end{aligned}$$

Therefore, the right-hand side of (5.7) equals

$$-\sum_{k=0}^{m-1} \binom{m-1}{k} B_k\frac{N^{m-k}}{m-k} - \frac{B_m}{m}$$

(since $B_0 = 1$, $B_1 = 1/2$). On the other hand, from the formula for the sum of powers (1.1) on p. 1, we have

$$\sum_{n=1}^{N} n^{m-1} = \sum_{k=0}^{m-1} \binom{m-1}{k} B_k \frac{N^{m-k}}{m-k}. \tag{5.8}$$

Thus, from (5.7) we have the formula

$$\zeta(1-m) = -\frac{B_m}{m}.$$

□

Remark 5.5. The value in (2) was first given by Euler ([32], see also Weil[5] [103]). The values of the Riemann zeta function at positive even integers in Corollary 4.12 (p. 61)

$$\zeta(2k) = \frac{(-1)^{k-1}}{2} \frac{B_{2k}}{(2k)!} (2\pi)^{2k} \quad (k \in \mathbf{N})$$

were also computed first by Euler (ibid.). There is not much known about the values at positive odd integers. It is conjectured that all $\zeta(2k+1)$ are transcendental numbers.[6] Apéry[7] [5] proved that $\zeta(3)$ is an irrational number, and Rivoal [80] proved that there are infinitely many irrational numbers among $\zeta(2k+1)$ $(k = 1, 2, 3, \ldots)$ (For more recent results, see [99].)

Remark 5.6. The formula (5.8) for the sum of powers can also be obtained using the Euler–Maclaurin summation formula (Theorem 5.1). Indeed, the sum $\sum_{n=1}^{N} n^{m-1}$ can be obtained by letting $f(x) = x^{m-1}$, $a = 1$, and $b = N$ in the summation formula.

Exercise 5.7. Use (5.6) ($N = M = 1$) and (5.3) to prove

$$\lim_{s \to 1} \left(\zeta(s) - \frac{1}{s-1} \right) = \gamma.$$

Exercise 5.8. Use (5.6) ($N = M = 1$) to deduce the formula

$$\zeta(s) = -s \int_0^{\infty} \frac{x - [x] - \frac{1}{2}}{x^{s+1}} \, dx$$

---

[5] André Weil (born on May 6, 1906 in Paris, France—died on August 6, 1998 in Princeton, USA).

[6] Since π is a transcendental number [71], $\zeta(2k)$ are all transcendental numbers.

[7] Roger Apéry (born on November 14, 1916 in Ruen, France—died on December 18, 1994 in Caen, France).

which is valid for $-1 < \mathrm{Re}(s) < 0$.

Exercise 5.9. Use the previous exercise and the formula (proved in Chap. 4, (4.5))

$$x - [x] - \frac{1}{2} = -\sum_{n=1}^{\infty} \frac{\sin 2\pi n x}{\pi n}$$

to deduce the functional equation of the Riemann zeta function

$$\zeta(s) = -2^s s \pi^{s-1} \Gamma(-s) \sin \frac{\pi s}{2} \zeta(1-s).$$

(This way of proof is rigorously described in Chap. 2.1 in [90]. We give a different proof of (the symmetric form of) the functional equation in Chap. 9.)

Exercise 5.10. Carry out the computation suggested in Remark 5.6.

Chapter 6
Quadratic Forms and Ideal Theory of Quadratic Fields

There are close relations between Bernoulli numbers and quadratic fields or quadratic forms. In order to explain those, we give a survey of the ideal theory of quadratic fields and quadratic forms. Since Gauss, it is well known that there is a deep relation between the ideal theory of quadratic fields (i.e. quadratic extensions of the rational number field) and integral quadratic forms. This is obvious for specialists, but textbooks which explain this in detail are rare. In most textbooks, they treat the correspondence of ideals only in the case of maximal orders (the ring of all integers) for which the description is simple and easy. This is very disappointing. We cannot see the whole picture of the theory of quadratic forms by such a restricted treatment and sometimes it causes misunderstanding. We need the full description of this kind of relation when we explain later the relation between L-functions of prehomogeneous vector spaces and the Bernoulli numbers. So it would be a good chance to try to explain the full relation here.

6.1 Quadratic Forms

For integers a, b, c and variables x, y, we call the homogeneous polynomial

$$Q(x, y) = ax^2 + bxy + cy^2$$

of degree two an integral binary quadratic form (or a quadratic form for short, since we treat only the integral and binary case in this book). In particular, in case a, b, c are mutually coprime (namely the greatest common divisor, g.c.d., of a, b, c is one), we say $Q(x, y)$ is primitive. If we define the matrix S by

$$S = \begin{pmatrix} a & b/2 \\ b/2 & c \end{pmatrix},$$

T. Arakawa et al., *Bernoulli Numbers and Zeta Functions*, Springer Monographs in Mathematics, DOI 10.1007/978-4-431-54919-2_6, © Springer Japan 2014

then we can write

$$Q(x, y) = (x, y)S \begin{pmatrix} x \\ y \end{pmatrix}.$$

Often this expression helps easy understanding. A symmetric matrix like S whose diagonal components are integers and the other components are half-integers[1] is called a half-integral symmetric matrix. We denote by L^* the set of all half-integral symmetric matrices of order two:

$$L^* = \left\{ \begin{pmatrix} a & b/2 \\ b/2 & c \end{pmatrix} ; \ a, b, c \in \mathbf{Z} \right\}.$$

This set corresponds bijectively to the set of all integral (binary) quadratic forms. When the corresponding quadratic form $(x, y)S \begin{pmatrix} x \\ y \end{pmatrix}$ is primitive, we say that S is primitive. We consider the following groups of integral matrices:

$$GL_2(\mathbf{Z}) = \{g \in M_2(\mathbf{Z}); \det(g) = \pm 1\},$$
$$SL_2(\mathbf{Z}) = \{g \in M_2(\mathbf{Z}); \det(g) = 1\},$$

where $M_2(\mathbf{Z})$ is the set of all integral 2×2 matrices. The group $SL_2(\mathbf{Z})$ is called the (full) modular group and this is the subgroup of $GL_2(\mathbf{Z})$ of index two. The group $GL_2(\mathbf{Z})$ (and hence also $SL_2(\mathbf{Z})$) acts on L^* from the right by

$$L^* \ni S \mapsto \det(\gamma)\,{}^t\gamma S \gamma \in L^* \qquad (\gamma \in GL_2(\mathbf{Z})).$$

Two elements of L^* or the corresponding quadratic forms are said to be $SL_2(\mathbf{Z})$-equivalent, or just equivalent, when they belong to the same $SL_2(\mathbf{Z})$-orbit under this action. Two elements of L^* are said to be $GL_2(\mathbf{Z})$-equivalent when they belong to the same $GL_2(\mathbf{Z})$ orbit by the above action. (In some books, $SL_2(\mathbf{Z})$-equivalence is called a proper equivalence, but we do not add the word "proper" for this in this book). Usually we call the number $b^2 - 4ac = -\det(2S)$ the discriminant of the quadratic form. We denote this number by $D(S)$. It is obvious that $D(S) \equiv 0$ or 1 mod 4. On the other hand, if D is an integer such that $D \equiv 0$ or 1 mod 4, there exists $S \in L^*$ such that $D(S) = D$. The discriminant is invariant by the action of $GL_2(\mathbf{Z})$. The cases where D is a square (i.e. a square of integers, e.g. 0, 1, 4, 9 ...) are exceptional, so in this chapter, we assume that D is not a square unless otherwise stated. We fix such a non-square integer $D \equiv 0$ or 1 mod 4 and put $L^*(D) = \{S \in L^*; D(S) = D, S \text{ primitive}\}$. Then it is known that both $SL_2(\mathbf{Z})$-equivalence classes and $GL_2(\mathbf{Z})$-equivalence classes in $L^*(D)$ are finite. (The proof

[1]Here half-integers mean their doubles are integers.

of the finiteness of the class number and an explicit formula for the class number will be given later.) When $D(S) > 0$, we call these numbers of equivalence classes the class number and the class number in the wide sense of quadratic forms of discriminant D for $SL_2(\mathbf{Z})$-equivalence and for $GL_2(\mathbf{Z})$ equivalence, respectively.
When $D(S) < 0$, we have $ac > 0$, and this means that S is positive definite or negative definite. We denote by $L_+^*(D)$ the subset of positive definite quadratic forms of $L^*(D)$ and by $L_-^*(D)$ negative definite forms. When $D(S) < 0$, we say that the number of $SL_2(\mathbf{Z})$ equivalence classes in $L_+^*(D)$ the class number of the quadratic forms of discriminant D. In this case, the number of $GL_2(\mathbf{Z})$ equivalence classes in $L^*(D)$ and the number of $SL_2(\mathbf{Z})$ equivalence classes in $L_+^*(D)$ are the same. Indeed this is proved as follows. Any $S \in L_-^*(D)$ is $GL_2(\mathbf{Z})$ equivalent to an element of $L_+^*(D)$ since for $\gamma = \begin{pmatrix} 1 & 0 \\ 0 & -1 \end{pmatrix}$, we have $\det(\gamma)^t\gamma S\gamma = -{}^t\gamma S\gamma \in L_+^*(D)$. If S_1 and $S_2 \in L_+^*(S)$ are $GL_2(\mathbf{Z})$ equivalent and $S_2 = \det(\gamma)^t\gamma S_1\gamma$ for $\gamma \in GL_2(\mathbf{Z})$, then since S_2 and ${}^t\gamma S_1\gamma$ are positive definite, we have $\det(\gamma) > 0$, hence $\det(\gamma) = 1$, and S_1 and S_2 are $SL_2(\mathbf{Z})$ equivalent.

6.2 Orders of Quadratic Fields

Now we review the theory of quadratic fields, but there are many books describing this theory, so here we write only briefly. Also, we describe the theory mostly from the viewpoint of quadratic forms. In some parts this makes our explanation more complicated. But this is still useful since this kind of explanation is seldom found in books nowadays.

When m is a non-square integer, we say that $K = \mathbf{Q}(\sqrt{m}) = \mathbf{Q} + \mathbf{Q}\sqrt{m}$ is a quadratic field. When $m > 0$, we call K a real quadratic field and when $m < 0$, we call it an imaginary quadratic field. If m has a square factor, we can bring this factor outside the square root, so we can assume that m is an integer without square factor. We assume this from now on. An element α in this field K satisfying $\alpha^2+b\alpha+c = 0$ for some rational integers b, c (i.e. integers in the usual sense) is called an integer of K. The set $\mathfrak{O}_{max}$ of all integers in K form a ring. If we put

$$\omega = \begin{cases} \frac{1+\sqrt{m}}{2} & \text{if } m \equiv 1 \bmod 4, \\ \sqrt{m} & \text{otherwise,} \end{cases} \tag{6.1}$$

then one can show $\mathfrak{O}_{max} = \mathbf{Z} + \mathbf{Z}\omega$. The ring $\mathfrak{O}_{max}$ is called the maximal order (or the ring of all integers) of $\mathbf{Q}(\sqrt{m})$.

If we define the linear map $\alpha \to \overline{\alpha}$ of $\alpha \in K$ to $\overline{\alpha} \in K$ as $\mathbf{Q}$-vector space by $\overline{a + b\sqrt{m}} = a - b\sqrt{m}$ $(a, b \in \mathbf{Q})$, which is called a conjugation then this is an automorphism as a field. For any element α of K, we write $N(\alpha) = \alpha\overline{\alpha}$, $Tr(\alpha) = \alpha + \overline{\alpha}$, the former is called the norm and the latter the trace of the element α.

We put $D_K = (\omega - \overline{\omega})^2 = Tr(\omega)^2 - 4N(\omega)$ and call it the discriminant, or the fundamental discriminant of K. More concretely we have $D_K = m$ if $m \equiv 1 \bmod 4$ and $D_K = 4m$ otherwise.

The maximal order $\mathfrak{O}_{max}$ is neither a principal ideal domain, nor a unique factorization domain in general, so it is known that we cannot define a useful prime factor decomposition of elements of $\mathfrak{O}_{max}$. But if we consider ideals of $\mathfrak{O}_{max}$ instead of elements, any ideal of $\mathfrak{O}_{max}$ is factored uniquely into a product of prime ideals of $\mathfrak{O}_{max}$. This fact is a direct consequence of the fact that $\mathfrak{O}_{max}$ is a Dedekind domain. But sometimes it is useful to treat subrings of $\mathfrak{O}_{max}$ consisting only of a part of integers. For example, it is inevitable to treat such rings when we consider the correspondence with quadratic forms. But such subrings are not Dedekind domains in general, and we must keep in mind that the prime ideal decomposition does not hold for such rings in general. Now any subring $\mathfrak{O}$ of $\mathfrak{O}_{max}$ containing 1 is a free $\mathbf{Z}$-module since any subgroup of a finitely generated free abelian group is free by the fundamental theorem of finitely generated abelian groups. When the rank of $\mathfrak{O}$ is two, we call it simply an order. Let us describe all such rings.

Lemma 6.1. *Let ω be as in* (6.1). *For an arbitrary natural number f, we put*

$$\mathfrak{O}_f = \mathbf{Z} + f\omega\mathbf{Z}.$$

Then this is an order of $\mathbf{Q}(\sqrt{m})$. Conversely, for any order $\mathfrak{O}$ of $\mathbf{Q}(\sqrt{m})$, there exists a natural number f such that $\mathfrak{O} = \mathfrak{O}_f$.

Proof. To show that $\mathbf{Z} + f\omega\mathbf{Z}$ is an order, we must show that this is a subring. This is shown by the following calculation.

$$(f\omega)^2 = \begin{cases} f(f\omega) + \dfrac{m-1}{4} f^2 & \text{if } m \equiv 1 \bmod 4, \\ f^2 m & \text{otherwise.} \end{cases}$$

Conversely, if $\mathfrak{O}$ is an order, then since it is of rank 2, the ring $\mathfrak{O}$ contains an element which does not belong to $\mathbf{Z}$. So there is an integer a and non-zero integer b such that $a + b\omega \in \mathfrak{O}$. Since we have $1 \in \mathfrak{O}$ by definition, we have $a \in \mathfrak{O}$ so $b\omega \in \mathfrak{O}$. We take the smallest positive integer f among those b such that $b\omega \in \mathfrak{O}$. All the other such b are divisible by f. Indeed, for b such that $b\omega \in \mathfrak{O}$, we divide b by f and write $b = fx + b_0$, $x \in \mathbf{Z}$, $0 \leq b_0 < f$. Then we have $b_0\omega = (b - fx)\omega = b\omega - x(f\omega) \in \mathfrak{O}$. By our choice of f, we must have $b_0 = 0$, and b is divisible by f. Hence we have $\mathfrak{O} = \mathbf{Z} + f\omega\mathbf{Z} = \mathfrak{O}_f$. □

Orders $\mathfrak{O}_f$ for different f are different rings. If f and f' are natural numbers, then it is easy to see that $\mathfrak{O}_{f'} \subset \mathfrak{O}_f$ if and only if $f \mid f'$. This ring can be also written as

$$\mathfrak{O}_f = \mathbf{Z} + f\mathfrak{O}_{max}.$$

We call f the conductor of the order $\mathfrak{O}_f$. The conductor of an order is 1 if and only if it is maximal. We define the discriminant of the order $\mathfrak{O}_f$ by $D_K f^2$. It is conventional to call non-zero ideal $\mathfrak{a}$ of $\mathfrak{O}_f$ simply an ideal of $\mathfrak{O}_f$. Since an ideal $\mathfrak{a}$ is a submodule of $\mathfrak{O}_f$, it is a free $\mathbf{Z}$-module, but since it is also a module over $\mathfrak{O}_f$, the rank is two. So the ring $\mathfrak{O}_f/\mathfrak{a}$ is a finite ring. The size of this finite ring is called the norm of $\mathfrak{a}$ and denoted by $\mathcal{N}(\mathfrak{a})$. We describe the free basis over $\mathbf{Z}$ of the ideal $\mathfrak{a}$ of $\mathfrak{O}_f$ more concretely. We say that an ideal $\mathfrak{a}$ is primitive if it cannot be written as $\mathfrak{a} = l\mathfrak{b}$ for any integer $l > 1$ and any ideal $\mathfrak{b}$ of $\mathfrak{O}_f$.

Lemma 6.2. *For any ideal $\mathfrak{a}$ of $\mathfrak{O}_f$, there exist positive integers a, e, d with $-a/2 \le d < a/2$ such that*

$$\mathfrak{a} = \mathbf{Z}a + \mathbf{Z}(d + ef\omega).$$

Here the integers a, b, d are uniquely determined by $\mathfrak{a}$. If $\mathfrak{a}$ is a primitive ideal, then we have $e = 1$ and the norm of $\mathfrak{a}$ is a.

Proof. For a while, we take an arbitrary ideal $\mathfrak{a}$ of $\mathfrak{O}_f$ and do not assume that it is primitive. Since the module $\mathfrak{O}_f/\mathfrak{a}$ is a finite module, there is some natural number m such that $m1 \in \mathfrak{a}$ and we have $\mathfrak{a} \cap \mathbf{Z} \neq 0$. So we take the smallest positive rational integer a in $\mathfrak{a}$. Then it is clear that $a\mathbf{Z} = \mathfrak{a} \cap \mathbf{Z}$. On the other hand, we denote by e the smallest positive integer among those y such that $x + yf\omega \in \mathfrak{a}$ for some integer x. We take $d \in \mathbf{Z}$ such that $d + ef\omega \in \mathfrak{a}$. Then we can conclude that

$$\mathfrak{a} = \mathbf{Z}a + \mathbf{Z}(d + ef\omega).$$

This can be seen as follows. First, if $x + yf\omega \in \mathfrak{a}$ for some x, $y \in \mathbf{Z}$, then y is divisible by e. Indeed, if $y = eq + r$ with $0 \le r < e$ and $0 \le a$, then we have $x + yf\omega - q(d + ef\omega) = (x - qd) + rf\omega \in \mathfrak{a}$. Here if $r \neq 0$, it contradicts the definition of e, so we have $r = 0$. Hence there is an integer z such that $(x + yf\omega) - z(d + ef\omega) = x - zd \in \mathbf{Z} \cap \mathfrak{a}$, which is a multiple of a. Hence we have $\mathfrak{a} = \mathbf{Z}a + \mathbf{Z}(d + ef\omega)$. By adding a multiple of a to $d + ef\omega$ if necessary, we may assume that d satisfies $-a/2 \le d < a/2$. If we take a, e and d as above, then it is clear by our choice that the integers a, d, e are uniquely determined by the ideal $\mathfrak{a}$. Now we show that if $\mathfrak{a}$ is primitive in addition, then we have $e = 1$ in the above. Indeed, since $\mathfrak{a}$ is an ideal of $\mathfrak{O}_f$, we have $f\omega\mathfrak{a} \subset \mathfrak{a}$ and hence $af\omega \in \mathfrak{a}$. So we have $af\omega = xa + y(d + fe\omega)$ for some integers x, y. This means $a = ye$ and $xa + yd = 0$. So $a/e = y \in \mathbf{Z}$ and $d/e = -x \in \mathbf{Z}$. So if we put $\mathfrak{b} = \mathbf{Z}(a/e) + \mathbf{Z}(d/e + f\omega)$, then we have $\mathfrak{a} = e\mathfrak{b}$. Since $\mathfrak{a}$ is an ideal of $\mathfrak{O}_f$, we have $e(\alpha\mathfrak{b}) = \alpha\mathfrak{a} \subset \mathfrak{a} = e\mathfrak{b}$, and hence we have $\alpha\mathfrak{b} \subset \mathfrak{b}$. So $\mathfrak{b}$ is also an ideal of $\mathfrak{O}_f$. Since we assumed that $\mathfrak{a}$ is primitive, we have $e = 1$. Since $\mathfrak{O}_f = \mathbf{Z} + \mathbf{Z}f\omega = \mathbf{Z} + \mathbf{Z}(d + f\omega)$ and $\mathfrak{a} = \mathbf{Z}a + \mathbf{Z}(d + f\omega)$, we have $\mathfrak{O}_f/\mathfrak{a} \cong \mathbf{Z}/a\mathbf{Z}$ and the norm of the primitive ideal $\mathfrak{a}$ is a. □

We call $(a, d + ef\omega)$ in the above lemma the standard basis of the ideal $\mathfrak{a}$. Although we described the standard basis of an ideal above, we are not claiming that the module $M = \mathbf{Z}a + \mathbf{Z}(d + f\omega)$ is always an ideal for any random choice of $a, d \in \mathbf{Z}$. The module M is an $\mathfrak{O}_f$-ideal only when a, d, f satisfy some special conditions. We shall describe these conditions below.

Lemma 6.3. *The module* $M = \mathbf{Z}a + \mathbf{Z}(d + f\omega)$ *is an ideal of* $\mathfrak{O}_f$ *if and only if*

$$a | N(d + f\omega).$$

This is equivalent to saying that there exists an integer c such that

$$b^2 - f^2 D_K = 4ac,$$

where we put $b = Tr(d + f\omega)$.

Proof. All we need to do is to examine the condition $f\omega M \subset M$. We have $f\omega \times a = (-d)a + a(d + f\omega) \in M$. As for $f\omega(d + f\omega)$, we have

$$f\omega(d + f\omega) = (d + Tr(f\omega))(d + f\omega) - N(d + f\omega). \tag{6.2}$$

So this belongs to M if and only if $N(d + f\omega)$ is divisible by a. Since we have

$$\begin{aligned} b^2 - f^2 D_K &= (2d + fTr(\omega))^2 - f^2(Tr(\omega)^2 - 4N(\omega)) \\ &= 4(d^2 + dfTr(\omega) + f^2 N(\omega)) \\ &= 4N(d + f\omega), \end{aligned}$$

this is equivalent to say that

$$b^2 - f^2 D_K = 4ac$$

for some $c \in \mathbf{Z}$. □

Remark 6.4. (1) Notation being the same as in the last lemma, we have

$$d + f\omega = \frac{b + \sqrt{D}}{2},$$

where $D = f^2 D_K$: the discriminant of $\mathfrak{O}_f$.

(2) The norm a of a primitive ideal and the conductor f might not be coprime. For example, the module $\mathbf{Z}5 + \mathbf{Z}5\sqrt{2}$ is a primitive ideal of the order $\mathfrak{O}_5$ of the quadratic field $K = \mathbf{Q}(\sqrt{2})$ with conductor 5 and its norm is 5. We can write $\mathbf{Z}5 + \mathbf{Z}5\sqrt{2} = 5(\mathbf{Z} + \mathbf{Z}\sqrt{2})$, but $\mathbf{Z} + \mathbf{Z}\sqrt{2}$ is not an ideal (nor a subset) of $\mathfrak{O}_5$.

For an ideal $\mathfrak{a}$ of $\mathfrak{O}_f$, the ring $\{x \in K; x\mathfrak{a} \subset \mathfrak{a}\}$ is a subring of $\mathfrak{O}_{max}$ which contains $\mathfrak{O}_f$, so it is $\mathfrak{O}_{f'}$ for some divisor f' of f. When this coincides with $\mathfrak{O}_f$, that is, when $f = f'$, we say that $\mathfrak{a}$ is a proper $\mathfrak{O}_f$-ideal. (In the usual ring theory, the term *proper ideal* is often used to indicate the ideal which is not the whole ring itself. Note that our usage here is different from that.) Next, we shall see the condition that a primitive ideal of $\mathfrak{O}_f$ is a proper $\mathfrak{O}_f$-ideal.

Lemma 6.5. *For a primitive ideal $\mathfrak{a}$ of $\mathfrak{O}_f$, we write $\mathfrak{a} = \mathbf{Z}a + \mathbf{Z}(d + f\omega)$ by the standard basis and define integers b, c as in Lemma 6.3. Then the ideal $\mathfrak{a}$ is a proper $\mathfrak{O}_f$-ideal if and only if gcd of a, b, c is one.*

Proof. Assume that $\mathfrak{a}$ is an $\mathfrak{O}_{f'}$ module for a divisor f' of f If $\mathfrak{a}$ is an $\mathfrak{O}_{f'}$ module, then $af'\omega \in \mathfrak{a}$, hence we have $af'\omega = xa + y(d + f\omega)$. Comparing the coefficients, we see $a = y(f/f')$ and $xa + yd = 0$. But since $a > 0$, we have $y \neq 0$ and so $d = -x(f/f')$. Hence we have $a, d \in (f/f')\mathbf{Z}$. So $bf'/f = df'/f + f'Tr(w) \in \mathbf{Z}$ and $b \in (f/f')\mathbf{Z}$. Multiplying the Eq. (6.2) in the proof of Lemma 6.3 by f'/f, we have

$$f'\omega(d + f\omega) = (f'd/f + f'Tr(\omega))(d + f\omega) - acf'/f.$$

Since $f'\omega(d + f\omega) \in \mathfrak{a}$, we have $cf'/f \in \mathbf{Z}$, i.e. $c \in f/f'\mathbf{Z}$ So if $f' < f$, then the g.c.d. of a, b, c is bigger than one. Conversely, assume that n is the greatest common divisor of a, b, c. We first show that n always divides f. Indeed, write $a = na_0, b = nb_0, c = nc_0$. Then $b^2 - 4ac = n^2(b_0^2 - 4a_0c_0) = f^2D_K$. If D_K is odd, then D_K does not contain any square factor, so we have $n|f$. If D_K is even, then $D_K = 4m$, where $m \equiv 2$ or $3 \bmod 4$ and m is square-free. If n is odd, then by the same reason as before, we have $n|f$. Assume n is even and $n = 2n_0$. Then we have $n_0|f$. If f/n_0 is odd, then $(f/n_0)^2m \equiv 2$ or $3 \bmod 4$ but this is equal to $b_0^2 - 4a_0c_0$ which is 0 or 1 mod 4 and we have a contradiction. So f/n_0 is even and we have $n|f$. So if we assume that $n > 1$ and define f' by $nf' = f$, then $f' < f$ and $a, b, c \in (f/f')\mathbf{Z}$. Then we also have $d \in (f/f')\mathbf{Z}$. Indeed we have $(f/f')|b = Tr(d + f\omega) = 2d + fTr(\omega)$ and $(f/f')^2|ac = N(d + f\omega) = d^2 + 2dfTr(\omega) + f^2N(\omega)$. So we see that $(f/f')^2$ divides

$$(2d + fTr(\omega))^2 - 2(d^2 + 2dfTr(\omega) + f^2N(\omega)) = 2d^2 + f^2(Tr(\omega)^2 - 2N(\omega)).$$

Since f/f' divides f, we have $(f/f')^2|2d^2$. So we have $(f/f')|d$. So we have $a, b, c, d \in (f/f')\mathbf{Z}$. Hence we see that $af'\omega = (af'/f)f\omega = -a(df'/f) + (af'/f)(d + f\omega) \in \mathfrak{a}$ and $f'\omega(d + f\omega) = (df'/f) + f'Tr(\omega))(d + f\omega)f\omega - ac(f'/f) \in \mathfrak{a}$, so $\mathfrak{a}$ is an ideal of $\mathfrak{O}_{f'}$. □

Lemma 6.6. *If the discriminant of a quadratic form $ax^2 + bxy + cy^2$ is not a square, then there exists a quadratic field K and a natural number f such that $b^2 - 4ac = f^2D_K$.*

Proof. We have $b^2 - 4ac \equiv 0 \bmod 4$ or $1 \bmod 4$. So we can write $b^2 - 4ac = 2^e f_0^2 m_0$ for some odd natural number f_0, some odd number m_0 with no square factor and some non-negative integer e. If $e = 0$ then $m_0 \equiv 1 \bmod 4$. Since we assumed that $b^2 - 4ac$ is not a square, the field $K = \mathbf{Q}(\sqrt{m_0})$ is quadratic and $D_K = m_0$. Then D_K and $f = f_0$ satisfy the demand. If $e \neq 0$, then

$b^2 - 4ac \equiv 0 \bmod 4$, so we have $e \geq 2$. If e is odd, then put $K = \mathbf{Q}(\sqrt{2m_0})$. Then we have $D_K = 8m_0$ and $f = 2^{(e-3)/2} f_0$ satisfies the demand. If e is even, put $K = \mathbf{Q}(\sqrt{m_0})$. If $m_0 \equiv 1 \bmod 4$ then we have $D_K = m_0$, and then put $f = 2^{e/2} f_0$. If $m_0 \equiv 3 \bmod 4$ then we have $D_K = 4m_0$ and put $f = 2^{(e-2)/2} f_0$. These satisfy the demand of the lemma. □

This lemma suggests that there is a deep relation between binary quadratic forms and quadratic fields. For each proper primitive ideal $\mathfrak{a}$ of an order of a quadratic field, using a, b, c of Lemma 6.3, we define a quadratic form $Q_{\mathfrak{a}}$ by

$$Q_{\mathfrak{a}}(x, y) = ax^2 + bxy + cy^2.$$

We can also write this as

$$Q_{\mathfrak{a}}(x, y) = N(\mathfrak{a})^{-1} N\left(xa + y\frac{b + \sqrt{D}}{2}\right) = aN\left(x + y\frac{b + \sqrt{D}}{2a}\right).$$

Here we put $D = b^2 - 4ac$, and a and $\frac{b+\sqrt{D}}{2} = d + f\omega$ form a basis of $\mathfrak{a}$.

Now in order to see relations between equivalence classes of quadratic forms with some equivalent classes of ideals of a quadratic order, we give the following definitions. We say that two proper $\mathfrak{O}_f$-ideals $\mathfrak{a}$ and $\mathfrak{b}$ are equivalent in the wide sense if there exists $\alpha \in K^{\times}$ such that $\mathfrak{a} = (\alpha)\mathfrak{b}$. If $N(\alpha) > 0$ in the above in addition, we say that $\mathfrak{a}$ and $\mathfrak{b}$ are equivalent in the narrow sense. Here the condition $N(\alpha) > 0$ is satisfied automatically if K is an imaginary quadratic field. If K is a real quadratic field, then taking $-\alpha$ instead of α if necessary, we may replace the condition $N(\alpha) > 0$ by the condition $\alpha > 0$ and $\overline{\alpha} > 0$. Actually this is the usual definition which fits into the general theory. The number of equivalence classes in the wide (narrow) sense of proper $\mathfrak{O}_f$-ideals is called the class number in the wide (narrow) sense of $\mathfrak{O}_f$. It is common to call the number of equivalence classes in the wide sense just the class number. If K is an imaginary quadratic field, the definition in the narrow sense and that in the wide sense are the same, as we saw already.

Now let K be a quadratic field and let D_K be the fundamental discriminant of K. Let f be a natural number. We define a mapping from the set of proper primitive ideals $\mathfrak{a}$ of $\mathfrak{O}_f$ to the set of quadratic forms by setting $Q_{\mathfrak{a}}(x, y) = ax^2 + bxy + cy^2$ by using the standard basis such that $\mathfrak{a} = \mathbf{Z}a + \mathbf{Z}(d + f\omega)$, where we define b, c by $b = 2d + Tr(\omega)$, $N(d + f\omega) = ac$. Since the standard basis is unique, this is a well-defined mapping from ideals.

This mapping does not give a bijective mapping of proper primitive ideals to primitive quadratic forms. But we can prove the following claim.

Theorem 6.7. *When K is an imaginary quadratic field, that is, if $D_K < 0$, then the above mapping induces a bijective correspondence between the set of equivalence classes in the narrow sense of proper $\mathfrak{O}_f$ ideals to the set of $SL_2(\mathbf{Z})$-equivalence classes of positive definite primitive quadratic forms with discriminant $f^2 D_K$. If K is a real quadratic field, that is, if $D_K > 0$, then this gives a bijection from* *the set of equivalence classes in the narrow (wide) sense of proper $\mathfrak{O}_f$ ideals to the set of $SL_2(\mathbf{Z})$-equivalence ($GL_2(\mathbf{Z})$-equivalence) classes of indefinite primitive quadratic forms with discriminant $f^2 D_K$.*

Proof. For simplicity, we write $L(D) = L^*(D)$ if $D > 0$ and $L(D) = L^*_+(D)$ if $D < 0$. We denote by $L(D)/\sim$ the set of $GL_2(\mathbf{Z})$ equivalence classes in $L(D)$ and by $L(D)/\approx$ the set of $SL_2(\mathbf{Z})$ equivalence classes. Since any $\mathfrak{O}_f$ ideal is equivalent (both in the wide sense and narrow sense) to the primitive ideal, it is sufficient to consider primitive ideals. For a given proper primitive $\mathfrak{O}_f$ ideal, we define a, b, c as before. Then since $(a, b, c) = 1$, the quadratic form $Q_\mathfrak{a}(x, y) = ax^2 + bxy + cy^2$ is primitive and the discriminant of $Q_\mathfrak{a}$ is $f^2 D_K$. By our choice of the standard basis of $\mathfrak{a}$, we have $a > 0$. So if $D(S) < 0$, then $Q_\mathfrak{a}(x, y)$ is positive definite. So the mapping $\mathfrak{a}$ to $Q_\mathfrak{a}(x, y)$ induces mappings to $L(D)/\sim$ and to $L(D)/\approx$. We first show that this mapping is surjective. Take a primitive quadratic form $Q(x, y) = ax^2 + bxy + cy^2$. If this is positive definite, then we have $a > 0$ automatically. If this is indefinite, then we might have $a < 0$, but changing the quadratic form by $SL_2(\mathbf{Z})$ equivalence, we may assume that $a > 0$. This is proved as follows. Assume that $a < 0$. For a given x, y, we have $4a(ax^2 + bxy + cy^2) = (2ax + by)^2 + (4ac - b^2)y^2$. We have $4ac - b^2 < 0$ by the assumption that the quadratic form is indefinite. We fix an integer y such that $|a|/\sqrt{b^2 - 4ac} < y$. Choosing x suitably, we can assume that $2ax + by$ is in any set of representatives modulo $2a$. So we can assume that $|2ax + by| \leq |a|$. So we have $(2ax + by)^2 - (b^2 - 4ac)y^2 < 0$ for some x. This means that there exists $(x, y) \in \mathbf{Z}^2$ such that $ax^2 + bxy + cy^2 > 0$. This property still holds if we divide x, y by the g.c.d. of x and y, so we may assume that x and y are coprime. We can take an element A of $SL_2(\mathbf{Z})$ such that (x, y) is the first row of A. So the (1,1) component of $A \begin{pmatrix} a & b/2 \\ b/2 & c \end{pmatrix} {}^tA$ is positive. This proves the assertion. We assume $a > 0$ from now on. Let D be the discriminant of Q, that is, $D = b^2 - 4ac = f^2 D_K$. Since we have $(fTr(\omega))^2 = f^2(D_K + 4N(\omega))$, we have $(fTr(\omega))^2 \equiv f^2 D_K \equiv b^2 \bmod 4$ and we have $b \equiv fTr(\omega) \bmod 2$. So if we define d by $b = 2d + fTr(\omega)$, then

$$\mathfrak{a} = \mathbf{Z}a + \mathbf{Z}(d + f\omega) = \mathbf{Z}a + \mathbf{Z}\frac{b + \sqrt{D}}{2}$$

is a proper primitive ideal of $\mathfrak{O}_f$ since we assumed $gcd(a, b, c) = 1$. This d might not satisfy the condition $-a/2 \leq d < a/2$, but changing b to $b_0 = b - 2ak$ for some k, we see that $d_0 = d - ak$ satisfies the condition. So we take $k \in \mathbf{Z}$ such that $-a/2 \leq d_0 = d - ak < a/2$. If we define $b_0 = b - 2ak$ and $c_0 = c - bk + ak^2$, then $b_0^2 - 4ac_0 = f^2 D_k$ and we have $ac_0 = N(d_0 + f\omega)$, $b_0 = 2d_0 + fTr(\omega)$. So we have $\mathfrak{a} = \mathbf{Z}a + \mathbf{Z}(d_0 + f\omega)$. We have

$$\begin{aligned} Q_\mathfrak{a}(x, y) &= a_0x^2 + b_0xy + c_0y^2 \\ &= a(x - ky)^2 + b(x - ky)y + cy^2, \end{aligned}$$

so this is $SL_2(\mathbf{Z})$ equivalent (a fortiori, $GL_2(\mathbf{Z})$-equivalent) to $Q(x, y)$. So the mapping to $L(D)/\sim$ or $L(D)/\approx$ is surjective. Next we see that these mappings to $L(D)/\sim$ and $L(D)/\approx$ induce well defined mappings from ideal classes in the wide sense and in the narrow sense, respectively. Write two proper primitive $\mathfrak{O}_f$ ideals by standard bases as $\mathfrak{a}_i = \mathbf{Z}a_i + \mathbf{Z}(b_i + \sqrt{D})/2$, where $D = b_i^2 - 4a_ic_i$ $(i = 1, 2)$, and assume that $\mathfrak{a}_2 = \mathfrak{a}_1\alpha$ for some $\alpha \in \mathbf{Q}(\sqrt{D})^\times$. Then there exists $A \in GL_2(\mathbf{Z})$ such that $(a_2, (b_2 + \sqrt{D})/2) = (a_1\alpha, (b_1 + \sqrt{D})\alpha/2)A$. Since we have

$$\begin{pmatrix} a_2 & \frac{b_2+\sqrt{D}}{2} \\ a_2 & \frac{b_2-\sqrt{D}}{2} \end{pmatrix} = \begin{pmatrix} a_1 & \frac{b_1+\sqrt{D}}{2} \\ a_1 & \frac{b_1-\sqrt{D}}{2} \end{pmatrix} \begin{pmatrix} \alpha & 0 \\ 0 & \overline{\alpha} \end{pmatrix} A,$$

we have $a_2 = a_1N(\alpha)\det(A)$. Writing $(X, Y) = (x, y)\,{}^tA$, we also have

$$\begin{aligned} a_2(a_2x^2 + b_2xy + c_2y^2) &= N\left(a_2x + \frac{b_2 + \sqrt{D}}{2}y\right) \\ &= a_1N(\alpha)(a_1X^2 + b_1XY + c_1Y^2) \\ &= a_2\det(A)(a_1X^2 + b_1XY + c_1Y^2). \end{aligned}$$

Since $a_i > 0$ for $i = 1, 2$, we have $N(\alpha)\det(A) > 0$. So we are done. Next we see that the mapping is injective. For two primitive proper O_f ideals $\mathfrak{a}_1 = \mathbf{Z}a_1 + \mathbf{Z}(d_1 + f\omega)$ and $\mathfrak{a}_2 = \mathbf{Z}a_2 + \mathbf{Z}(d_2 + f\omega)$, we define

$$\begin{aligned} Q_{\mathfrak{a}_1}(x, y) &= a_1x^2 + b_1xy + c_1y^2 = a_1^{-1}N(a_1x + (d_1 + f\omega)y) \quad \text{and} \\ Q_{\mathfrak{a}_2}(x, y) &= a_2x^2 + b_2xy + c_2y^2 = a_2^{-1}N(a_2x + (d_2 + f\omega)y), \end{aligned}$$

as before. Assume that $Q_{\mathfrak{a}_2}(x, y) = \det(A)Q_{\mathfrak{a}_1}((x, y)A)$ for some $A \in GL_2(\mathbf{Z})$. We define $\omega_1, \omega_2 \in K = \mathbf{Q}(\sqrt{D_K})$ by

$$(\omega_1, \omega_2) = (a_1, d_1 + f\omega)\,{}^tA = \left(a_1, \frac{b_1 + \sqrt{D}}{2}\right){}^tA.$$

and put $\alpha = a_2/\omega_1$. We will show that $\mathfrak{a}_2 = \alpha\mathfrak{a}_1$ and $\det(A)N(\alpha) > 0$. Since $a_1x + (d_1 + f\omega)y = (x, y)\begin{pmatrix} a_1 \\ d_1 + f\omega \end{pmatrix}$, and $A\begin{pmatrix} a_1 \\ d_1 + f\omega \end{pmatrix} = \begin{pmatrix} \omega_1 \\ \omega_2 \end{pmatrix}$, we have

$$\begin{aligned} Q_{\mathfrak{a}_1}((x, y)A) &= a_1^{-1}N\left((x, y)A\begin{pmatrix} a_1 \\ d_1 + f\omega \end{pmatrix}\right) = a_1^{-1}N(x\omega_1 + y\omega_2) \\ &= a_1^{-1}(x^2N(\omega_1) + xyTr(\omega_1\overline{\omega_2}) + y^2N(\omega_2)). \end{aligned}$$

Since this is equal to $\det(A)Q_{\mathfrak{a}_1}(x, y)$, we have

$$N(\omega_1) = \det(A)a_1a_2 \quad \text{and} \quad Tr(\overline{\omega_1}\omega_2) = \det(A)a\_b_2.$$

So we have

$$N(\alpha)\det(A) = \det(A)a_2^2/N(\omega_1) = a_2/a_1 > 0. \tag{6.3}$$

Now, since $A \in GL_2(\mathbf{Z})$, we have $\mathfrak{a}_1 = \mathbf{Z}\omega_1 + \mathbf{Z}\omega_2$. So we have $\mathfrak{a}_1\alpha = \mathbf{Z}a_2 + \mathbf{Z}(a_2\omega_2/\omega_1)$. So it is sufficient to show that $a_2\omega_2/\omega_1 = d_2 + f\omega$. We have $a_2\omega_2/\omega_1 = a_2\omega_2\overline{\omega_1}/N(\omega_1) = \det(A)\omega_2\overline{\omega_1}/a_1$. We must calculate $\omega_2\overline{\omega_1}$. Taking the conjugate of the definition of ω_i, we have

$$\begin{pmatrix} \omega_1, \omega_2 \\ \overline{\omega}_1, \overline{\omega}_2 \end{pmatrix} = \begin{pmatrix} a_1, d_1 + f\omega \\ a_1, d_1 + f\overline{\omega} \end{pmatrix} {}^tA.$$

So taking the determinant, we have

$$-\omega_1\overline{\omega}_2 + \omega_2\overline{\omega}_1 = \det(A)a_1 f(\omega - \overline{\omega}).$$

So adding $\omega_1\overline{\omega}_2 + \overline{\omega_1}\omega_2 = \det(A)b_2a_1 = \det(A)a_1(2d_2 + Tr(f\omega))$ to the both sides and dividing by 2, we have $\omega_2\overline{\omega}_1 = \det(A)a_1(d_2 + f\omega) = (N(\omega_1)/a_2)(d_2 + f\omega)$. So we have $\mathfrak{a}_2 = \mathfrak{a}_1\alpha$. Hence $\mathfrak{a}_1$ and $\mathfrak{a}_2$ are equivalent in the wide sense, and if $\det(A) = 1$, then equivalent in the narrow sense. Hence the mapping is injective.

□

When $D_K > 0$, the equivalences of proper O_f-ideals in the wide sense and in the narrow sense are the same if and only if there exists $\varepsilon \in O_f^\times$ with $N(\varepsilon) = -1$, as we can see easily. We give the corresponding fact in the case of symmetric matrices (or quadratic forms) in the following proposition.

Proposition 6.8. *Assume that $D = f^2D_K$ is positive and is not a square. Assume that for $S_1, S_2 \in L^*(D)$ we have $S_2 = -{}^tBS_1B$ for some $B \in GL_2(\mathbf{Z})$ with* $\det(B) = -1$. *Then there exists $C \in SL_2(\mathbf{Z})$ such that $S_2 = {}^tCS_1C$ if and only if there exists $\varepsilon \in \mathfrak{O}_f^\times$ such that $N(\varepsilon) = -1$.*

Proof. We may assume that S_1 and S_2 are primitive. First we assume the existence of C. Then if we write $A = BC^{-1} \in GL_2(\mathbf{Z})$, then we have $\det(A) = -1$ and $S_1 = \det(A)\,{}^tAS_1A$. We have shown in the proof of the last theorem that S_1 is $SL_2(\mathbf{Z})$ equivalent to a matrix S in $L(D)$ corresponding to a quadratic form $Q_{\mathfrak{a}}(x, y)$ for some primitive O_f proper ideal $\mathfrak{a} = \mathbf{Z}a + \mathbf{Z}(d + f\omega)$. Now we put $(\omega_1, \omega_2) = (a, d + f\omega)A$ and $\alpha = a/\omega_1$, and apply the argument of the proof of the last theorem for $\mathfrak{a}_1 = \mathfrak{a}_2 = \mathfrak{a}$. Then by (6.3) we have $N(\alpha) < 0$ and we have $\mathfrak{a}\alpha = \mathfrak{a}$. Since $\mathfrak{a}$ is a proper O_f-ideal, we have $\alpha \in O_f$. We also have $\mathfrak{a} = \mathfrak{a}\alpha^{-1}$, so $\alpha^{-1} \in O_f$ and $N(\alpha)$ and $N(\alpha^{-1})$ are integers. So we have $N(\alpha) = -1$ and $\alpha \in O_f^\times$. Conversely, if there exists $\epsilon \in \mathfrak{O}_f^\times$ with $N(\epsilon) = -1$, then we have $\mathfrak{a}\epsilon = \mathfrak{a}$. If we define $A \in GL_2(\mathbf{Z})$ by

$$(a, d + f\omega)\epsilon = (a, d + f\omega)A,$$

then $\det(A) = N(\epsilon) = -1$ and $Q_{\mathfrak{a}}((x, y)A) = \det(A)Q_{\mathfrak{a}}(x, y)$, that is, $AS^t A = \det(A)S$. Assuming that $S_1 = PS^t P$ for $P \in SL_2(\mathbf{Z})$, $C = PAP^{-1}$ satisfies the condition in the proposition. □

Example 6.9. (1) If we put $K = \mathbf{Q}(\sqrt{m})$ and $\mathfrak{a} = \mathbf{Z}m + \mathbf{Z}\sqrt{m}$, then $Q_{\mathfrak{a}}(x, y) = |m|^{-1}N(|m|x + y\sqrt{m}) = |m|x^2 - sgn(m)y^2$.
(2) We take $K = \mathbf{Q}(\sqrt{-5})$. For an ideal $\mathfrak{a} = 3\mathbf{Z} + (1 + \sqrt{-5})\mathbf{Z}$ of $\mathfrak{O}_{max} = \mathbf{Z} + \mathbf{Z}\sqrt{-5}$, we have $Q_{\mathfrak{a}}(x, y) = 3x^2 + 2xy + 2y^2$. On the other hand, for $\mathfrak{O}_{max} = (1)$ itself, we have $Q_{(1)} = x^2 + 5y^2$. Any primitive quadratic form with discriminant -20 is equivalent to one of these two.

Remark 6.10. In the above, we did not treat negative definite quadratic forms. But this causes no problem since all these can be obtained by multiplying -1 to the positive definite ones. On the other hand, we excluded primitive quadratic forms with square discriminant. These quadratic forms are (as far as discriminant is not zero) very special quadratic forms in various senses. (For example, the volume of the fundamental domain with respect to the unit group (the automorphism group) is not finite etc.) We would like to omit such details, but at least we try to give a classification of equivalence classes below.

Let $Q(x, y) = ax^2 + bxy + cy^2$ be a primitive quadratic form and assume that $b^2 - 4ac = f^2$ for a positive integer f. First we see that we may assume $a = 0$ by replacing Q by an equivalent quadratic form. Indeed, if $a \neq 0$, then noting that the equation $ax^2 + bx + c = 0$ has a solution $x = (-b \pm f)/2a$, we denote by e the g.c.d. of $-b + f$ and $2a$ and define x_0, y_0 by $-b + f = x_0 e$ and $2a = y_0 e$. Then x_0 and y_0 are coprime and there exist w_0, $z_0 \in \mathbf{Z}$ such that $x_0 w_0 - y_0 z_0 = 1$. Since we have $a\left(\frac{x_0}{y_0}\right)^2 + b\left(\frac{x_0}{y_0}\right) + c = 0$, we get $ax_0^2 + bx_0 y_0 + cy_0^2 = 0$ and

$$\begin{pmatrix} x_0 & y_0 \\ z_0 & w_0 \end{pmatrix}\begin{pmatrix} a & b/2 \\ b/2 & c \end{pmatrix}\begin{pmatrix} x_0 & z_0 \\ y_0 & w_0 \end{pmatrix} = \begin{pmatrix} ax_0^2 + bx_0 y_0 + cy_0^2 & * \\ * & * \end{pmatrix} = \begin{pmatrix} 0 & * \\ * & * \end{pmatrix}.$$

So we can assume that $Q(x, y) = bxy + cy^2$ up to $SL_2(\mathbf{Z})$-equivalence. If the discriminant of Q is 0, then b is also 0, and by the assumption that it is primitive, we have $c = \pm 1$. It is easy to see that the quadratic forms y^2 and $-y^2$ are not $SL_2(\mathbf{Z})$-equivalent but and $GL_2(\mathbf{Z})$-equivalent. So we classified the case where the discriminant is zero. From now on, we assume that the discriminant is not zero. We consider an element $A = \begin{pmatrix} \alpha & \beta \\ \gamma & \delta \end{pmatrix} \in GL_2(\mathbf{Z})$ which maps $bxy + cy^2$ to quadratic forms whose coefficient of x^2 is 0. Since

$${}^t A \begin{pmatrix} 0 & b/2 \\ b/2 & c \end{pmatrix} A = \begin{pmatrix} b\alpha\gamma + c\gamma^2 & * \\ * & * \end{pmatrix},$$

we should have $\gamma = 0$ or $b\alpha + c\gamma = 0$. Since b and c are coprime, we have z, $w \in \mathbf{Z}$ such that $bz + cw = \pm 1$. So such an A is given either by

$$A = \pm \begin{pmatrix} c & z \\ -b & w \end{pmatrix} \text{ or } \pm \begin{pmatrix} 1 & z \\ 0 & \pm 1 \end{pmatrix}.$$

with $\det(A) = \pm 1$. For the former one, if we take $bz + cw = 1$, then we have $\det(A) = 1$ and

$${}^t A \begin{pmatrix} 0 & b/2 \\ b/2 & c \end{pmatrix} A = \begin{pmatrix} 0 & -b(bz+cw)/2 \\ -b(bz+cw)/2 & bzw + cw^2 \end{pmatrix} = \det(A) \begin{pmatrix} 0 & -b/2 \\ -b/2 & w \end{pmatrix},$$

so we can choose a representative of $SL_2(\mathbf{Z})$ equivalence classes such that $b > 0$. On the other hand, we have

$$\begin{pmatrix} 1 & 0 \\ z & 1 \end{pmatrix} \begin{pmatrix} 0 & b/2 \\ b/2 & c \end{pmatrix} \begin{pmatrix} 1 & z \\ 0 & 1 \end{pmatrix} = \begin{pmatrix} 0 & b/2 \\ b/2 & c + bz \end{pmatrix}$$

so we can take c among representatives modulo b. We have

$$-\begin{pmatrix} 1 & 0 \\ 0 & -1 \end{pmatrix} \begin{pmatrix} 0 & b/2 \\ b/2 & c \end{pmatrix} \begin{pmatrix} 1 & 0 \\ 0 & -1 \end{pmatrix} = \begin{pmatrix} 0 & b/2 \\ b/2 & -c \end{pmatrix}.$$

So representatives of $SL_2(\mathbf{Z})$-equivalence classes and $GL_2(\mathbf{Z})$-equivalence classes with discriminant $f^2 \neq 0$ ($f > 0$) can be taken in the set of the following matrices:

$$\begin{pmatrix} 0 & f/2 \\ f/2 & c \end{pmatrix}$$

with $0 \leq c \leq f - 1$ and $0 \leq c \leq f/2$, respectively. The fact that these are not equivalent with each other can be seen easily by the above calculation. In particular, the number of $SL_2(\mathbf{Z})$- or $GL_2(\mathbf{Z})$- equivalence classes of quadratic forms with a fixed square discriminant is finite. The number of $SL_2(\mathbf{Z})$ equivalence classes is f, and if we restrict to the primitive classes, the number is $\varphi(f)$, where $\varphi(f)$ is the Euler function. The case where the discriminant is not a square will be explained in the next section.

6.3 Class Number Formula of Quadratic Forms

In this section, first we prove the finiteness of the class number (the number of equivalence classes) of primitive quadratic forms. As we mentioned in the last section, this is equivalent to the finiteness of the class number (in the narrow sense or in the wide sense) of quadratic orders $\mathfrak{O}_f$, which are not necessarily maximal. Next we state a formula of the class numbers (in the wide sense) of quadratic orders. (The proof is postponed until Chap. 10.) As we saw in the last section, the class number in the narrow sense is easily obtained from the one in the wide sense. Actually, once we know the class number formula for the maximal order, the class number of an arbitrary order is determined. The reason for this can be explained if we use group-theoretic language, but in Chap. 10, we will give a different direct proof using zeta functions.

Proposition 6.11. *The number of $SL_2(\mathbf{Z})$ equivalence classes of binary primitive quadratic forms of a given discriminant is finite. In particular, the class number of binary primitive quadratic forms with discriminant f^2D_K is finite.*

Proof. We already explained the case where the discriminant is square. So, here we treat only those with discriminant f^2D_K, where K is a quadratic field and D_K is the fundamental discriminant of K. We fix f^2D_K now and we consider only those quadratic forms which have this number as discriminant. We say first that any primitive quadratic form is equivalent to $ax^2+bxy+cy^2$ with $|b| \le |a| \le |c|$. Indeed, we take a quadratic form $ax^2+bxy+cy^2$ so that $|a|$, the absolute value of the coefficient of x^2, is the smallest among those equivalent to the original one. Then by the transformation $(x,y) \to (-y,x)$, the coefficient x^2 becomes c, so by our choice of a, we have $|a| \le |c|$. For some integer $m \in \mathbf{Z}$, we have $-|a| \le b-2ma \le |a|$. So, changing the quadratic form to the equivalent one by the transformation $(x,y) \to (x-my,y)$, we have $a(x-my)^2+b(x-my)y+cy^2 = ax^2+(b-2am)xy+(c-bm+am^2)y^2$, we may assume that $-|a| \le b \le |a|$ from the first. So we can assume that $|b| \le |a| \le |c|$. (In particular, if $D_K<0$, we may assume that the quadratic form is positive definite, so we have $|b| \le a \le c$.) On the other hand, we have $b^2-4ac = f^2D_K$ by definition, so we have

$$4|ac|-b^2 \le |4ac-b^2| = f^2|D_K|.$$

But since $b^2 \le a^2$, $a^2 \le |ac|$, we have $3a^2 \le 4|ac|-b^2 \le f^2|D_K|$. So $|a| \le \sqrt{f^2|D_K|/3}$. It is obvious that there is only a finite number of such integers a. Hence the number of b is also finite by $|b| \le |a|$ and $c=(b^2-f^2D_K)/4a$ is also. Hence the assertion is proved. □

Let D_K be the fundamental discriminant of a quadratic field and put $D=f^2D_K$. We denote by $h(D)$ the class number in the wide sense of the order $\mathfrak{O}_f$ of the quadratic field $K=\mathbf{Q}+\mathbf{Q}\omega$. To describe the class number formula, we introduce the quadratic residue symbol. For an odd prime p and an integer a, the quadratic residue symbol $\left(\frac{a}{p}\right)$ with respect to p is defined as follows.

$$\left(\frac{a}{p}\right) = \begin{cases} 1 & \text{if } a \notin p\mathbf{Z} \text{ and there exists } x \in \mathbf{Z} \text{ such that } x^2 \equiv a \bmod p, \\ -1 & \text{if } a \notin p\mathbf{Z} \text{ and there is no } x \in \mathbf{Z} \text{ such that } x^2 \equiv a \bmod p, \\ 0 & \text{if } a \in p\mathbf{Z}. \end{cases}$$

The following three relations are known as the reciprocity law of the quadratic residue symbols.

(1) For odd primes p, q, we have

$$\left(\frac{q}{p}\right)\left(\frac{p}{q}\right) = (-1)^{(p-1)(q-1)/4}.$$

(2) For an odd prime p, we have

$$\left(\frac{-1}{p}\right) = (-1)^{(p-1)/2}.$$

(3) For an odd prime p, we have

$$\left(\frac{2}{p}\right) = (-1)^{(p^2-1)/8}.$$

The proof will be omitted here. (See e.g. [83].)

For the fundamental discriminant D_K of a quadratic field and a prime p, we define the notation $\chi_K(p)$ by

$$\chi_K(p) = \begin{cases} \left(\frac{D_K}{p}\right) & \text{if } p \text{ is an odd prime,} \\ 1 & \text{if } p = 2 \text{ and } D_K \equiv 1 \bmod 8, \\ -1 & \text{if } p = 2 \text{ and } D_K \equiv 5 \bmod 8, \\ 0 & \text{if } p = 2 \text{ and } D_K \equiv 0 \bmod 4. \end{cases}$$

We also define $\chi_K(1) = 1$ and

$$\chi_K(-1) = \begin{cases} 1 & \text{if } D_K > 0, \\ -1 & \text{if } D_K < 0. \end{cases}$$

Moreover, if u is an integer and if $u = \pm p_1^{e_1} \cdots p_r^{e_r}$ is a prime factor decomposition of u, we put $\chi_K(u) = \chi_K(\pm 1)\chi_K(p_1)^{e_1} \cdots \chi_K(p_r)^{e_r}$. Actually, by using the reciprocity law and other things, one can show that $\chi_K(u)$ depends only on $u \bmod D_K$. We omit the details of the proofs of this fact (cf. [107, §5]). The function χ_K prolonged in this way is a primitive Dirichlet character modulo $|D_K|$. It is common to write this character as $\left(\frac{D_K}{u}\right)$. So in this book, we sometimes use this notation instead of χ_K.

Now we can write down the class number formula by using the notation $\chi_K(u) = \left(\frac{D_K}{u}\right)$. When $D_K > 0$, by Dirichlet, it is known that $O_{max}^\times = \{\pm 1\} \times \{\varepsilon^n; n \in \mathbf{Z}\}$ for some $\varepsilon \in O_{max}^\times$ with $\varepsilon \neq \pm 1$. This ε is called the fundamental unit of K.

Theorem 6.12. (1) *The class number of the quadratic field K with discriminant D_K is given by*

$$h(D_K) = \begin{cases} -\dfrac{w}{D_K} \displaystyle\sum_{u=1}^{|D_K|-1} \left(\frac{D_K}{u}\right) u & \text{if } D_K < 0, \\ \log(\varepsilon) \displaystyle\sum_{u=1}^{D_K-1} \left(\frac{D_K}{u}\right) \log(\sin(u\pi/D_K)) & \text{if } D_K > 0. \end{cases}$$

Here w is the half of the number of units of $\mathfrak{O}_{max}$ and when $D_K > 0$, we denote by ε the fundamental unit of K.

(2) *The class number of the order $\mathfrak{O}_f$ of K with discriminant $D = f^2 D_K$ is given by*

$$h(D) = \frac{h(D_K) f}{[\mathfrak{O}_{max}^\times : \mathfrak{O}_f^\times]} \prod_{p|f} \left(1 - \frac{1}{p}\left(\frac{D_K}{p}\right)\right).$$

Here for any ring R, the notation $R^\times$ means the group of all units of R.

As we mentioned before, the class number in the narrow sense can be easily obtained by this formula, and the class number of primitive quadratic forms also. The proof of the class number formula will be given in Chap. 10 when K is an imaginary quadratic field. When K is a real quadratic field, the proof will be omitted from this book.

The class numbers of imaginary quadratic fields have a relation to the Bernoulli numbers. Indeed, if we put $\chi(u) = \left(\frac{D_K}{u}\right)$, then by Sect. 4.2, we have

$$B_{1,\chi} = \frac{1}{D_K} \sum_{u=1}^{|D_K|-1} \chi(u) u, \tag{6.4}$$

so if $D_K < 0$, we see

$$h(D_K) = -wB_{1,\chi}. \tag{6.5}$$

In particular, if p is a prime such that $p \equiv 3 \bmod 4$ and $p > 3$, then it is shown by using (6.4) and the reciprocity law of quadratic residues that the class number $h(-p)$ of $\mathbf{Q}(\sqrt{-p})$ is given by

$$h(-p) = -\frac{1}{p} \sum_{u=1}^{p-1} \left(\frac{u}{p}\right) u. \tag{6.6}$$

By the above formula, we see that $h(D) = 1$ for $D = -3, -4, -7, -8, -11, -19, -43, -67$ and -163. It is known that there are no other (fundamental discriminants of) imaginary quadratic fields such that $h(D_K) = 1$. Even if we take the non-maximal orders of imaginary quadratic fields into account, there are only four more orders with class number one, namely $D = -12, -16, -27$ and -28. In contrast to this, it is conjectured that there are infinitely many real quadratic fields with class number one,[2] but no proof is known.

Remark 6.13. For S_1 and $S_2 \in L^*(D)$, we can also define an equivalence class by $S_2 = {}^tAS_1A$ for some $A \in GL_2(\mathbf{Z})$. If S_1 and S_2 are $SL_2(\mathbf{Z})$ equivalent in the sense we defined before, then of course these are equivalent in the above sense. But the equivalences $S_2 = {}^tA_1S_1A_1$ and $S_2 = -{}^tA_2S_1A_2$ for some A_1 and $A_2 \in GL_2(\mathbf{Z})$ with $\det(A_1) = \det(A_2) = -1$ are different conditions in general. For example, for two symmetric matrices $S_1, S_2 \in L^*(12)$ given by

$$S_1 = \begin{pmatrix} 3 & 0 \\ 0 & -1 \end{pmatrix} \text{ and } S_2 = \begin{pmatrix} 1 & 0 \\ 0 & -3 \end{pmatrix},$$

there do not exist $A \in GL_2(\mathbf{Z})$ such that $S_2 = {}^tAS_1A$. This can be seen from the fact that the diophantine equation $a^2 - 3b^2 = -1$ does not have integer solutions. On the other hand, each corresponds to the symmetric matrix obtained from the standard basis of $3\mathbf{Z} + \sqrt{3}\mathbf{Z}$ or $\mathbf{Z} + \mathbf{Z}\sqrt{3}$. Since $\sqrt{3}(\mathbf{Z} + \mathbf{Z}\sqrt{3}) = 3\mathbf{Z} + \sqrt{3}\mathbf{Z}$, these ideals are equivalent in the wide sense. In fact, we have

$$-\begin{pmatrix} 0 & 1 \\ 1 & 0 \end{pmatrix}\begin{pmatrix} 3 & 0 \\ 0 & -1 \end{pmatrix}\begin{pmatrix} 0 & 1 \\ 1 & 0 \end{pmatrix} = \begin{pmatrix} 1 & 0 \\ 0 & -3 \end{pmatrix},$$

so we have $S_2 = \det(A){}^tASA$ for $A = \begin{pmatrix} 0 & 1 \\ 1 & 0 \end{pmatrix} \in GL_2(\mathbf{Z})$. We will give one more example. Take $S_3, S_4 \in L^*(23)$ given by

$$S_3 = \begin{pmatrix} 3 & 1/2 \\ 1/2 & 2 \end{pmatrix} \text{ and } S_4 = \begin{pmatrix} 3 & -1/2 \\ -1/2 & 2 \end{pmatrix}.$$

Then we have

$$S_4 = \begin{pmatrix} 1 & 0 \\ 0 & -1 \end{pmatrix} S_3 \begin{pmatrix} 1 & 0 \\ 0 & -1 \end{pmatrix}$$

but there exists no $A \in GL_2(\mathbf{Z})$ such that $S_4 = \det(A)\,{}^tAS_3A$. Indeed if there exists such A, then since S_3 and S_4 are positive definite, we should have $\det(A) > 0$. Then comparing diagonal components of both sides, we see that we should have $A = \pm 1_2$. This is a contradiction since $(1, 2)$ components do not coincide for $A = \pm 1_2$. We explain shortly the difference between $SL_2(\mathbf{Z})$ equivalence and the equivalence in the above sense. For any $S_1 \in L^*(D)$, define

[2] In Section 304 of his book on number theory, *Disquisitiones Arithmeticae* [35] Gauss stated his speculation on this and also his reasonable insight on the relation to the size of fundamental units.

$$S_2 = \begin{pmatrix} 1 & 0 \\ 0 & -1 \end{pmatrix} S_1 \begin{pmatrix} 1 & 0 \\ 0 & -1 \end{pmatrix}.$$

If S_2 is $SL_2(\mathbf{Z})$ equivalent to S_1, then S_1 (or the corresponding $SL_2(\mathbf{Z})$ equivalence class, or ideal or class in the narrow sense) is called ambig. In this case, for any $S \in L^*(D)$ such that $S = {}^t A S_1 A$ for some $A \in GL_2(\mathbf{Z})$, we see that S is $SL_2(\mathbf{Z})$ equivalent to S_1. So there is no difference between two equivalences. In the case when S_2 is not $SL_2(\mathbf{Z})$ equivalent to S_1, then the $GL_2(\mathbf{Z})$ equivalence class which contains S_1 contains two different $SL_2(\mathbf{Z})$ equivalent classes. To count the equivalence classes of $L^*(D)$ up to the equivalence that $S_2 = {}^t A S_1 A$ for $A \in GL_2(\mathbf{Z})$, we must count ambig classes, which is a part of the genus theory of Gauss. Since this is complicated in general, we do not explain this theory in this book. (See [29]. See also Exercises 6.21 and 10.25.)

Exercise 6.14. For a square-free integer m, show that 1 and ω in (6.1) gives a basis of the maximal order $\mathfrak{O}_{max}$ of $\mathbf{Q}(\sqrt{m})$.

Exercise 6.15. Show that the quadratic form $Q(x, y) = x^2 + bxy + cy^2$ with discriminant D is equivalent to the following quadratic form.

$$\begin{cases} x^2 - \frac{D}{4}y^2 & \text{if } b \text{ is even,} \\ x^2 + xy + \frac{1-D}{4}y^2 & \text{if } b \text{ is odd.} \end{cases}$$

Exercise 6.16. (1) Let $\mathfrak{a} = \mathbf{Z}a + \mathbf{Z}\frac{b+\sqrt{D}}{2}$ be a proper primitive ideal of order $\mathfrak{O}_f$ of discriminant $D = f^2 D_K$, where D_K is a discriminant of a quadratic field K. We assume $a > 0$. Show that $\mathfrak{a}$ is a principal ideal of $\mathfrak{O}_f$ (i.e. $\mathfrak{a} = \mathfrak{O}_f\alpha$ for some element $\alpha \in \mathfrak{O}_f$) if and only if there exists integers x and y such that $ax^2 + bxy + cy^2 = \pm 1$.

(2) Show that the class of quadratic forms in the wide sense corresponding to a principal ideal $\mathfrak{a}$ contains the one given in Exercise 6.15.

(3) Notation being as above, give an example of a principal ideal $\mathfrak{a}$ such that there exists no $x, y \in \mathbf{Z}$ with $ax^2 + bxy + cy^2 = 1$.

(4) Give an example of an equivalence class of quadratic forms $Q(x, y)$ in the narrow sense corresponding to a principal ideal (in the wide sense) such that $Q(x, y)$ is not $SL_2(\mathbf{Z})$ equivalent to any form given in Exercise 6.15.

Exercise 6.17. Fix a positive discriminant $D = f^2 D_K$, where K is a real quadratic field, and consider an integral quadratic form $Q(x, y) = Ax^2 + Bxy + Cy^2$ whose discriminant is D.

(1) Let $|a|$ be minimum among coefficients of x^2 of quadratic forms which are equivalent to Q. Show that there exists a quadratic form $Q_2(x, y) = ax^2 + bxy + y^2$ equivalent to Q such that $\sqrt{D} - 2|a| < b < \sqrt{D}$ and $|a| \leq |c|$.

(2) Taking Q_2 as in (1), show that $ac < 0$ and $0 < b$. (Use $(\sqrt{D} + b)(\sqrt{D} - b) = -4ac$.) In particular, show that $0 < b < \sqrt{D}$, $\sqrt{D} - b < 2|a| < \sqrt{D} + b$.

Exercise 6.18. Show that for any integral quadratic form Q, there exists a quadratic form $SL_2(\mathbf{Z})$ equivalent to Q such that the coefficient of x^2 is positive. (See the proof of Theorem 6.7, for example.)

Exercise 6.19. (1) Show that the following pairs of quadratic forms in (i) and (ii) are $SL_2(\mathbf{Z})$ equivalent respectively.

$$\text{(i)}\quad ax^2 + bxy + ay^2 \quad \text{and} \quad ax^2 - bxy + ay^2.$$

$$\text{(ii)}\quad ax^2 + axy + cy^2 \quad \text{and} \quad ax^2 - axy + cy^2.$$

(2) Assume that $D = f^2 D_K < 0$ and a positive definite quadratic form $Q(x, y) = ax^2 + bxy + cy^2$ has discriminant D. As shown in the proof of Proposition 6.11, by replacing $SL_2(\mathbf{Z})$ equivalence, we may assume that $|b| \le a \le c$. Such quadratic forms are called reduced. Conversely, assume that there are two $SL_2(\mathbf{Z})$ equivalent positive definite quadratic forms $Q_i(x, y) = a_i x^2 + b_i xy + c_i y^2$ with discriminant D with $|b_i| \le a_i \le c_i$ $(1 \le i \le 2)$. Show that $(a_1, b_1, c_1) = (a_2, b_2, c_2)$ except for the following pairs.

(i) $a_1 = c_1$ and $(a_2, b_2, c_2) = (a_1, -b_1, a_1)$.
(ii) $a_1 = b_1$ and $(a_2, b_2, c_2) = (a_1, -a_1, c_1)$.

Hint: Write down the transformation explicitly as $a_2 = a_1\alpha^2 + b_1\alpha\gamma + c_1\gamma^2$ and assuming that $a_2 \ge a_1$ and using the condition of a_1, b_1, c_1, show that $|\alpha\gamma| \le 1$.

Exercise 6.20. (1) Show that $h(D) = 1$ for $D = -3, -4, -7, -8, -11, -12, -16, -19, -27, -28, -43, -67$ and -163.
(2) Show that $h(-20) = 2$ and give a complete set of representatives of the ideal classes.
(3) Show that $h(40) = 2$ and give a complete set of representatives of the ideal classes.

Exercise 6.21. Fix a discriminant $D = f^2 D_K$ and consider $S = \begin{pmatrix} a & b/2 \\ b/2 & c \end{pmatrix} \in L^*(D)$. Show that the following conditions on S are equivalent (the class to which such an S belongs is called an ambig class).

(1) There exists $A \in GL_2(\mathbf{Z})$ with $\det(A) = -1$ such that ${}^tASA = S$.
(2) The matrix S is $SL_2(\mathbf{Z})$ equivalent to

$$\begin{pmatrix} a & -b/2 \\ -b/2 & c \end{pmatrix}.$$

(3) There exists $S' = \begin{pmatrix} a' & b'/2 \\ b'/2 & c' \end{pmatrix} \in L^*(D)$ which is $SL_2(\mathbf{Z})$ equivalent to S such that b' is divisible by a'. (Here we may take $b' = 0$ or $b' = a'$.)

Hint: If ${}^tASA = S$ with $\det(A) = -1$, then show that $Tr(A) = 0$ and $A^2 = \begin{pmatrix} 1 & 0 \\ 0 & 1 \end{pmatrix}$. Then reduce the case when A is upper triangular.

Chapter 7
Congruence Between Bernoulli Numbers and Class Numbers of Imaginary Quadratic Fields

In this chapter, we prove a congruence relation between Bernoulli numbers and class numbers of imaginary quadratic fields (Theorem 7.1). For that purpose, we study in Sect. 7.2 a certain type of power series which Hurwitz[1] introduced (we call this "Hurwitz-integral" series). Although one may prove Theorem 7.2 without using this, we nevertheless introduce this notion because of its elegance and usefulness.

7.1 Congruence Between Bernoulli Numbers and Class Numbers

For a prime p satisfying $p \equiv 3 \bmod 4$, recall that $h(-p)$ denotes the class number of the imaginary quadratic field of discriminant $-p$.

Theorem 7.1. *Let $p > 3$ be a prime, $p \equiv 3 \bmod 4$. We have the congruence*[2]

$$h(-p) \equiv -2B_{\frac{p+1}{2}} \mod p.$$

Remark 7.2. (1) By the theorem of Clausen and von Staudt (Theorem 3.1), the denominator of $B_{\frac{p+1}{2}}$ is prime to p.

(2) Using Euler's formula $\zeta(1-k) = -B_k/k$ (Theorem 5.4 (2), p. 72), we have

[1]Adolf Hurwitz (born on March 26, 1859 in Hildesheim, Germany—died on November 18, 1919 in Zürich, Switzerland).

[2]This is due to Augustin Louis Cauchy (born on August 21, 1789 in Paris, France—died on May 23, 1857 in Sceaux, France). He expressed the left-hand side by a difference of numbers of quadratic residues and non-residues in the interval $(0, p/2)$. That this is equal to the class number is nothing but the class number formula of Dirichlet.

T. Arakawa et al., *Bernoulli Numbers and Zeta Functions*, Springer Monographs in Mathematics, DOI 10.1007/978-4-431-54919-2_7, © Springer Japan 2014

$$\zeta\left(\frac{1-p}{2}\right) = -\frac{B_{\frac{p+1}{2}}}{\frac{p+1}{2}} \equiv -2B_{\frac{p+1}{2}} \mod p$$

and hence the theorem can be written as

$$h(-p) \equiv \zeta\left(\frac{1-p}{2}\right) \mod p.$$

The class number formula (6.6) at the end of the last chapter gives

$$1 \le h(-p) = \frac{1}{p}\left|\sum_{u=1}^{p-1}\left(\frac{u}{p}\right)u\right| \le \frac{1}{p}\sum_{u=1}^{p-1} u = \frac{p-1}{2}.$$

This shows that the class number $h(-p)$ is uniquely determined by its value mod p. Therefore, the above congruence asserts that the value $\zeta((1-p)/2)$ of the Riemann zeta function "knows completely" the class number $h(-p)$.

(3) For a more advanced interpretation of this theorem using p-adic modular forms, see [82].

(4) An analogous result in the real quadratic case is known as the following Ankeny[3]–Artin[4]–Chowla[5] congruence [4]: Let p be a prime number congruent to 1 modulo 4, and $\varepsilon = (t + u\sqrt{p})/2$ be the fundamental unit. Then, we have

$$\frac{u}{t}h(p) \equiv B_{\frac{p-1}{2}} \mod p.$$

Example 7.3. Let us compute several class numbers by using the theorem.

(1) For $p = 7$,

$$-2B_{\frac{7+1}{2}} = -2B_4 = -2\left(-\frac{1}{30}\right) = \frac{1}{15} \equiv 1 \mod 7,$$

so $h(-7) = 1$.

(2) For $p = 11$,

$$-2B_{\frac{11+1}{2}} = -2B_6 = -2\left(\frac{1}{42}\right) = -\frac{1}{21} = \frac{1}{1-22} \equiv 1 \mod 11,$$

so $h(-11) = 1$.

[3]Nesmith Cornett Ankeny (born in 1927 in Walla Walla, USA—died on August 4, 1993 in Seattle, USA).

[4]Emil Artin (born on March 3, 1898 in Vienna, Austria—died on December 20, 1962 in Hamburg, Germany).

[5]Sarvadaman D. S. Chowla (born on October 22, 1907 in London, England—died on December 10, 1995 in Laramie, USA).

(3) For $p = 19$,

$$-2B_{\frac{19+1}{2}} = -2B_{10} = -2\left(\frac{5}{66}\right) = -\frac{5}{33} = \frac{5}{5-38} \equiv 1 \mod 19,$$

so $h(-19) = 1$.

(4) For $p = 23$,

$$-2B_{\frac{23+1}{2}} = -2B_{12} = -2\left(-\frac{691}{2730}\right) = \frac{691}{1365} \equiv \frac{1}{8} \equiv 3 \mod 23,$$

so $h(-23) = 3$.

7.2 "Hurwitz-integral" Series

Definition 7.4. A formal power series with rational number coefficients of the form $\sum_{n=0}^{\infty} c_n\, t^n/n!$ such that all c_n are integers is said to be a Hurwitz-integral series. The set of all Hurwitz-integral series is denoted by $\mathcal{H}$.

Note that $P(t) \in \mathcal{H}$ if and only if $P(0), P'(0), P''(0), P^{(3)}(0), \ldots$ are all in $\mathbf{Z}$. Typical examples of elements of $\mathcal{H}$ are e^t and $\log(1+t)$.

Proposition 7.5. (1) *$\mathcal{H}$ is an integral domain.*

(2) *$\mathcal{H}$ is closed under term-by-term differentiation $\dfrac{d}{dt}$ and integration $\displaystyle\int_0^t$.*

(3) *The set of invertible elements of $\mathcal{H}$ is $\mathcal{H}^\times = \{P(t) \in \mathcal{H} \mid P(0) = \pm 1\}$.*

Proof. (1) We only need to show that the set $\mathcal{H}$ is closed under multiplication, other properties being trivial to verify. For two series $P(t) = \sum_{n=0}^{\infty} c_n\, t^n/n!$, $Q(t) = \sum_{n=0}^{\infty} d_n\, t^n/n!$, write the product $P(t)Q(t)$ as $\sum_{n=0}^{\infty} e_n\, t^n/n!$. The numbers e_n are given by

$$e_n = \sum_{i=0}^{n} \binom{n}{i} c_i\, d_{n-i}$$

and hence are integers if c_i, d_{n-i} are all integers.

(2) For $P(t) = \sum_{n=0}^{\infty} c_n\, t^n/n! \in \mathcal{H}$, we have

$$P'(t) = \sum_{n=1}^{\infty} c_n \frac{t^{n-1}}{(n-1)!} = \sum_{n=0}^{\infty} c_{n+1} \frac{t^n}{n!},$$

$$\int_0^t P(t)dt = \sum_{n=0}^{\infty} c_n \frac{t^{n+1}}{(n+1)!} = \sum_{n=1}^{\infty} c_{n-1} \frac{t^n}{n!}.$$

Both of these are clearly in $\mathcal{H}$.

(3) In order that $P(t) = \sum_{n=0}^{\infty} c_n\, t^n/n!$ is invertible in $\mathbf{Q}[[t]]$, the condition $c_0 \neq 0$ is necessary (Proposition 1.9 on p. 15). This being so, write the reciprocal $P(t)^{-1}$ in $\mathbf{Q}[[t]]$ as $P(t)^{-1} = \sum_{n=0}^{\infty} b_n\, t^n/n!$. To have $P(t)^{-1} \in \mathcal{H}$, we need $b_0 = c_0^{-1} \in \mathbf{Z}$ and thus $c_0 = \pm 1$. If this is satisfied, other coefficients b_n automatically belong to $\mathbf{Z}$, because b_n is the constant term of the nth derivative $\left(P(t)^{-1}\right)^{(n)}$ of $P(t)^{-1}$ and the derivative is of the form

$$\left(P(t)^{-1}\right)^{(n)} = \frac{\text{polynomial in } P(t), P'(t), P''(t), \ldots \text{ with integer coefficients}}{\text{power of } P(t)},$$

and so, if $c_0 = P(0) = \pm 1$, then the value of this at $t = 0$ is an integer since $P(0), P'(0), P''(0), \cdots \in \mathbf{Z}$. We therefore have $P(t) \in \mathcal{H}^{\times}$. Conversely, if $P(t) \in \mathcal{H}^{\times}$, then from $P(t)^{-1}|_{t=0} = c_0^{-1} \in \mathbf{Z}$ we have $c_0 = \pm 1$.

□

Remark 7.6. The series $\sin(t), \cos(t)$ (Taylor expansions of $\sin(x), \cos(x)$ at the origin, viewed as formal power series in $x = t$) are both in $\mathcal{H}$, and $\cos(t) = 1 - t^2/2! + t^4/4! + \cdots$ is invertible by proposition (3) above. Hence the series $\tan(t) = \sin(t)/\cos(t)$ is Hurwitz-integral. This shows that the tangent numbers T_n in Remark 1.18 (p. 24) are integers.

Proposition 7.7. *Suppose $P(t) \in \mathcal{H}$, $P(0) = 0$. Then for any $N \in \mathbf{N}$ we have*

$$\frac{P(t)^N}{N!} \in \mathcal{H}.$$

Proof. We proceed by induction on N. The case $N = 1$ is trivial. Assume the proposition is valid up to $N - 1$. Using $\left(P(t)^N/N!\right)' = P(t)^{N-1}P'(t)/(N-1)!$ and $P(0) = 0$, we have

$$\frac{P(t)^N}{N!} = \int_0^t \frac{P(t)^{N-1}}{(N-1)!} P'(t)dt.$$

The induction hypothesis and Proposition 7.5 (1), (2) show the right-hand side belongs to $\mathcal{H}$. □

Definition 7.8. For series $P(t)(\neq 0), Q(t) \in \mathcal{H}$, we write $P(t)|_H Q(t)$ or $Q(t) \equiv_H 0 \mod P(t)$ if the condition $Q(t)/P(t) \in \mathcal{H}$ holds. Also we write $Q_1(t) \equiv_H Q_2(t) \mod P(t)$ if $P(t)|_H(Q_1(t) - Q_2(t))$.

In particular, when m is an integer (this is also an element in $\mathcal{H}$), the congruence $Q_1(t) \equiv_H Q_2(t) \mod m$ means, if we write $Q_1(t) = \sum_{n=0}^{\infty} c_n^{(1)}\, t^n/n!$, $Q_2(t) = \sum_{n=0}^{\infty} c_n^{(2)}\, t^n/n!$, that the congruence $c_n^{(1)} \equiv c_n^{(2)} \mod m$ holds for every n.

7.3 Proof of Theorem 7.1

Put $(p-1)/2 = m$. We have $m > 1$ because of our assumption $p > 3$. In the class number formula ((6.6) on p. 90)

$$h(-p) = -\frac{1}{p}\sum_{u=1}^{p-1}\left(\frac{u}{p}\right)u,$$

divide the sum into two parts according to $u < p/2$ and $u > p/2$, and note $\left(\frac{-1}{p}\right) = -1$ because $p \equiv 3 \mod 4$. Then we obtain

$$\begin{aligned} ph(-p) &= -\sum_{0<u<p/2}\left(\frac{u}{p}\right)u - \sum_{0<u<p/2}\left(\frac{p-u}{p}\right)(p-u) \\ &= -2\sum_{0<u<p/2}\left(\frac{u}{p}\right)u + p\sum_{0<u<p/2}\left(\frac{u}{p}\right). \end{aligned} \tag{7.1}$$

On the other hand, if we divide the sum according to the parity of u, then we have

$$\begin{aligned} ph(-p) &= -\sum_{0<u<p/2}\left(\frac{2u}{p}\right)(2u) - \sum_{0<u<p/2}\left(\frac{p-2u}{p}\right)(p-2u) \\ &= -4\sum_{0<u<p/2}\left(\frac{2u}{p}\right)u + p\sum_{0<u<p/2}\left(\frac{2u}{p}\right). \end{aligned}$$

From this we have

$$\left(\frac{2}{p}\right)ph(-p) = -4\sum_{0<u<p/2}\left(\frac{u}{p}\right)u + p\sum_{0<u<p/2}\left(\frac{u}{p}\right). \tag{7.2}$$

Subtract (7.2) from two times (7.1) to obtain

$$\left(2-\left(\frac{2}{p}\right)\right)ph(-p) = p\sum_{0<u<p/2}\left(\frac{u}{p}\right)$$

and hence

$$h(-p) = \frac{1}{2-\left(\frac{2}{p}\right)}\sum_{a=1}^{m}\left(\frac{a}{p}\right).$$

Applying Euler's criterion $\left(\frac{a}{p}\right) \equiv a^m \mod p$ (cf. [50, Proposition 5.1.2]) here, we have

$$h(-p) \equiv \frac{1}{2-2^m}\sum_{a=1}^{m} a^m \mod p.$$

Hence, the proof of the theorem $h(-p) \equiv -2B_{m+1} \mod p$ boils down to showing

$$\sum_{a=1}^{m} a^m \equiv -2(2-2^m)B_{m+1} \mod p.$$

Further, by $2 \equiv \frac{1}{m+1} \mod p$ and $-(2-2^m) \equiv -(2^{2m+1}-2^m) \equiv 2^m(1-2^{m+1}) \mod p$, this reduces to showing

$$\sum_{a=1}^{m} a^m \equiv 2^m(1-2^{m+1})\frac{B_{m+1}}{m+1} \mod p.$$

First, by $\frac{t}{e^t-1} = 1 - \frac{t}{2} + \sum_{n=2}^{\infty} B_n \frac{t^n}{n!}$, we have

$$\begin{aligned}
\frac{1}{e^t+1} &= \frac{1}{e^t-1} - \frac{2}{e^{2t}-1} \\
&= \frac{1}{t}\left(\left(1-\frac{t}{2}+\sum_{n=2}^{\infty} B_n \frac{t^n}{n!}\right) - \left(1-\frac{2t}{2}+\sum_{n=2}^{\infty} B_n \frac{(2t)^n}{n!}\right)\right) \\
&= \frac{1}{2} + \sum_{n=2}^{\infty}(1-2^n)B_n \frac{t^{n-1}}{n!} \\
&= \frac{1}{2} + \sum_{n=1}^{\infty}(1-2^{n+1})\frac{B_{n+1}}{n+1}\cdot\frac{t^n}{n!}.
\end{aligned}$$

Replacing $t \mapsto 2t$ and multiplying both sides by 2, we obtain

$$\frac{2}{e^{2t}+1} = 1 + \sum_{n=1}^{\infty} 2^{n+1}(1-2^{n+1})\frac{B_{n+1}}{n+1}\frac{t^n}{n!}. \tag{7.3}$$

Now since

$$\frac{2}{e^{2t}+1} = \frac{1}{1-\frac{1-e^{2t}}{2}} = 1 + \left(\frac{1-e^{2t}}{2}\right) + \left(\frac{1-e^{2t}}{2}\right)^2 + \left(\frac{1-e^{2t}}{2}\right)^3 + \cdots$$

and $\left(\frac{1-e^{2t}}{2}\right) = -\sum_{n=1}^{\infty} 2^{n-1}\frac{t^n}{n!} \in \mathcal{H}$, we have $\frac{2}{e^{2t}+1} \in \mathcal{H}$. Also, by Proposition 7.7, we have $\left(\frac{1-e^{2t}}{2}\right)^N \equiv_H 0 \mod N!$. In particular, if $N \geq 2$, we have $\left(\frac{1-e^{2t}}{2}\right)^N \equiv_H 0 \mod 2$. By this, combined with the fact that the coefficient of $\frac{t^n}{n!}$ in $\frac{1-e^{2t}}{2}$ is divisible by 2 if $n \geq 2$, we have

$$\frac{2}{e^{2t}+1} \equiv_H 1+t \mod 2. \tag{7.4}$$

Therefore, by (7.3), we have

$$2^n(1-2^{n+1})\frac{B_{n+1}}{n+1} \in \mathbf{Z}$$

if $n \geq 2$. Next observe

$$\begin{aligned}(1+e^{2t})\sum_{a=0}^{m} e^{4at} &= \sum_{a=0}^{m}\left(e^{4at}+e^{(4a+2)t}\right) = \sum_{j=0}^{2m+1} e^{2jt} \\ &= \sum_{j=0}^{p}\sum_{n=0}^{\infty}\frac{(2jt)^n}{n!} = \sum_{n=0}^{\infty}\left(\sum_{j=0}^{p}(2j)^n\right)\frac{t^n}{n!} \\ &= p+1+\sum_{n=1}^{\infty}2^n\left(\sum_{j=1}^{p}j^n\right)\frac{t^n}{n!} \\ &\equiv_H 1-\sum_{\substack{n\geq 1\\ p-1|n}} 2^n\frac{t^n}{n!} \mod p.\end{aligned}$$

(Here we have used (3.1) on p. 42, i.e., $\sum_{j=1}^{p} j^n \equiv -1 \bmod p$ if $p-1|n$ and $\sum_{j=1}^{p} j^n \equiv 0 \bmod p$ otherwise.) Multiplying both sides by $\frac{2}{e^{2t}+1}$ $(\in \mathcal{H})$, we have

$$2\sum_{a=0}^{m} e^{4at} \equiv_H \frac{2}{e^{2t}+1}\left(1-\sum_{\substack{n\geq 1\\ p-1|n}} 2^n\frac{t^n}{n!}\right) \mod p.$$

Comparing the coefficients of $\frac{t^m}{m!}$ on both sides, we obtain by (7.3) (note $m=\frac{p-1}{2}<p-1$)

$$2\sum_{a=0}^{m}(4a)^m \equiv 2^{m+1}(1-2^{m+1})\frac{B_{m+1}}{m+1} \mod p.$$

Dividing both sides by 2 and using $4^m = 2^{p-1} \equiv 1 \mod p$, we have

$$\sum_{a=0}^{m} a^m \equiv 2^m(1-2^{m+1})\frac{B_{m+1}}{m+1} \mod p.$$

This is what we wanted to show. □

Exercise 7.9. Compute the class number $h(-31)$ by using Theorem 7.1 and Table 1.1 in Chap. 1.

Exercise 7.10. Let p be a prime congruent to 3 modulo 4. Prove that the numerator of $B_{\frac{p+1}{2}}$ can never be divisible by p.

Exercise 7.11. The Euler number E_n is defined by the generating series

$$\frac{2}{e^t + e^{-t}} = \sum_{n=0}^{\infty} E_n \frac{t^n}{n!}.$$

Prove that all E_n are integers. Moreover, prove that E_{2n} is odd for all $n \geq 1$. (E_{2n+1} is easily seen to be 0.) Hint: Use (7.4).

Chapter 8
Character Sums and Bernoulli Numbers

We would like to explain arithmetic identities between Bernoulli numbers and the root of unity. Namely we shall explain certain small collections of formulas between exponential sums or character sums and Bernoulli numbers. We often encounter such formulas when we compare the dimension formulas of modular forms obtained by the Riemann–Roch theorem and by the trace formula. Often, the exponential sums appear in the first method and the Bernoulli numbers appear in the second method. But sometimes it is not easy to prove these relations in an elementary way, and there are several cases that no elementary proofs are known. It seems that there is no general elementary way to study these relations. Our elementary method in this section is very restrictive and clearly not enough to attack difficult problems, but still, it should give us some general feeling of what are easy exponential sums.

Before going to easy things treated in this section, we give here one example for which no elementary proof is known.

Let p be a prime such that $p \equiv 3 \bmod 4$. We put $\zeta = e^{2\pi i/p}$. Then the following formula holds.

$$\sum_{(a,b,c)\in S} \frac{\psi(abc)}{(1-\zeta^a)(1-\zeta^b)(1-\zeta^c)} = -\sqrt{-p}\left(\frac{p+1}{4}B_{1,\psi} + \frac{1}{5}B_{3,\psi}\right).$$

Here we put $S = \left\{(a,b,c) \in (\mathbf{F}_p^\times)^3; ab+bc+ca = 0\right\}$, $\mathbf{F}_p$ is the prime field $\mathbf{Z}/p\mathbf{Z}$ with characteristic p, ψ is the quadratic residue character defined by $\psi(x) = \left(\frac{x}{p}\right)$, and $B_{n,\psi}$ is the generalized Bernoulli number.

Motivated by dimension formulas, this relation is conjectured in [69], and later proved in [47]. The proof there is a remote indirect proof. It is natural to ask if there exists a more elementary and direct proof. But no elementary alternative proof is currently known. Assuming that such a direct proof exists, we have no idea if such a proof is simple or complicated before we really prove it.

T. Arakawa et al., *Bernoulli Numbers and Zeta Functions*, Springer Monographs in Mathematics, DOI 10.1007/978-4-431-54919-2_8, © Springer Japan 2014

The results we shall give below are much easier compared with this example. But we see that even researchers sometimes fail to recognize that these are easy, so we think it is of some use to explain such elementary formulas here. To fix our standpoint, we assume here that we are satisfied if an exponential sum or a character sum can be written by the generalized Bernoulli numbers.

8.1 Simplest Examples

We start from simple examples which can be treated easily. For a prime p, we put $\zeta_p = e^{2\pi i/p}$. As one of our examples, we consider the sum $\sum_{a=1}^{p-1}(1-\zeta_p^a)^{-1}$. In fact this sum is equal to

$$\sum_{a=1}^{p-1} \frac{1}{1-\zeta_p^a} = \frac{p-1}{2}. \tag{8.1}$$

Similar sums obtained by attaching the quadratic residue symbols or character values of a to the numerator of $1/(1-\zeta_p^a)$ can be expressed by using Bernoulli numbers and Gaussian sums. We call loosely this kind of sum obtained by combination of roots of unity and characters *an exponential sum* or *a character sum.* As for the first example (8.1) we gave above, the argument is very simple. For similar sums with some characters too, this kind of formula can be obtained by fairly easy calculation.

Proof 1. We give the first proof. Since ζ_p^a $(1 \le a \le p-1)$ are mutually different $p-1$ roots of the equation $X^p - 1 = 0$ which is not equal to 1, we have

$$\frac{X^p - 1}{X - 1} = X^{p-1} + X^{p-2} + \cdots + 1 = \prod_{a=1}^{p-1}(X - \zeta_p^a).$$

Take the logarithm of both sides and compare their derivatives with respect to X. Then the left-hand side is

$$\frac{d}{dX}\log(X^{p-1} + X^{p-2} + \cdots + 1) = \frac{\sum_{a=1}^{p-1} aX^a}{X^{p-1} + \cdots + 1},$$

and the right-hand side is

$$\frac{d}{dX}\sum_{a=1}^{p-1}\log(X - \zeta_p^a) = \sum_{a=1}^{p-1}\frac{1}{X - \zeta_p^a}.$$

Here if we put $X = 1$, then we get the relation to be proved since $p^{-1}\sum_{a=1}^{p-1} a = (p-1)/2$. □

Proof 2. The next proof might appear in an exercise of the elementary algebra. Here we use a little tool of algebra (Galois theory of cyclotomic fields). We never use this argument later, so if the readers are not familiar with this tool, they are recommended to skip this and to go directly to Proof 3. We call the field generated by the nth roots of unity over the rational number field a *cyclotomic field*. Gauss studied cyclotomic fields deeply [35, Chapter 7]. The cyclotomic field $\mathbf{Q}(\zeta_p)$ is a Galois extension of $\mathbf{Q}$, and the set of the mappings $\sigma_a\colon \zeta_p \mapsto \zeta_p^a$ for integers a with $1 \le a \le p-1$ gives the Galois group of $\mathbf{Q}(\zeta_p)$ over $\mathbf{Q}$. So the sum which appears in (8.1) is equal to the trace of $x = 1/(1-\zeta_p) \in \mathbf{Q}(\zeta_p)$ over $\mathbf{Q}$, that is, the sum $\sum_{a=1}^{p-1} \sigma_a(x)$, where $\sigma_a(x)$ is the action of the element σ_a in the Galois group on x. This trace is invariant by each element σ_a of the Galois group, so by the general Galois theory, we see this is a rational number. The degree of extension of $\mathbf{Q}(\zeta_p)$ over $\mathbf{Q}$ (namely the dimension as a vector space over $\mathbf{Q}$) is $p-1$, but since we have $\zeta_p = 1 - 1/x$, x also generates $\mathbf{Q}(\zeta_p)$, so the minimal polynomial of x over $\mathbf{Q}$ is also of degree $p-1$. The minimal polynomial of ζ_p is $1 + X + X^2 + \cdots + X^{p-1} = (X^p - 1)/(X - 1)$, so the minimal polynomial of $-1/x = \zeta_p - 1$ is given by $((X + 1)^p - 1)/X$. If we put $X = -1/Y$, then this is written as $-Y^{1-p}((Y - 1)^p - Y^p)$, so the minimal polynomial of x whose coefficient of the highest degree is one is given by

$$\frac{1}{p}(Y^p - (Y - 1)^p).$$

The trace of x is -1 times the coefficient of Y^{p-2}, so by the binomial theorem, we get

$$\frac{1}{p} \times \frac{p(p-1)}{2} = \frac{p-1}{2}.$$

Hence we prove (8.1). □

Proof 3. Proof 2 is somewhat clumsy and similar methods cannot be used so easily in other cases as imagined. Here we explain a more general method. Let t be an indeterminate (i.e. a variable, or an element transcendental over the field in question), and consider the rational function

$$\frac{1}{1-\zeta_p t}.$$

By expanding this as a formal power series along $t = 0$, we have

$$\frac{1}{1-\zeta_p t} = \sum_{i=0}^{\infty} \zeta_p^i t^i = \sum_{j=0}^{\infty} \sum_{c=0}^{p-1} \zeta_p^c t^{c+pj} = \frac{\sum_{c=0}^{p-1} \zeta_p^c t^c}{1 - t^p}.$$

If $|t| < 1$, this calculation is valid also as a convergent power series. If we take here $t \to 1$, then by the definition of differentiation (or the theorem of l'Hôpital[1]), we easily see that

$$\frac{1}{1-\zeta_p} = -\frac{1}{p}\sum_{c=1}^{p-1} c\zeta_p^c.$$

In the above calculation, we used only the properties that $\zeta_p^p = 1$ and $\zeta_p \neq 1$, so for any a prime to p, we can apply the same calculation to ζ_p^a instead of ζ_p, so we get also

$$\frac{1}{1-\zeta_p^a} = -\frac{1}{p}\sum_{c=1}^{p-1} c\zeta_p^{ac}.$$

Here if we take the summation over a, we have

$$\sum_{a=1}^{p-1} \zeta_p^{ca} = \sum_{a=0}^{p-1} \zeta_p^{ca} - 1 = \begin{cases} -1 & \text{if } c \not\equiv 0 \bmod p, \\ p-1 & \text{if } c \equiv 0 \bmod p, \end{cases}$$

so we get

$$\sum_{a=1}^{p-1} \frac{1}{1-\zeta_p^a} = -\frac{1}{p}\sum_{c=1}^{p-1}(-c) = \frac{1}{p} \times \frac{p(p-1)}{2} = \frac{p-1}{2}.$$

Hence we prove (8.1). □

An advantage of Proof 3 is that we can calculate mechanically without any special insight on the result beforehand.

An alternative proof for the expression for $(1-\zeta^a)^{-1}$ that we obtained in Proof 3 above will be given also in Lemma 8.5. By the way, we see of course without any calculation that the element $(1-\zeta^a)^{-1}$ of $\mathbf{Q}(\zeta_p)$ is a linear combination of the basis $\zeta_p, \ldots, \zeta_p^{p-1}$ of $\mathbf{Q}(\zeta_p)$ over $\mathbf{Q}$. In principle, for any element of a finite field extension K over k, a concrete expression of the inverse by a basis can be obtained by Euclidean algorithm on polynomials. But we cannot expect in general that such a linear combination is written by a simple formula. Since we have an accidentally easy expression in the above case, we could use it.

[1] Guillaume François Antoine de l'Hôpital, (born in 1661 in Paris, France—died on February 2, 1704 in Paris, France).

8.2 Gaussian Sum

Before we start explaining the relation between exponential sums and the generalized Bernoulli numbers, we explain about Gaussian sums. The simplest definition of a Gaussian sum is given by

$$g(p) = \sum_{x=0}^{p-1} \zeta_p^{x^2} \qquad \zeta_p = e^{\frac{2\pi i}{p}}$$

for an odd prime p. This number was calculated by Gauss and the result is very beautiful. Namely we have

$$g(p) = \sqrt{(-1)^{(p-1)/2}\, p}\ .$$

Here the choice of the square root is given as follows. If $p \equiv 1 \bmod 4$, we take $\sqrt{p}$ as a positive number and if $p \equiv 3 \bmod 4$ we take $\sqrt{-p}$ so that its argument is $\pi/2$. As we shall see later, it is very easy to prove that $g(p)^2 = (-1)^{(p-1)/2}p$, but it is considerably difficult to determine the sign of the square root in $g(p)$. It is said that Gauss spent 4 years getting the proof. [2]

At the present time, there are many known ways to determine this sign, but it would be beyond the scope or different from the aim of this book to carry those proofs here. So we would like to ask readers to refer to some other books on the determination of the sign. (For example, many proofs are introduced in [15].)

Now, this Gaussian sum can be also expressed in a slightly different way. Let ψ be the quadratic residue symbol modulo p defined on p. 88 (i.e. $\psi(n) = \left(\frac{n}{p}\right)$) and put

[2]Gauss wrote in his letter (*Collected Works X*, p. 24) to Olbers dated September 3, 1805: "Since four years ago, a week has seldom passed when I had not made one or another attempt to solve this difficulty, especially lively now also again in the latest period. But all pondering, all search have been in vain, and each time sadly I must have laid down my pen again. At last a couple of days ago it succeeded, not for my wearisome search, but only through the grace of God, I would like to say. As the lightning strikes, the problem has been solved; ⋯ Strangely enough, the solution of this problem appears now easier than many others which did not well keep me so many days as these years, and certainly no one will, when I give a lecture on this matter some day, get suspicious of the long dilemma in which it brought me." Incidentally, Olbers (Heinrich Wilhelm Matthias Olbers, 1758–1840) was a medical doctor and an astronomer. The biggest asteroid, Ceres, which made Gauss leap to fame, was discovered by Italian astronomer Giuseppe Piazzi (1746–1826) on New Year's day in 1801, but became lost through the difficulty of observation, then was accurately located by Gauss by calculation. It was rediscovered by Olbers on New Year's day in 1802. Also, he newly discovered the asteroid Pallas in 1802 and Vesta in 1807. Olbers was a friend of Gauss whom he visited most often.

$$g(\psi) = \sum_{n=0}^{p-1} \psi(n)\zeta_p^n.$$

For each n, the number of representatives mod p of $x \in \mathbf{Z}$ such that $n \equiv x^2 \bmod p$ is given by $1 + \psi(n)$, so we get

$$g(p) = \sum_{n=0}^{p-1}(1 + \psi(n))\zeta_p^n = \sum_{n=0}^{p-1} \zeta_p^n + \sum_{n=0}^{p-1} \psi(n)\zeta_p^n.$$

Here the first sum is obviously 0, so this coincides with $g(\psi)$.

Here we shall explain more general Gaussian sums which will be needed later. If we change x^2 to a quadratic form of several variables and integral vectors in the above definition of Gaussian sum $g(p)$, we can define the "Gaussian sum of quadratic forms". This plays an important role in the transformation formula of theta functions for example, and it is possible to calculate this as the above formula of Gauss, though we do not go into this direction here. On the other hand, if we change ψ into a general Dirichlet character in the definition of $g(\psi)$, then we can define a Gaussian sum associated with a character. We treat this in this book.

Definition 8.1. Let χ be a primitive Dirichlet character with conductor f and put $\zeta_f = e^{2\pi i/f}$. We write

$$g(\chi) = \sum_{n=0}^{f-1} \chi(n)\zeta_f^n$$

and we call this the Gaussian sum associated with character χ.

In such a general case too, the absolute value of $g(\chi)$ is a simple quantity and given by $g(\chi)\overline{g(\chi)} = f$. (The proof will be given later.) But for general χ, there is no formula of the Gaussian sum itself which is simple and explicit. When $\chi^2 = 1$, there are some formulas which will be explained later.

Lemma 8.2. *Let χ be a primitive Dirichlet character with conductor f and let a be an arbitrary integer. Then the following formula holds.*

$$\overline{\chi(a)}g(\chi) = \sum_{n=1}^{f} \chi(n)\zeta_f^{an}.$$

In particular, we have $\overline{g(\chi)} = \chi(-1)g(\overline{\chi})$.

A good point of this relation is that we are not assuming that a is coprime to f and this is often useful for calculation of exponential sums.

Proofs of Lemma. If a is coprime to f, the relation is obtained trivially by replacing n by an in the definition of the Gaussian sum $g(\chi)$. If a is not coprime to f, then the left-hand side is of course 0 by definition of a Dirichlet character, but the right-hand side is also 0 by the following reasoning. If $(a, f) = d > 1$, then ζ_f^{an} depends only on n mod f/d, but elements $1 \le n \le f$ such that 1 mod f/d form a subgroup in $(\mathbf{Z}/f\mathbf{Z})^\times$ and so for each fixed n_0 coprime to f, we have

$$\sum_{n \equiv n_0 \bmod f/d} \chi(n) = \chi(n_0) \sum_{n \equiv 1 \bmod f/d} \chi(n) = 0.$$

The second relation is obtained by putting $a = -1$ and taking the conjugate of both sides in the first relation. Namely we have $\chi(-1)\overline{g(\chi)} = g(\overline{\chi})$ and since $\chi(-1)^2 = \chi(1) = 1$, we have $\chi(-1)^{-1} = \chi(-1)$. □

Since the Gaussian sum is never 0 as we see below, the above formula can be written also as

$$\chi(a) = g(\overline{\chi})^{-1} \sum_{n=1}^{f} \overline{\chi(n)} \zeta_f^{an}. \tag{8.2}$$

Lemma 8.3. *We have*

$$g(\chi)\overline{g(\chi)} = f.$$

Proof. By Lemma 8.2, for any integer a, we have

$$\chi(a)\overline{\chi(a)} g(\chi)\overline{g(\chi)} = \sum_{n,m=1}^{f-1} \chi(n)\overline{\chi(m)} \zeta_f^{a(n-m)}.$$

Summing up both sides from $a = 0, \ldots, f-1$, the left-hand side becomes $\varphi(f) g(\chi)\overline{g(\chi)}$ since $\chi(a)\overline{\chi(a)} = 1$ if $(a, f) = 1$ and 0 otherwise. Here $\varphi(f)$ is the Euler function, namely the number of positive integers not more than f which are coprime to f. By using

$$\sum_{a=0}^{f-1} \zeta_f^{a(n-m)} = \begin{cases} 0 & \text{if } n \ne m, \\ f & \text{if } n = m, \end{cases}$$

the right-hand side becomes

$$f \times \sum_{n=1}^{f-1} \chi(n)\overline{\chi(n)} = f\varphi(f).$$

Comparing both sides, we prove the lemma. □

Next we will give a formula for the Gaussian sum for a primitive Dirichlet character with $\chi^2 = 1$.

Proposition 8.4. *Let χ be a non-trivial primitive Dirichlet character such that $\chi^2 = 1$. Then there exists a quadratic field K such that $\chi(u) = \chi_K(u) = \left(\frac{D_K}{u}\right)$, where D_K is the fundamental discriminant of K. Furthermore, for such χ, we have*

$$g(\chi_K) = \sqrt{D_K}.$$

By using the Chinese remainder theorem, we decompose the sum in the definition of $g(\chi_K)$ into powers of primes and we reduce the formula to those in the case of primes. In the middle of the proof, we need the reciprocity law of quadratic residues. We omit the details here. We refer to [15]. See also Exercises 8.26 and 8.27.

8.3 Exponential Sums and Generalized Bernoulli Numbers

In this section, we consider sums similar to those in Sect. 8.1, attaching characters to them. In fact, these new sums are far more interesting. In the previous sections, we took only the prime power root of unity, but in this section we consider the fth root $\zeta_f = e^{2\pi i/f}$ of unity. Since it is troublesome to write the suffix f always, we fix a natural number f once and for all and we denote ζ_f simply by ζ. We will explain below how to obtain formulas for sums like those in the last section by generalized Bernoulli numbers and simple combinatorial numbers.

First of all, when a is not divisible by f, we get the formula given below. (This formula will be used also in Chap. 11, Sect. 11.2.)

Lemma 8.5. *If $f \nmid a$, then we have*

$$\frac{1}{1-\zeta^a} = -\frac{1}{f}\sum_{c=1}^{f-1} c\zeta^{ac}.$$

Proof. This is obtained by the same calculation as in Proof 3 in Sect. 8.1, but here we give more direct proof. By expanding polynomials, we have

$$\begin{aligned}(1-X)\times\sum_{c=1}^{f-1} cX^c &= -(f-1)X^f + \sum_{c=1}^{f-1}(c-(c-1))X^c \\ &= -fX^f + \sum_{c=1}^{f} X^c.\end{aligned}$$

If we substitute as $X = \zeta^a$ in the above, then the second term of the right-hand side is zero since $\sum_{c=1}^{f} \zeta^{ac} = \zeta^a(1-\zeta^{af})/(1-\zeta^a) = 0$, and we have

$$(1-\zeta^a)\sum_{c=1}^{f-1} c\zeta^{ac} = -f\zeta^{af} = -f,$$

Dividing both sides by $-f(1-\zeta^a)$, we have the equality in the lemma. □

Now we consider a Dirichlet character χ whose conductor is f (i.e. a primitive Dirichlet character modulo f). First, we put

$$P_k(\chi) = \sum_{a=1}^{f} \frac{\chi(a)}{(1-\zeta^a)^k}$$

and try to describe this. Of course if $f = 1$, then the denominator becomes zero so this has no meaning. So we assume that $f > 1$ from now on. As in the last section, we define the Gaussian sum with respect to χ by

$$g(\chi) = \sum_{a=1}^{f} \chi(a)\zeta^a.$$

To treat the exponential sums slightly more generally, we put

$$P_{k,\chi}(t) = \sum_{a=1}^{f} \frac{\chi(a)}{(1-\zeta^a e^t)^k}.$$

We have $P_{k,\chi}(0) = P_k(\chi)$. The next relation holds.

$$P_{1,\chi}(0) = -g(\chi)B_{1,\overline{\chi}}. \tag{8.3}$$

Proof. By the last lemma and Lemma 8.2, we have

$$P_{1,\chi}(0) = -\frac{1}{f}\sum_{a=1}^{f}\chi(a)\sum_{c=1}^{f-1} c\zeta^{ac} = -\frac{1}{f}g(\chi)\sum_{c=1}^{f}\overline{\chi(c)}c = -g(\chi)B_{1,\overline{\chi}}.$$

In the last equality, we used the formula (4.1) for $B_{1,\chi}$ in Sect. 4.2 (p. 54). □

In fact, $P_{1,\chi}(t)$ is almost identical to the definition of the generalized Bernoulli numbers. Indeed, calculating the development, we have

$$P_{1,\chi}(t) = \sum_{a=1}^{f}\sum_{j=0}^{\infty}\chi(a)\zeta^{aj}e^{jt} = g(\chi)\sum_{j=0}^{\infty}\overline{\chi(j)}e^{jt} = -g(\chi)\frac{\sum_{j=0}^{f-1}\overline{\chi(j)}e^{jt}}{e^{ft}-1}.$$

(Here we are regarding e^t as a formal variable and expanding it as a formal power series. Or we may say we are expanding it in the neighborhood of $t = -\infty$. Anyway, in the final stage, it is an equality between rational functions of e^t, and the calculation is justified regardless of any interpretation.) Since we are taking $f > 1$, we have $\chi(0) = \chi(f) = 0$. So by definition, we have

$$P_{1,\chi}(t) = -g(\chi)\sum_{n=1}^{\infty} B_{n,\overline{\chi}}\frac{t^{n-1}}{n!}.$$

Hence we have

$$\frac{d^l P_{1,\chi}}{dt^l}(0) = P_{1,\chi}^{(l)}(0) = -g(\chi)\frac{1}{l+1}B_{l+1,\overline{\chi}}. \tag{8.4}$$

Next we want to give a formula for $P_{k,\chi}(0)$. Now we put

$$Q_{k,a}(t) = \frac{1}{(1-\zeta^a e^t)^k}.$$

Then we have

$$P_{k,\chi}(t) = \sum_{a=1}^{f} \chi(a) Q_{k,a}(t),$$

and

$$\frac{d}{dt}Q_{k,a}(t) = k(Q_{k+1,a}(t) - Q_{k,a}(t)),$$

so inductively $Q_{k,a}(t)$ is expressed by higher-order derivatives of $Q_{1,a}(t)$, so $P_{k,\chi}(t)$ is also expressed by derivatives of higher order of $P_{1,\chi}(t)$. More explicitly, by using the Stirling number of the first kind (as for the definition, see Sect. 2.1), it is expressed as follows.

Proposition 8.6.

$$Q_{k,a}(t) = \frac{1}{(k-1)!}\sum_{l=0}^{k-1}\begin{bmatrix} k \\ l+1 \end{bmatrix} Q_{1,a}^{(l)}(t),$$

$$P_{k,\chi}(t) = \frac{1}{(k-1)!}\sum_{l=0}^{k-1}\begin{bmatrix} k \\ l+1 \end{bmatrix} P_{1,\chi}^{(l)}(t).$$

Proof. This is proved by induction with respect to k. The assertion is true for $k = 1$. We assume it is true up to k. Since $Q_{k+1,a}(t) = Q_{k,a}(t) + \frac{1}{k}\frac{d}{dt}Q_{k,a}(t)$, we can express $Q_{k+1,a}(t)$ by the linear combination of derivatives of $Q_{k,1}(t)$ by the inductive assumption, and the coefficient of $Q_{1,a}^{(l)}(t)$ in $Q_{k+1,a}(t)$ in this expression is given by

$$\frac{1}{(k-1)!}\left\{\left[\begin{matrix} k \\ l+1 \end{matrix}\right] + \frac{1}{k}\left[\begin{matrix} k \\ l \end{matrix}\right]\right\}$$

for $0 \le l \le k$. By the recurrence relation of the Stirling number of the first kind (p. 28, Eq. (2.2)), this is equal to

$$\frac{1}{k!}\left[\begin{matrix} k+1 \\ l+1 \end{matrix}\right].$$

Hence we prove the assertion. □

Corollary 8.7.

$$P_{k,\chi}(0) = \sum_{a=1}^{f} \frac{\chi(a)}{(1-\zeta^a)^k} = -\frac{g(\chi)}{(k-1)!}\sum_{l=0}^{k-1}\left[\begin{matrix} k \\ l+1 \end{matrix}\right]\frac{B_{l+1,\overline{\chi}}}{l+1}. \tag{8.5}$$

Proof. In the second formula in the proposition, put $t = 0$ and then use (8.4). □

Example 8.8. For example, we have

$$\sum_{a=1}^{f-1} \frac{\chi(a)}{(1-\zeta^a)} = -g(\chi)B_{1,\overline{\chi}},$$

$$\sum_{a=1}^{f-1} \frac{\chi(a)}{(1-\zeta^a)^2} = -g(\chi)\left(B_{1,\overline{\chi}} + \frac{1}{2}B_{2,\overline{\chi}}\right),$$

$$\sum_{a=1}^{f-1} \frac{\chi(a)}{(1-\zeta^a)^3} = -g(\chi)\left(B_{1,\overline{\chi}} + \frac{3}{4}B_{2,\overline{\chi}} + \frac{1}{6}B_{3,\overline{\chi}}\right).$$

In the above, we assumed that the character χ is primitive in order to evaluate $P_{1,\chi}^{(l)}(0)$ by Bernoulli numbers in the final stage, but we did not use this condition at any other place. So, in the above, even if we take 1 instead of $\chi(a)$ for any a, we can execute almost the same calculation. For example, we get

$$\begin{aligned}\sum_{a=1}^{f-1} Q_{1,a}(t) = \sum_{a=1}^{f-1}\frac{1}{1-\zeta^a e^t} &= \frac{f}{1-e^{ft}} - \frac{1}{1-e^t} \\ &= f - \frac{fe^{ft}}{e^{ft}-1} - \left(1 - \frac{e^t}{e^t-1}\right) \\ &= f - 1 + \sum_{n=1}^{\infty} B_n \frac{(1-f^n)t^{n-1}}{n!}.\end{aligned}$$

Hence if we put $t = 0$ here, we have

$$\sum_{a=1}^{f-1} \frac{1}{1-\zeta^a} = \frac{f-1}{2} \tag{8.6}$$

since $B_1 = 1/2$. The case $f = p$ is the formula in Sect. 8.1.

We give similar formulas in more general cases. Noting the relation

$$\sum_{a=1}^{f-1} Q_{1,a}^{(l)}(0) = \frac{1}{l+1}(1-f^{l+1})B_{l+1} + \delta_{l0}(f-1)$$

(where δ_{l0} is Kronecker's delta, that is, 1 for $l = 0$ and 0 otherwise), we get the following formula.

Lemma 8.9.

$$\begin{aligned}\sum_{a=1}^{f-1} \frac{1}{(1-\zeta^a)^k} &= \sum_{a=1}^{f-1} Q_{k,a}(0) \\ &= \frac{1}{(k-1)!}\sum_{l=0}^{k-1} \left[{k \atop l+1}\right] \left\{\frac{1}{l+1}(1-f^{l+1})B_{l+1} + (f-1)\delta_{l0}\right\}.\end{aligned}$$

We give examples of this formula.

Example 8.10.

$$\sum_{a=1}^{f-1} \frac{1}{1-\zeta^a} = \frac{f-1}{2},$$

$$\sum_{a=1}^{f-1} \frac{1}{(1-\zeta^a)^2} = -\frac{(f-1)(f-5)}{12},$$

$$\sum_{a=1}^{f-1} \frac{1}{(1-\zeta^a)^3} = -\frac{(f-1)(f-3)}{8},$$

$$\sum_{a=1}^{f-1} \frac{1}{(1-\zeta^a)^4} = \frac{(f-1)(f^3+f^2-109f+251)}{720}.$$

Using similar methods, we will give formulas for the following sums:

$$\sum_{c=0}^{f-1} c^k \zeta^{ca}.$$

When f divides a, then we have $\zeta^{ac} = 1$, so this should give the formula to describe sums of powers. Also under the assumption that f does not divide a, we give formulas to express these sums by $Q_{k,a}(0)$. This is a generalization of the formula to express the right-hand side of Lemma 8.5 by the left-hand side. (These formulas will be applied to calculations of exponential sums later.) As usual, for natural numbers k, j, we denote by $\binom{k}{j}$ the binomial coefficient

$$\binom{k}{j} = \frac{k(k-1)\cdots(k-j+1)}{j!}.$$

Here for $j = 0$, we put $\binom{k}{0} = 1$.

Lemma 8.11. (1) *When a is divisible by f, we have*

$$\sum_{c=0}^{f} c^k \zeta^{ac} = \sum_{c=0}^{f} c^k = \sum_{j=0}^{k} \binom{k}{j} B_j \frac{f^{k-j+1}}{k-j+1}.$$

(2) *When a is not divisible by f, we have*

$$\sum_{c=0}^{f} c^k \zeta^{ac} = \sum_{m=1}^{k} (-1)^m (m-1)! Q_{m,a}(0) \left(\sum_{j=1}^{k-m+1} (-1)^{k-j} f^j \binom{k}{j} \left\{ {k-j+1 \atop m} \right\} \right).$$

The first formula is nothing but the formula (1.1) in Chap. 1.

Proof. First we assume that a is an arbitrary integer. To extract c^k by differentiating an easily handled function, we consider the following transformation, taking t as a variable.

$$\begin{aligned} \sum_{c=0}^{f-1} c^k \zeta^{ca} e^{ct} &= \frac{d^k}{dt^k} \left(\sum_{c=0}^{f-1} \zeta^{ac} e^{ct} \right) \\ &= \frac{d^k}{dt^k} \left(\frac{1-e^{ft}}{1-\zeta^a e^t} \right) \\ &= Q_{1,a}^{(k)}(t) - \sum_{j=0}^{k} \binom{k}{j} f^{k-j} e^{ft} Q_{1,a}^{(j)}(t). \end{aligned} \tag{8.7}$$

To show the last equality, we used the Leibniz rule[3] on the higher-order derivatives of products.

[3]Gottfried Wilhelm Freiherr von Leibniz (born on July 1, 1646 in Leipzig, Saxony (now Germany)—died on November 14, 1716 in Hannover, Hanover (now Germany))

If $f|a$, then we have

$$Q_{1,a}(t) = Q_{1,0}(t) = -\frac{1}{e^t - 1} = 1 - \frac{e^t}{e^t - 1} = 1 - \sum_{n=0}^{\infty} \frac{B_n}{n!} t^{n-1}.$$

This has a pole at $t = 0$. So comparing the constant terms of both sides of (8.7) as a power series expansion with respect to t and noting that

$$Q_{1,a}^{(0)}(t) = -\frac{1}{t} + (1 - B_1) - \frac{B_2}{2} t - \cdots,$$

$$Q_{1,a}^{(j)}(t) = -\frac{(-1)^j j!}{t^{j+1}} - \frac{B_{j+1}}{j+1} - \frac{B_{j+2}}{(j+2)} t - \cdots \quad (j \geq 1),$$

$$e^{ft} = \sum_{n=0}^{\infty} \frac{f^n t^n}{n!},$$

we get

$$\begin{aligned}
\sum_{c=0}^{f-1} c^k &= -\frac{B_{k+1}}{k+1} + \sum_{j=0}^{k} \binom{k}{j} f^{k-j} \frac{B_{j+1}}{j+1} + \sum_{j=0}^{k} \binom{k}{j} f^{k-j} \frac{f^{j+1}}{(j+1)!} \times (-1)^j j! - f^k \\
&= \sum_{j=0}^{k-1} \binom{k}{j} f^{k-j} \frac{B_{j+1}}{j+1} + \frac{f^{k+1}}{k+1} - f^k \\
&= \sum_{j=0}^{k-1} \binom{k}{j+1} f^{k-j} \frac{B_{j+1}}{k-j} + \frac{f^{k+1}}{k+1} - f^k \\
&= \sum_{j=1}^{k} \binom{k}{j} f^{k-j+1} \frac{B_j}{k-j+1} + \frac{f^{k+1}}{k+1} - f^k.
\end{aligned}$$

(In the second equality, we used the fact $\sum_{j=0}^{k} (-1)^j \binom{k}{j} \frac{1}{j+1} = \frac{1}{k+1}$.) So we get

$$\sum_{c=0}^{f} c^k = \sum_{j=0}^{k} \binom{k}{j} B_j \frac{f^{k-j+1}}{k-j+1}.$$

In this way, we obtained the formula for the sum of powers again.

Next, we assume that a is not divisible by f. Here we want to express $Q_{1,a}^{(n)}(t)$ by $Q_{k,a}(t)$. This is a converse of the formula in Proposition 8.6 and this time, it is described by using the Stirling number of the second kind. (As for definition, see Sect. 2.1.)

In the formula (5.1) of Proposition 2.6 p. 28, if we substitute as $m \to m+1$, $n \to k+1$, we get

$$\delta_{m+1,k+1}(=\delta_{m,k}) = (-1)^{k+1}\sum_{l=m+1}^{k+1}(-1)^l \left\{ {k+1 \atop l} \right\} \left[{l \atop m+1} \right].$$

By virtue of this equality and Proposition 8.6, we have

$$\begin{aligned}
\mathcal{Q}_{1,a}^{(k)}(t) &= \sum_{m=0}^{k}\delta_{m,k}\mathcal{Q}_{1,a}^{(m)}(t) \\
&= \sum_{m=0}^{k}(-1)^{k+1}\sum_{l=m+1}^{k+1}(-1)^l \left\{ {k+1 \atop l} \right\} \left[{l \atop m+1} \right] \mathcal{Q}_{1,a}^{(m)}(t) \\
&= \sum_{l=1}^{k+1}(-1)^{k+l+1}\left\{ {k+1 \atop l} \right\} \sum_{m=0}^{l-1}\left[{l \atop m+1} \right] \mathcal{Q}_{1,a}^{(m)}(t) \\
&= \sum_{l=1}^{k+1}(-1)^{k+l+1}(l-1)!\left\{ {k+1 \atop l} \right\} \mathcal{Q}_{l,a}(t).
\end{aligned}$$

So putting $t=0$ in (8.7), we have

$$\begin{aligned}
\sum_{c=0}^{f-1}c^k\zeta^{ac} &= \sum_{m=1}^{k+1}(-1)^{k+m+1}(m-1)!\left\{ {k+1 \atop m} \right\} \mathcal{Q}_{m,a}(0) \\
&\quad + \sum_{j=0}^{k}\binom{k}{j} f^j \sum_{m=1}^{k-j+1}(-1)^{k-j+m}(m-1)!\left\{ {k-j+1 \atop m} \right\} \mathcal{Q}_{m,a}(0) \\
&= \sum_{j=1}^{k}\binom{k}{j} f^j \sum_{m=1}^{k-j+1}(-1)^{k-j+m}(m-1)!\left\{ {k-j+1 \atop m} \right\} \mathcal{Q}_{m,a}(0). \\
&= \sum_{m=1}^{k}(-1)^m(m-1)!\mathcal{Q}_{m,a}(0)\left(\sum_{j=1}^{k-m+1}(-1)^{k-j} f^j \binom{k}{j} \left\{ {k-j+1 \atop m} \right\} \right).
\end{aligned}$$

□

By the above calculation, we get the following examples.

Example 8.12.

$$\sum_{c=0}^{f-1} c\zeta^{ac} = -\frac{f}{1-\zeta^a},$$

$$\sum_{c=0}^{f-1} c^2\zeta^{ac} = \frac{-f^2+2f}{1-\zeta^a} - \frac{2f}{(1-\zeta^a)^2},$$

$$\sum_{c=0}^{f-1} c^3\zeta^{ac} = \frac{-f^3+3f^2-3f}{1-\zeta^a} + \frac{-3f^2+9f}{(1-\zeta^a)^2} + \frac{-6f}{(1-\zeta^a)^3}.$$

Application of these formulas will be given in the next section.

8.4 Various Examples of Sums

In this section, we treat several sporadic examples. Sometimes we can use the similar method for sums on character values as for sums on roots of unity. We shall see this. Let χ be a primitive Dirichlet character modulo f. We consider the following sum.

$$S_k(\chi) = \sum_{a_1,\dots,a_k=0}^{f-1} \chi(a_1+\cdots+a_k)a_1\cdots a_k.$$

Here we should note that the sum depends on the range of natural numbers a_i (namely, if we change them to other representatives mod f, then the value of the sum would change). Even for $k = 2$, it is not an efficient method to calculate this kind of sum by dividing the sum into various pieces and to seek the answers by case-by-case analysis. First of all, let us recall the following relation (cf. (8.2)).

$$\chi(a) = g(\overline{\chi})^{-1}\sum_{n=1}^{f}\overline{\chi(n)}\zeta^{an}.$$

By virtue of this relation, we get

$$S_k(\chi) = g(\overline{\chi})^{-1}\sum_{a_1,\dots,a_k=0}^{f-1}\sum_{n=1}^{f}\overline{\chi(n)}\zeta^{n(a_1+\cdots+a_k)}a_1\cdots a_k. \tag{8.8}$$

But since we have

$$\sum_{a=0}^{f-1} a\zeta^{an} = -\frac{f}{1-\zeta^n},$$

we obtain

$$\begin{aligned} S_k(\chi) &= (-f)^k g(\overline{\chi})^{-1} \sum_{n=1}^{f} \frac{\overline{\chi(n)}}{(1-\zeta^n)^k} \\ &= (-f)^k g(\overline{\chi})^{-1} P_{k,\overline{\chi}}(0) \\ &= \frac{(-1)^{k+1} f^k}{(k-1)!} \sum_{l=0}^{k-1} \frac{1}{l+1} \begin{bmatrix} k \\ l+1 \end{bmatrix} B_{l+1,\chi}. \end{aligned}$$

For example, we have

$$\begin{aligned} S_2(\chi) &= -f^2 \left(B_{1,\chi} + \frac{1}{2} B_{2,\chi} \right), \\ S_3(\chi) &= f^3 \left(B_{1,\chi} + \frac{3}{4} B_{2,\chi} + \frac{1}{6} B_{3,\chi} \right). \end{aligned}$$

We also consider the following similar examples.

$$S(\chi, e_1, \ldots, e_k) = \sum_{a_1,\ldots,a_k=1}^{f-1} \chi(a_1 + \cdots + a_k) a_1^{e_1} \cdots a_k^{e_k}.$$

The formula to describe this by Bernoulli numbers can be obtained by almost the same method as before. Indeed, as before, writing the character values by Gaussian sum and roots of unity, we have

$$\begin{aligned} S(\chi, e_1, \ldots, e_k) &= g(\overline{\chi})^{-1} \sum_{n=1}^{f} \overline{\chi(n)} \prod_{j=1}^{k} \left(\sum_{c=1}^{f-1} c^{e_j} \zeta^{cn} \right) \\ &= g(\overline{\chi})^{-1} \sum_{n=1}^{f} \overline{\chi(n)} \sum_{\substack{1 \le l_j \le e_j \\ 1 \le m_j \le e_j - l_j + 1 \\ 1 \le j \le k}} (-1)^{e_1 + \cdots + e_k - l_1 - \cdots - l_k + m_1 + \cdots + m_k} f^{l_1 + \cdots + l_k} \\ &\quad \times \binom{e_1}{l_1} \cdots \binom{e_k}{l_k} Q_{m_1+\cdots+m_k,n}(0)(m_1-1)! \cdots (m_k-1)! \\ &\quad \times \left\{ {e_1 - l_1 + 1 \atop m_1} \right\} \cdots \left\{ {e_k - l_k + 1 \atop m_k} \right\} \\ &= - \sum_{\substack{1 \le l_j \le e_j \\ 1 \le m_j \le e_j - l_j + 1 \\ 1 \le j \le k}} (-1)^{e_1 + \cdots + e_k - l_1 - \cdots - l_k} f^{l_1 + \cdots + l_k} \binom{e_1}{l_1} \cdots \binom{e_k}{l_k} \end{aligned}$$

$$\times \frac{(-1)^{m_1+\cdots+m_k}}{(m_1+\cdots+m_k-1)!}(m_1-1)!\cdots(m_k-1)!$$
$$\times \left\{ {e_1 - l_1 + 1 \atop m_1} \right\} \cdots \left\{ {e_k - l_k + 1 \atop m_k} \right\}$$
$$\times \left(\sum_{l=0}^{m_1+\cdots+m_k-1} \frac{1}{l+1} \left[{m_1+\cdots+m_k \atop l+1} \right] B_{l+1,\chi} \right).$$

Below, we sum up the above formula again.

Proposition 8.13. *We have the following formula.*

$$S(\chi, e_1, \ldots, e_k) = (-1)^{e_1+\cdots+e_k+1} \sum_{l_1=1}^{e_1} \cdots \sum_{l_k=1}^{e_k} (-f)^{l_1+\cdots+l_k} \prod_{j=1}^{k} \binom{e_j}{l_j}$$
$$\times \sum_{m_1=1}^{e_1-l_1+1} \cdots \sum_{m_k=1}^{e_k-l_k+1} \frac{(-1)^{m_1+\cdots+m_k}}{(m_1+\cdots+m_k-1)!} \prod_{j=1}^{k} (m_j-1)! \left\{ {e_j - l_j + 1 \atop m_j} \right\}$$
$$\times \left(\sum_{l=0}^{m_1+\cdots+m_k-1} \frac{1}{l+1} \left[{m_1+\cdots+m_k \atop l+1} \right] B_{l+1,\chi} \right).$$

Example 8.14. We give examples of applications of the above lemma below.

$$S_k(\chi) = S(\chi, \overbrace{1, \ldots, 1}^{k}) = \frac{(-1)^{k+1} f^k}{(k-1)!} \sum_{l=0}^{k-1} \frac{1}{l+1} \left[{k \atop l+1} \right] B_{l+1,\chi},$$

$$S(\chi, e) = \sum_{n=0}^{f-1} \chi(n) n^e = \sum_{l=0}^{e-1} \frac{f^{e-l}}{l+1} \binom{e}{l} B_{l+1,\chi},$$

$$S(\chi, 2, 1) = \sum_{m,n=0}^{f-1} \chi(m+n) m^2 n = -f^2 \left(f B_{1,\chi} + \frac{(f+1)}{2} B_{2,\chi} + \frac{1}{3} B_{3,\chi} \right).$$

We also give another similar example. Let m be a natural number which is coprime to a fixed f and smaller than f. We consider the following sum.

$$D_m(\chi) = \sum_{a=1}^{f} \chi(a+m) a.$$

Then we have

$$D_m(\chi) = g(\bar{\chi})^{-1} \sum_{a,n=1}^{f} a \bar{\chi}(n) \zeta^{(a+m)n}$$

$$= g(\bar{\chi})^{-1}\left(\sum_{n=1,(n,f)=1}^{f}\bar{\chi}(n)\zeta^{mn}\left(-\frac{f}{(1-\zeta^n)}+f\right)\right),$$

but since we have

$$\sum_{n=1,(n,f)=1}^{f}\bar{\chi}(n)(1-\zeta^{mn})(1-\zeta^n)^{-1}=\sum_{l=0}^{m-1}\sum_{n=1,(n,f)=1}^{f}\bar{\chi}(n)\zeta^{nl}=g(\bar{\chi})\sum_{l=1}^{m-1}\chi(l),$$

taking the relation $\sum_{n=1}^{f}\bar{\chi}(n)(1-\zeta^n)^{-1} = -g(\bar{\chi})B_{1,\chi}$ in the last section into account, we get

$$D_m(\chi)=f\left(\sum_{l=1}^{m}\chi(l)+B_{1,\chi}\right).$$

We will give several similar examples in the exercise of this chapter.

8.5 Sporadic Examples: Using Functions

Lemma 8.15. *Let $f>1$ be an odd number and let χ be a primitive character with conductor f. Then we have*

$$\sum_{a=1}^{f-1}\frac{\chi(a)}{(1-\zeta^a)(1-\zeta^{2a})^2}$$
$$=-g(\chi)\left(\frac{1}{24}B_{3,\bar{\chi}}+\frac{1}{4}(1+\overline{\chi(2)})B_{2,\bar{\chi}}+\frac{1}{8}(3+5\overline{\chi(2)})B_{1,\bar{\chi}}\right).$$

Proof. If we put

$$Q_{k,\chi}(t)=\sum_{a=1}^{f}\frac{\chi(a)(1+\zeta^a e^t)}{(1-\zeta^{2a}e^{2t})^k},$$

$$P_{k,\chi}(t)=\sum_{a=1}^{f}\frac{\chi(a)}{(1-\zeta^a e^t)^k},$$

then we get

$$Q_{1,\chi}(t)=P_{1,\chi}(t)=-g(\chi)\sum_{n=1}^{\infty}B_{n,\bar{\chi}}\frac{t^{n-1}}{n!}.$$

Since

$$\frac{d}{dt}Q_{k,\chi}(t) = 2kQ_{k+1,\chi}(t) + (1-2k)Q_{k,\chi}(t) - \overline{\chi(2)}P_{k,\chi}(2t).$$

we can write $Q_{k,\chi}(0)$ inductively by $P_{k,\chi}(0)$, and then by Bernoulli numbers. For example, if we put $k = 1$ in the above relation and take the derivatives, we get

$$Q'_{1,\chi}(t) = 2Q_{2,\chi}(t) - Q_{1,\chi}(t) - \overline{\chi(2)}P_{1,\chi}(2t),$$
$$Q'_{2,\chi}(t) = \frac{1}{2}(Q''_{1,\chi}(t) + Q'_{1,\chi}(t)) + \overline{\chi(2)}P'_{1,\chi}(2t),$$

and

$$Q_{2,\chi}(0) = -\frac{g(\chi)}{2}\left(\frac{1}{2}B_{2,\overline{\chi}} + (1+\overline{\chi(2)})B_{1,\overline{\chi}}\right),$$
$$Q'_{2,\chi}(0) = -\frac{g(\chi)}{2}\left(\frac{1}{3}B_{3,\overline{\chi}} + \left(\frac{1}{2} + \overline{\chi(2)}\right)B_{2,\overline{\chi}}\right).$$

In the same way we get

$$Q'_{2,\chi}(t) = 4Q_{3,\chi}(t) - 3Q_{2,\chi}(t) - \overline{\chi(2)}P_{2,\chi}(2t)$$

and

$$Q_{3,\chi}(0) = \frac{1}{4}(Q'_{2,\chi}(0) + 3Q_{2,\chi}(0) + \overline{\chi(2)}P_{2,\chi}(0)).$$

Since the left-hand side of the lemma is equal to $Q_{3,\chi}(0)$, substituting this for the above value and using the relation $P_{2,\chi}(0) = -g(\chi)(B_{1,\overline{\chi}} + B_{2,\overline{\chi}}/2)$, we get the right-hand side. □

8.6 Sporadic Examples: Using the Symmetry

In the examples below, we denote by $\psi(n) = (\frac{n}{p})$ the quadratic residue character where p is an odd prime.

Example 8.16. We put

$$I = \sum_{a=1}^{p-1} \psi(a)a.$$

By definition, this is equal to $pB_{1,\psi}$ and we know already that $B_{1,\psi} = 0$ if $\psi(-1) = 1$ (i.e. if $p \equiv 1 \bmod 4$ by the quadratic reciprocity law). This is also shown directly as follows. We assume that $p \equiv 1 \bmod 4$. Substituting $p - a$ for a and noting $\psi(a) = \psi(-a) = \psi(p - a)$, we get

$$I = \sum_{a=1}^{p-1} \psi(p-a)(p-a) = \sum_{a=1}^{p-1} \psi(a)(p-a).$$

Hence adding this to the original expression, we get

$$2I = \sum_{a=1}^{p-1} \psi(a)(a + p - a) = p \sum_{a=1}^{p-1} \psi(a) = 0.$$

Of course this argument cannot be used when $p \equiv 3 \bmod 4$. Actually, if $p \equiv 3 \bmod 4$ and $p > 3$, then $-I/p$ is the class number of the imaginary quadratic field $\mathbf{Q}(\sqrt{-p})$, so it is a more complicated value.

Now there are many cases where the above kind of substitution is effective in our calculation. For example:

Example 8.17. The character ψ being as above, put

$$I = \sum_{l,m,n=1, 4ln \equiv m^2 \bmod p}^{p-1} \psi(l)lmn.$$

Then substituting $m \to p - m$, we get

$$\begin{aligned} 2I &= \sum_{l,m,n=1, 4ln \equiv m^2 \bmod p}^{p-1} \psi(l)ln(m + p - m) \\ &= p \sum_{l,n=1}^{p-1} \psi(l)l(\psi(ln) + 1)n \\ &= p \sum_{l,n=1}^{p-1} (l\psi(n)n + \psi(l)ln) \\ &= p^3(p-1)B_{1,\psi}. \end{aligned}$$

So we have

$$I = \frac{p^3(p-1)}{2} B_{1,\psi}.$$

Example 8.18. The next sum was first considered in connection with dimension formulas of automorphic forms [92]. We put $\psi(n) = \left(\frac{n}{p}\right)$, where p is a prime and

$$S = \{(s,t,r) \in \mathbf{F}_p^3;\ s(r+s)(s+t) \neq 0\}.$$

Put $\zeta = e^{2\pi i/p}$ and define a sum I by

$$I = \sum_{(s,t,r)\in S} \frac{\psi(s^2 - rt)}{(\zeta^{r+s}-1)(\zeta^{s+t}-1)(\zeta^{-s}-1)}.$$

At first glance, it might seem difficult to calculate this sum in an elementary way, but actually this can be easily calculated as follows. If we put $a = r+s$, $b = s+t$, $c = -s$, then $s^2 - rt = c^2 - (a+c)(b+c) = -ab - bc - ca$. Hence if we put $T = \mathbf{F}_p^{\times 3}$, then substitute $-a$, $-b$, $-c$ for a, b, c, we get

$$\begin{aligned}
I &= \sum_{(a,b,c)\in T} \frac{\psi(-ab-bc-ca)}{(\zeta^a-1)(\zeta^b-1)(\zeta^c-1)} \\
&= \sum_{(a,b,c)\in T} \frac{\psi(-ab-bc-ca)}{(\zeta^{-a}-1)(\zeta^{-b}-1)(\zeta^{-c}-1)} \\
&= \sum_{(a,b,c)\in T} \frac{\psi(-ab-bc-ca)\zeta^a\zeta^b\zeta^c}{(1-\zeta^a)(1-\zeta^b)(1-\zeta^c)} \\
&= -\sum_{(a,b,c)\in T} \frac{\psi(-ab-bc-ca)(\zeta^a-1+1)(\zeta^b-1+1)(\zeta^c-1+1)}{(\zeta^a-1)(\zeta^b-1)(\zeta^c-1)}.
\end{aligned}$$

So by expanding the numerator and transposing the same term to the left-hand side, we get

$$\begin{aligned}
2I = &-\sum_{(a,b,c)\in T} \psi(-ab-bc-ca) \\
&\times \left(\frac{1}{(1-\zeta^a)(1-\zeta^b)} + \frac{1}{(1-\zeta^b)(1-\zeta^c)} + \frac{1}{(1-\zeta^c)(1-\zeta^a)} \right. \\
&\left. -\frac{1}{1-\zeta^a} - \frac{1}{1-\zeta^b} - \frac{1}{1-\zeta^c} + 1 \right).
\end{aligned}$$

If we fix a and b, then we have

$$\sum_{c\in\mathbf{F}_p^\times}\psi(-ab-(a+b)c)=\begin{cases}-\psi(-ab) & \text{if } a+b\neq 0\,,\\ (p-1)\psi(-ab)=(p-1) & \text{if } a=-b\neq 0\,.\end{cases}$$

So if we fix $a\neq 0$, then we have

$$\sum_{b,c=1}^{p-1}\psi(-ab-bc-ca)=(p-1)-\sum_{b\neq -a,\ b\neq 0}\psi(-ab)=p.$$

Here by Examples 8.8 and 8.10, we have

$$\sum_{a,b\in\mathbf{F}_p^\times}\frac{\psi(-ab)}{(1-\zeta^a)(1-\zeta^b)}=\psi(-1)g(\psi)^2B_{1,\psi}^2=pB_{1,\psi}^2,$$

$$p\sum_{a\in\mathbf{F}_p^\times}\frac{1}{(1-\zeta^a)(1-\zeta^{-a})}=\frac{p(p-1)}{2}+\frac{p(p-1)(p-5)}{12}.$$

So noting $p-1=p-\psi(a^2)$, we get

$$\sum_{(a,b,c)\in T}\psi(-ab-bc-ca)\frac{1}{(1-\zeta^a)(1-\zeta^b)}=-pB_{1,\psi}^2+\frac{p(p-1)(p-5)}{12}+\frac{p(p-1)}{2}.$$

We have also

$$\sum_{(a,b,c)\in T}\frac{\psi(-ab-bc-ca)}{1-\zeta^a}=\frac{p(p-1)}{2}.$$

$$\sum_{(a,b,c)\in T}\psi(-ab-bc-ca)=p(p-1).$$

So as a total, we have

$$I=\frac{3}{2}pB_{1,\psi}^2-\frac{p(p-1)(p-5)}{8}-\frac{p(p-1)}{2}=\frac{3}{2}pB_{1,\psi}^2-\frac{p(p-1)^2}{8}.$$

In particular, if $p\equiv 1 \bmod 4$ then $B_{1,\psi}=0$, and if $p\equiv 3 \bmod 4$ and $p>3$, then $B_{1,\psi}=-h(-p)$ ($h(-p)$ is the class number of the imaginary quadratic field $\mathbf{Q}(\sqrt{-p})$), so there exists a formula for I by class numbers.

Example 8.19. We put $S=\{(a,b,c)\in\mathbf{Z}^3; 1\le a,b,c\le p-1, ab+bc+ca\equiv 0 \bmod p\}$ and

$$I = \sum_{(a,b,c)\in S} \psi(abc)abc.$$

The condition in the sum I is complicated and it is not easy to calculate I. But if we assume that $\psi(-1) = 1$, then I can be calculated as follows. By the relation

$$I = \sum_{(a,b,c)\in S} \psi(-abc)(p-a)(p-b)(p-c)$$

we get

$$\begin{aligned} 2I &= \sum_{(a,b,c)\in S} \psi(abc)((p-a)(p-b)(p-c) + abc) \\ &= \sum_{(a,b,c)\in S} \psi(abc)(p^3 - p^2(a+b+c) + p(ab+bc+ca)). \end{aligned}$$

But we have $\psi(abc) = \psi(-c^2(a+b)) = \psi(a+b)$ and since the condition of the definition of S is symmetric with respect to a, b, c, we have

$$\begin{aligned} \sum_{(a,b,c)\in S} \psi(abc)(ab+bc+ca) &= 3 \sum_{(a,b,c)\in S} \psi(a+b)ab \\ &= 3 \sum_{a,b,a+b\not\equiv 0 \bmod p} \psi(a+b)ab \\ &= 3 \sum_{a,b=1}^{p-1} \psi(a+b)ab = -3p^2\left(B_{1,\psi} + \frac{1}{2}B_{2,\psi}\right). \end{aligned}$$

Since we assumed $\psi(-1) = 1$, we have $B_{1,\psi} = 0$ now. Moreover we have

$$\begin{aligned} \sum_{(a,b,c)\in S} \psi(a+b)(a+b+c) &= 3\sum_{a=1}^{p-1}\left(\sum_{b=1}^{p-1} \psi(a+b)\right) a \\ &= -3\sum_{a=1}^{p-1} \psi(a)a = -3pB_{1,\psi} \end{aligned}$$

and this is equal to 0. We have

$$\sum_{a,b,c\in S} \psi(a+b) = \sum_{a,b=1}^{p-1} \psi(a+b) = \sum_{a=1}^{p-1}(-\psi(a)) = 0.$$

So we have

$$I = -\frac{3p^3}{4} B_{2,\psi}.$$

Remark 8.20. When $\psi(-1) = -1$, a simple formula to write down I by generalized Bernoulli numbers is known only conjecturally.

8.7 Sporadic Example: Symmetrize Asymmetry

Let $\psi(x)$ be the quadratic residue character modulo p (p is a prime) and consider the following special sum.

$$I = \sum_{m,n=1}^{p-1} \psi(m-4n)\psi(m)mn.$$

A formula for this sum was first given in [91] and [39] by comparing the trace formula and algebraic geometry, but this can be calculated in an elementary way as well. We illustrate this calculation below.

In the above definition, if we take n instead of $m - 4n$, then the sum is very simple and given by

$$\sum_{m,n=1}^{p-1} \psi(mn)mn = \left(\sum_{m=1}^{p-1} \psi(m)m\right)^2 = p^2 B_{1,\psi}^2.$$

But the sum I we actually defined above is subtly breaking its symmetry and that obstructs an easy calculation. We get over this by changing it to an exponential sum and transforming it to a symmetric one. By the relation

$$\psi(m) = g(\psi)^{-1} \sum_{a=1}^{p-1} \psi(a)\zeta^{am} \qquad \psi(m-4n) = g(\psi)^{-1} \sum_{b=1}^{p-1} \psi(b)\zeta^{b(m-4n)},$$

we have

$$I = g(\psi)^{-2} \sum_{a,b=1}^{p-1} \psi(ab) \sum_{m,n=1}^{p-1} mn\zeta^{am+b(m-4n)}.$$

But for $c \notin p\mathbf{Z}$, we have $\sum_{m=1}^{p-1} m\zeta^{cm} = -\frac{p}{1-\zeta^c}$, hence we have

$$I = g(\psi)^{-2}\left(p^2 \sum_{\substack{a,b=1\\ a\not\equiv -b \bmod p}}^{p-1} \frac{\psi(ab)}{(1-\zeta^{-4b})(1-\zeta^{a+b})} + \psi(-1)\sum_{m,n,b=1}^{p-1} mn\zeta^{-4nb}\right)$$
$$= p\psi(-1)\sum_{\substack{a,b=1\\ a+b\neq p}}^{p-1} \frac{\psi(ab)}{(1-\zeta^{-4b})(1-\zeta^{a+b})} - \frac{p(p-1)^2}{4},$$

using $g(\psi)^2 = \psi(-1)p$. In order to calculate I from the above relation, it is sufficient to calculate

$$J = \psi(-1)\sum_{\substack{a,b=1\\ a+b\neq p}}^{p-1} \frac{\psi(ab)}{(1-\zeta^{-4b})(1-\zeta^{a+b})}.$$

Now a direct calculation of J seems difficult partly because the sum in the definition of J is not symmetric with respect to a and b. So we try to rewrite this in a symmetric way. If we exchange a and b, this is just a change of notation, so we have

$$J = \psi(-1)\sum_{\substack{a,b=1\\ a+b\neq p}}^{p-1} \frac{\psi(ab)}{(1-\zeta^{-4a})(1-\zeta^{b+a})}.$$

Now we will add this to the original expression and take the average of both. To calculate this, we use the following equality.

$$\frac{1}{1-\zeta^{-4a}} + \frac{1}{1-\zeta^{-4b}} = \frac{2-\zeta^{-4a}-\zeta^{-4b}}{(1-\zeta^{-4a})(1-\zeta^{-4b})} = \frac{2\zeta^{4a+4b}-\zeta^{4a}-\zeta^{4b}}{(1-\zeta^{4a})(1-\zeta^{4b})}$$
$$= \frac{\zeta^{4a+4b}-1+(1-\zeta^{4a})(1-\zeta^{4b})}{(1-\zeta^{4a})(1-\zeta^{4b})}$$
$$= 1 - \frac{1-\zeta^{4a+4b}}{(1-\zeta^{4a})(1-\zeta^{4b})}.$$

After all, if we put

$$J_1 = \sum_{\substack{a,b=1\\ a+b\neq p}}^{p-1} \frac{\psi(ab)}{1-\zeta^{a+b}},$$

$$J_2 = \sum_{\substack{a,b=1\\ a+b\neq p}}^{p-1} \frac{\psi(ab)(1-\zeta^{4a+4b})}{(1-\zeta^{4a})(1-\zeta^{4b})(1-\zeta^{a+b})},$$

then we have

$$J = \frac{\psi(-1)}{2}(J_1 - J_2).$$

Here we have

$$\begin{aligned} J_1 &= -\frac{1}{p}\sum_{k=1}^{p-1}\sum_{\substack{a,b=1\\ a+b\neq p}}^{p-1} \psi(ab)k\zeta^{k(a+b)} \\ &= -\frac{1}{p}(g(\psi)^2 - \psi(-1)(p-1)) \times \frac{p(p-1)}{2} = -\frac{\psi(-1)(p-1)}{2}. \end{aligned}$$

Also we have

$$\begin{aligned} J_2 &= \sum_{\substack{a,b=1\\ a+b\neq p}}^{p-1} \psi(ab)\frac{1-\zeta^{4a+4b}}{(1-\zeta^{4a})(1-\zeta^{4b})(1-\zeta^{a+b})} \\ &= \sum_{a,b=1}^{p-1} \frac{\sum_{l=0}^{3}\psi(ab)\zeta^{l(a+b)}}{(1-\zeta^{4a})(1-\zeta^{4b})} - \sum_{a=1}^{p-1}\frac{4\psi(-1)}{(1-\zeta^{4a})(1-\zeta^{-4a})}. \end{aligned}$$

Now, by virtue of Example 8.10, we have

$$\begin{aligned} \sum_{a=1}^{p-1}\frac{1}{(1-\zeta^{4a})(1-\zeta^{-4a})} &= \sum_{a=1}^{p-1}\left(\frac{1}{1-\zeta^{4a}} - \frac{1}{(1-\zeta^{4a})^2}\right) \\ &= \frac{p-1}{2} + \frac{(p-1)(p-5)}{12} = \frac{p^2-1}{12}. \end{aligned}$$

Furthermore we have

$$\sum_{a=1}^{p-1}\frac{\psi(a)\zeta^{la}}{1-\zeta^{4a}} = -\frac{1}{p}\sum_{a,c=1}^{p-1}\psi(a)c\zeta^{4ac+al} = -\frac{1}{p}g(\psi)\sum_{c=1}^{p-1}\psi(4c+l)c.$$

Hence if we put $S(l) = \sum_{c=1}^{p-1}\psi(4c+l)c$, then we have

$$J_2 = \frac{\psi(-1)}{p}\sum_{l=0}^{3}S(l)^2 - \frac{\psi(-1)(p^2-1)}{3}.$$

Summing up, we have

$$J = -\frac{p-1}{4} + \frac{p^2-1}{6} - \frac{1}{2p}\sum_{l=0}^{3}S(l)^2 = \frac{(p-1)(2p-1)}{12} - \frac{1}{2p}\sum_{l=0}^{3}S(l)^2.$$

Next the following quantity should be evaluated.

$$\sum_{l=0}^{3} S(l)^2 = \left(\sum_{c=1}^{p-1} \psi(4c)c\right)^2 + \left(\sum_{c=1}^{p-1} \psi(4c+1)c\right)^2 + \left(\sum_{c=1}^{p-1} \psi(4c+2)c\right)^2 + \left(\sum_{c=1}^{p-1} \psi(4c+3)c\right)^2. \tag{8.9}$$

There are many ways to evaluate this, but here we calculate it just directly. (As for more general method of calculation, see [45].) First we have

$$\begin{aligned}
&\sum_{c=1}^{p-1}(\psi(2c)c + \psi(2c+1)c) \\
&\quad = \frac{1}{2}\left(\sum_{c=0}^{p-1}(\psi(2c)\times(2c) + \psi(2c+1)(2c+1)) - \sum_{c=0}^{p-1}\psi(2c+1)\right) \\
&\quad = \frac{1}{2}\left(\sum_{c=1}^{2p-1}\psi(c)c\right) \\
&\quad = \frac{1}{2}\sum_{c=1}^{p-1}\psi(c)(c+c+p) = \sum_{c=1}^{p-1}\psi(c)c.
\end{aligned}$$

Hence we have

$$\sum_{c=1}^{p-1}\psi(2c+1)c = (1-\psi(2))\sum_{c=1}^{p-1}\psi(c)c.$$

In the same way, we have

$$\begin{aligned}
&\sum_{c=1}^{p-1}(\psi(4c)c + \psi(4c+1)c + \psi(4c+2)c + \psi(4c+3)c) = \frac{1}{4}\sum_{c=1}^{4p-1}\psi(c)c \\
&\quad = \sum_{c=1}^{p-1}\psi(c)c.
\end{aligned}$$

This implies that

$$\sum_{c=1}^{p-1}(\psi(4c+1)+\psi(4c+3))c = (1-\psi(2))\sum_{c=1}^{p-1}\psi(c)c.$$

Now we consider the case $p \equiv 1 \bmod 4$ and $p \equiv 3 \bmod 4$ separately.

First if we assume $p \equiv 3 \bmod 4$, the $\psi(-1) = -1$. So we have $\psi(4c+3) = -\psi(4(p-c)-3) = -\psi(4(p-1-c)+1)$, and

$$\sum_{c=0}^{p-1} \psi(4c+3)c = -\sum_{c=0}^{p-1} \psi(4c+1)(p-1-c) = \sum_{c=0}^{p-1} \psi(4c+1)c.$$

Hence in this case we have

$$\sum_{c=1}^{p-1} \psi(4c+1)c = \sum_{c=1}^{p-1} \psi(4c+3)c = \frac{1-\psi(2)}{2} \sum_{c=1}^{p-1} \psi(c)c.$$

So we have

$$\begin{aligned}\sum_{l=0}^{3} S(l)^2 &= \left(1 + (1-\psi(2))^2 + \frac{(1-\psi(2))^2}{2}\right)\left(\sum_{c=1}^{p-1} \psi(c)c\right)^2 \\ &= (4 - 3\psi(2))p^2 B_{1,\psi}^2.\end{aligned}$$

By the assumption $p \equiv 3 \bmod 4$, if $p \neq 3$ then we have $B_{1,\psi} = -h(-p)$, where $h(-p)$ is the class number of the imaginary quadratic field $\mathbf{Q}(\sqrt{-p})$ (p. 90, (6.5)).

Next we assume that $p \equiv 1 \bmod 4$. In this case we have $B_{1,\psi} = 0$, so we cannot separate $\sum_{c=1}^{p-1} \psi(4c+1)c$ and $\sum_{c=1}^{p-1} \psi(4c+3)c$ in the above expression in the same way. In order to separate these two, we introduce a Dirichlet character δ with conductor 4. Here for $x \equiv 3 \bmod 4$ we have $\delta(x) = -1$ and we see that $\psi\delta$ is a primitive Dirichlet character with conductor $4p$. By definition, we have

$$\begin{aligned}4pB_{1,\psi\delta} &= \sum_{c=0}^{4p-1} \psi(c)\delta(c)c = \sum_{c=0}^{p-1} (\psi(4c+1)(4c+1) - \psi(4c+3)(4c+3)) \\ &= 4(S(1) - S(3)) = 8S(1) = -8S(3).\end{aligned}$$

By virtue of the fact $B_{1,\psi} = 0$, we have $S(0) = S(2) = 0$. So we get

$$S(1)^2 + S(3)^2 = \frac{p^2}{2}(B_{1,\psi\delta})^2.$$

Also we have $h(-p) = -B_{1,\psi\delta}$. Summing up all the above results, we get

$$I = -\frac{p(p-1)(p-2)}{12} - \begin{cases} 7/2 & \text{if } p=3\,, \\ p^2 \times h(-p)^2/4 & \text{if } p \equiv 1 \bmod 4, \\ (4-3\psi(2))p^2 \times h(-p)^2/2 & \text{if } p \equiv 3 \bmod 4 \text{ and } p \neq 3. \end{cases}$$

By the way, as for further generalization of the results in this section, see [45].

8.8 Quadratic Polynomials and Character Sums

The example given below is not related to Bernoulli numbers so much, but since the method is similar, we take this opportunity to explain it. Here too, we assume that p is an odd prime and $\psi(x) = \left(\frac{x}{p}\right)$ is the quadratic residue character modulo p. Let $P(x)$ be a polynomial in a variable x. If we consider the sum $\sum_{x=0}^{p-1} \psi(P(x))$, then this is related with the number of pairs $(x, y) \in \mathbf{F}_p^2$ of integral solutions of $y^2 = P(x)$, so related with the congruence zeta function of algebraic curves and a complicated quantity. There is no simple formula for general $P(x)$. But in the special case where $P(x)$ is a quadratic polynomial, it corresponds to a conic. In this case the structure of the curve itself is simple and there is a simple formula for the sum as above. Let us practice the calculation. We write a quadratic polynomial of x as $P(x) = ax^2 + bx + c$ ($a, b, c \in \mathbf{Z}$, $a \not\equiv 0 \bmod p$). It is not so difficult to calculate $\sum_{x \bmod p} \psi(P(x))$. Since we take p to be an odd prime, we consider the finite field $\mathbf{F}_p$ and rewrite $P(x) = a((x-d)^2 + e)$. Then we have

$$\sum_{x \bmod p} \psi(P(x)) = \psi(a) \sum_{x \bmod p} \psi(x^2 + e).$$

So this right-hand side is the point.

Lemma 8.21.

$$\sum_{x \bmod p} \psi(x^2 + e) = \begin{cases} -1 & \textit{if } e \not\equiv 0 \bmod p, \\ p-1 & \textit{if } e \equiv 0 \bmod p. \end{cases}$$

Proof. If we write $\zeta = e^{2\pi i/p}$, then we have

$$\psi(x^2 + e) = \frac{1}{g(\psi)} \sum_{k=0}^{p-1} \psi(k) \zeta^{k(x^2+e)},$$

where $g(\psi)$ is the Gaussian sum. So we have

$$\sum_{x=0}^{p-1} \psi(x^2 + e) = \frac{1}{g(\psi)} \sum_{k=0}^{p-1} \psi(k) \sum_{x=0}^{p-1} \zeta^{k(x^2+e)}.$$

But we have $\sum_{x=0}^{p-1} \zeta^{kx^2} = \psi(k) g(\psi)$, so we get the following result.

$$\sum_{k=0}^{\infty} \psi(k)^2 \zeta^{ke} = \sum_{k=1}^{p-1} \zeta^{ke} = \begin{cases} -1 & \text{if } e \not\equiv 0 \bmod p, \\ p-1 & \text{if } e \equiv 0 \bmod p. \end{cases}$$

□

By the way, the result above has of course a relation to the number of pairs $(x, y) \in \mathbf{F}_p^2$ such that $x^2+e = y^2$. Namely, for each $x \in \mathbf{F}_p$, there are $1+\psi(x^2+e)$ number of y which satisfy this equation. So $p-1$ pairs of (x, y) are solutions of this equation if $e \not\equiv 0 \bmod p$.

In general, also for a quadratic form $Q(x)$ of n variables, that is, a homogeneous polynomial of variables x_i of $x = (x_1, x_2, \ldots, x_n)$ of degree two, we sometimes consider the sum $\sum_{x \bmod p} \zeta^{Q(x)}$ and the results are well known, but this is remote from Bernoulli numbers and we omit it here.

Next we treat the sum $\sum_{x=0}^{p-1} \psi(x^2+e)x$. (Note that the sum $\sum_{x=0}^{p-1} \psi(P(x))x$ for a general quadratic polynomial $P(x)$ does not reduce to this sum.) Here we have

$$\sum_{x=1}^{p-1} \psi(x^2+e)x = \sum_{x=1}^{p-1} (p-x)\psi((p-x)^2+e) = \sum_{x=1}^{p-1} (p-x)\psi(x^2+e),$$

so we have

$$2\sum_{x=1}^{p-1} \psi(x^2+e)x = p\sum_{x=1}^{p-1} \psi(x^2+e) = p\left(\sum_{x=0}^{p-1} \psi(x^2+e) - \psi(e)\right).$$

Hence we have

$$\sum_{x=0}^{p-1} \psi(x^2+e)x = \begin{cases} -\frac{(1+\psi(e))p}{2} & \text{if } e \not\equiv 0 \bmod p, \\ \frac{p(p-1)}{2} & \text{if } e \equiv 0 \bmod p. \end{cases}$$

As an application, when $e \notin p\mathbf{Z}$, we can show the following formula.

$$\sum_{x,y=0}^{p-1} \psi(x^2+ey^2)xy = -\frac{(1+\psi(e))p^2(p-1)}{4}.$$

We omit the proof here.

8.9 A Sum with Quadratic Conditions

For a polynomial $f(x, y, z)$, we put

$$M(f) = \{(a, b, c) \in \mathbf{Z}^3; 1 \le a, b, c \le p-1; f(a, b, c) \equiv 0 \bmod p\}.$$

Proposition 8.22. *For an integer $e \not\equiv 0 \bmod p$, we write $h_e(x, y, z) = exz - y^2$. Then we have*

$$\sum_{(a,b,c)\in M(h_e)} \frac{\psi(a)}{(1-\zeta^a)(1-\zeta^b)(1-\zeta^c)} = -\frac{p-1}{4}(1+\psi(e))g(\psi)B_{1,\psi}.$$

Proof. Denote the left-hand side by I. In the definition of I, we may replace b by $-b$. Noting that

$$(1-\zeta^{-b})^{-1} + (1-\zeta^b)^{-1} = 1,$$

and taking integers $\bar{a}$ and $\bar{e}$ such that $a\bar{a} \equiv e\bar{e} \equiv 1 \bmod p$, we have

$$\begin{aligned} 2I &= \sum_{a,b=1}^{p-1} \frac{\psi(a)}{(1-\zeta^a)(1-\zeta^{\bar{a}\bar{e}b^2})} \\ &= \sum_{a,b=1}^{p-1} \frac{\psi(a)}{(1-\zeta^a)} \left(-\frac{1}{p}\sum_{m=1}^{p-1} m\zeta^{\bar{a}\bar{e}b^2 m}\right) \\ &= -\frac{1}{p}\sum_{a,m=1}^{p-1} \frac{\psi(a)}{(1-\zeta^a)} m(\psi(\bar{a}\bar{e}m)g(\psi)-1) \\ &= -\frac{p-1}{2}\psi(\bar{e})g(\psi)B_{1,\psi} - \frac{p-1}{2}g(\psi)B_{1,\psi} \\ &= -\frac{p-1}{2}(1+\psi(\bar{e}))g(\psi)B_{1,\psi}. \end{aligned}$$

□

As an application of this formula, we can show the following.

Corollary 8.23.

$$\sum_{(l,m,n)\in M(h_e)} \psi(n)lmn = \frac{p^3(p-1)}{4}(1+\psi(e))B_{1,\psi}.$$

Proof. For a while, for any rational number α whose denominator is not divisible by p, we understand that we define ζ^α by regarding α as an element of $\mathbf{F}_p$. Then we have

$$\begin{aligned} -p^3 I &= \sum_{l,m,n=1}^{p-1} lmn \sum_{(a,b,c)\in M(h_e)} \psi(a)\zeta^{al+bm+cn} \\ &= \sum_{l,m,n=1}^{p-1} lmn \sum_{a=1}^{p-1} \psi(a)\zeta^{al}\zeta^{-\frac{aem^2}{4n}} \sum_{b=1}^{p-1} \zeta^{\frac{n}{ae}(b+\frac{aem}{2n})^2} \end{aligned}$$

$$= \sum_{l,m,n,a=1}^{p-1} lmn\zeta^{al}\psi(a)\zeta^{-aem^2/4n}(\psi(n/(ae))g(\psi) - \zeta^{aem^2/4n})$$

$$= \sum_{l,m,n,a=1}^{p-1} lmn\zeta^{al}(\psi(ne)\zeta^{-aem^2/4n}g(\psi) - \psi(a))$$

$$= \sum_{l,m,n=1}^{p-1} (-g(\psi)\psi(ne) - \psi(l)g(\psi))lmn + pg(\psi) \sum_{4ln\equiv em^2 \bmod p} \psi(n)lmn.$$

Hence we have

$$I = \frac{g(\psi)}{p^3}\left((1+\psi(e))pB_{1,\psi}\frac{p^2(p-1)^2}{4} - p\sum_{(l,m,n)\in M(h_{4/e})} \psi(n)lmn\right).$$

In the above calculation, we used $\sum_{x=0}^{p-1}\zeta^{cx^2} = \psi(c)g(\psi)$. Here in the final expression of I and in the first term of the right-hand side above, the number e appears only as $\psi(e)$. Since we have $\psi(4/e) = \psi(e)$, we can replace the condition $M(h_{4/e})$ of the sum on the right-hand side by $M(h_e)$. By these considerations, we get the formula for $\sum_{(l,m,n)\in M(h_e)} \psi(n)lmn$. □

Several such sums with quadratic conditions have something to do with special values of zeta functions of prehomogeneous vector spaces. But there are a lot of mysteries around this in general. It has not been clarified when such formulas for exponential sums can exist or what essential reasons there are for that. For some special quadratic conditions, we have several conjectures but proofs are not known. It seems premature to treat this problem here, so we omit the concrete description of conjectures or other things, but it would be a very interesting problem to study these sums from the viewpoint of relations between exponential sums and special values of zeta functions of prehomogeneous vector spaces.

Exercise 8.24. Give an example of non-trivial primitive Dirichlet character χ such that χ^3 is trivial and for that χ, calculate the Gaussian sum $g(\chi)$ explicitly.

Exercise 8.25. (1) Assume that the cardinality of a finite group G is odd. Then show that any character χ such that $\chi^2 = 1$ (the identity character) is a trivial character.

(2) Assume that G is a finite cyclic group of even order. Then show that the non-trivial character χ of G such that $\chi^2 = 1$ exists uniquely.

Exercise 8.26. (1) Let p be an odd prime. For any natural number r, it is known that $(\mathbf{Z}/p^r\mathbf{Z})^\times \cong \mathbf{Z}/p^{r-1}\mathbf{Z} \times \mathbf{Z}/(p-1)\mathbf{Z}$. By using this, show that if a non-trivial Dirichlet character χ modulo p^r satisfies $\chi(a)^2 = 1$ for any a with $(a,p) = 1$, then the conductor of χ is p.

(2) Let r be a natural number with $r \geq 2$. It is known that $(\mathbf{Z}/2^r\mathbf{Z})^\times \cong \mathbf{Z}/2\mathbf{Z} \times \mathbf{Z}/2^{r-2}\mathbf{Z}$ (or $\cong ((1+2\mathbf{Z})/(1+4\mathbf{Z})) \times \big((1+4\mathbf{Z})/(1+2^{r-2}\mathbf{Z})\big)$, the latter is cyclic generated by 5.) Show that if a non-trivial Dirichlet character χ modulo 2^r satisfies $\chi(a)^2 = 1$ for any a with $(a,2) = 1$, then the conductor of χ is either 4 or 8.

(3) Show that if χ is any non-trivial Dirichlet character χ such that χ^2 is trivial, then the conductor of χ is $|D_K|$ for a fundamental discriminant of some quadratic field K.

(4) Show that if χ is any non-trivial primitive Dirichlet character χ such that χ^2 is trivial, then $\chi = \chi_K$ for some quadratic field K, where χ_K is defined as in Sect. 6.3.

Exercise 8.27. Let χ be a primitive Dirichlet character with conductor f. Assume that $f = m_1 m_2$ with $(m_1, m_2) = 1$. We regard χ as a character of $(\mathbf{Z}/f\mathbf{Z})^\times \cong (\mathbf{Z}/m_1\mathbf{Z})^\times \times (\mathbf{Z}/m_2\mathbf{Z})^\times$ and denote by χ_i the restriction to $\mathbf{Z}/m_i\mathbf{Z}$ for $i = 1, 2$. We also denote by χ_i the Dirichlet character modulo m_i associated with χ_i.

(1) Prove that the conductor of χ_i is m_i.

(2) For any n, put $\zeta_n = exp\left(\frac{2\pi i}{n}\right)$. We take integers x, y with $xm_1 + ym_2 = 1$. Show that for any $a \in \mathbf{Z}$, we have

$$\zeta_f^a = \zeta_{m_1}^{ay} \zeta_{m_2}^{ax}.$$

(3) Show the following relation between Gaussian sums.

$$g(\chi) = \chi_1(m_2)\chi_2(m_1)g(\chi_1)g(\chi_2).$$

(4) For each primitive Dirichlet character χ with conductor 4 or 8 defined in Sect. 4.1, calculate the Gaussian sum $g(\chi)$.

(5) Let χ_K be the Dirichlet character modulo D_K defined in Sect. 6.3. Show that $g(\chi_K) = \sqrt{D_K}$ by using $g(p) = \sqrt{(-1)^{(p-1)/2}p}$ for odd primes p and the reciprocity law of the quadratic residue symbol.

Exercise 8.28. Prove Example 8.14 using Proposition 8.13.

Exercise 8.29. Show that the following relations hold.

(1) For any primitive character χ modulo f, we have

$$\sum_{a=1}^{f-1} \frac{\chi(a)\zeta^a}{(1-\zeta^a)^2} = -\frac{1}{2} g(\chi) B_{2,\overline{\chi}}.$$

$$\sum_{a=1}^{f-1} \frac{\chi(a)\zeta^a(1+\zeta^a)}{(1-\zeta^a)^3} = -\frac{1}{3} g(\chi) B_{3,\overline{\chi}}.$$

(2) If we denote by $\psi(n) = (\frac{n}{p})$ the quadratic residue character, we have

$$\sum_{m,n=0}^{p-1} \psi(m+n)\psi(m)mn = \frac{1}{2}\left(p^2 B_{1,\psi}^2 - \frac{p(p-1)(7p+1)}{12}\right).$$

(3) We have

$$\sum_{l,m,n=0}^{p-1} \psi(lm+mn+nl) = \psi(-1)p(p-1).$$

$$\sum_{l,m,n=0}^{p-1} \psi(lm+mn+nl)l = \psi(-1)\frac{p^2(p-1)}{2}.$$

$$\sum_{l,m,n=0}^{p-1} \psi(lm+mn+nl)l^2 = \psi(-1)\frac{p^2(p-1)(2p-1)}{6}.$$

$$\sum_{l,m,n=0}^{p-1} \psi(lm+mn+nl)l^3 = \psi(-1)\frac{p^3(p-1)^2}{4}.$$

$$\sum_{l,m,n=0}^{p-1} \psi(lm+mn+nl)lm = \psi(-1)\frac{p^2(p^2-1)}{6}.$$

$$\sum_{l,m,n=0}^{p-1} \psi(lm+mn+nl)l^2m = \psi(-1)\frac{p^3(p^2-1)}{12}.$$

$$\sum_{l,m,n=0}^{p-1} \psi(lm+mn+nl)lmn = -\frac{3p^3}{2}(B_{1,\psi})^2 + \psi(-1)\frac{p^3(p-1)}{4}.$$

(Hint: $\sum_{n=0}^{p-1} \psi(n(l+m)+lm)$ is 0 if $l+m \not\equiv 0 \bmod p$. In the last relation, substitute l, m, n by $p-l$, $p-m$, $p-n$ and add that to the original expression.)

Chapter 9
Special Values and Complex Integral Representation of L-Functions

As a continuation of Chaps. 4 and 5, we study here properties of Hurwitz zeta functions and Dirichlet L-functions such as their analytic continuation and functional equation, and calculate their special values at negative integers. There are various proofs for the functional equation; here we explain the method using a contour integral. Although there would be a viewpoint that it would be too much to introduce a contour integral, it is interesting for its own sake and useful too, so we venture to derive the functional equation from a contour integral by a method to cut out the path of the integral.

9.1 The Hurwitz Zeta Function

For a positive real number a, we define a zeta function $\zeta(s,a)$ by

$$\zeta(s,a) = \sum_{n=0}^{\infty} (n+a)^{-s}.$$

Since the series on the right-hand side converges absolutely for $\mathrm{Re}(s) > 1$, $\zeta(s,a)$ is defined in this range. This function $\zeta(s,a)$ is called a Hurwitz zeta function. (Hurwitz [43], Whittaker[1] and Watson[2] [104, Part II]). In particular, when $a = 1$, $\zeta(s,1)$ is nothing but the Riemann zeta function $\zeta(s)$. In this section, we shall give an integral representation of $\zeta(s,a)$.

We denote by $\Gamma(s)$ the gamma function. This is defined by

---

[1]Edmund Taylor Whittaker (born on October 24, 1873 in Southport, England—died on March 24, 1956 in Edinburgh, Scotland).

[2]George Neville Watson (born on January 31, 1886 in Devon, England—died on February 2, 1965 in Warwickshire, England).

T. Arakawa et al., *Bernoulli Numbers and Zeta Functions*, Springer Monographs in Mathematics, DOI 10.1007/978-4-431-54919-2_9, © Springer Japan 2014

$$\Gamma(s) = \int_0^\infty e^{-t} t^{s-1}\, dt \tag{9.1}$$

for complex numbers s such that $\mathrm{Re}(s) > 0$. The integral on the right-hand side converges absolutely and uniformly on any compact sets in $\mathrm{Re}(s) > 0$, and so we see that $\Gamma(s)$ is a holomorphic function for $\mathrm{Re}(s) > 0$. By integration by parts, we get the most important property

$$\Gamma(s+1) = s\Gamma(s) \tag{9.2}$$

of the gamma function. Through this equality (9.2), $\Gamma(s)$ can be continued analytically to a meromorphic function on the whole s-plane. More concretely, if we rewrite (9.2) as

$$\Gamma(s) = \frac{\Gamma(s+1)}{s},$$

then since the right-hand side is a meromorphic function for $\mathrm{Re}(s) > -1$, $\Gamma(s)$ is continued analytically to $\mathrm{Re}(s) > -1$ and $s = 0$ is a pole of order 1 with residue 1. We also have

$$\Gamma(s) = \frac{\Gamma(s+2)}{s(s+1)},$$

and repeating this, we get

$$\Gamma(s) = \frac{\Gamma(s+M)}{s(s+1)\cdots(s+M-1)} \tag{9.3}$$

for arbitrary natural number M. Through (9.3), $\Gamma(s)$ is continued analytically to a meromorphic function on $\mathrm{Re}(s) > -M$. Since M is arbitrary, $\Gamma(s)$ is continued analytically to a meromorphic function on the whole s-plane and $s = 0, -1, -2, \ldots$ are poles of order 1, and the residues can be calculated by (9.3).

Changing the variable as $t \to (n+a)t$ $(n+a > 0)$ in (9.1), we get

$$\frac{1}{(n+a)^s} = \frac{1}{\Gamma(s)} \int_0^\infty e^{-(n+a)t} t^{s-1}\, dt.$$

Formal calculation leads us to

$$\sum_{n=0}^\infty \frac{1}{(n+a)^s} = \frac{1}{\Gamma(s)} \int_0^\infty \sum_{n=0}^\infty e^{-(n+a)t} t^{s-1}\, dt,$$

and for the range $\mathrm{Re}(s) > 1$, we see that the right-hand side converges absolutely, so the above calculation is justified. Hence, for the range $\mathrm{Re}(s) > 1$, we get an integral representation of $\zeta(s, a)$ as

$$\zeta(s,a) = \frac{1}{\Gamma(s)} \int_0^\infty \frac{e^{-at}}{1-e^{-t}} t^{s-1}\, dt = \frac{1}{\Gamma(s)} \int_0^\infty \frac{te^{(1-a)t}}{e^t - 1} t^{s-2}\, dt. \tag{9.4}$$

This integral representation suggests us that $\zeta(s, a)$ has something to do with Bernoulli polynomial $B_k(a)$.

9.2 Contour Integral

We transform the integral representation (9.4) to a complex integral in order to apply to it the residue theorem of complex analysis. We regard the interval $[0, \infty)$ of integration as a path of a complex integral, and then expanding this a little, we consider the following contour $I(\varepsilon, \infty)$ $(\varepsilon > 0)$. We define $I(\varepsilon, \infty)$ by a curve $\varphi : (-\infty, \infty) \longrightarrow \mathbf{C}$ given by

$$I(\varepsilon,\infty): \qquad \varphi(u) = \begin{cases} -u & u < -\varepsilon, \\ \varepsilon \exp\left(\pi i \frac{u+\varepsilon}{\varepsilon}\right) & -\varepsilon \le u \le \varepsilon, \\ u & u > \varepsilon. \end{cases}$$

In the definition of $I(\varepsilon, \infty)$, the parts for $u < -\varepsilon$ and $u > \varepsilon$ overlap, but we interpret it that for $u < -\varepsilon$ we take the path above the real axis and for $u > \varepsilon$ below the real axis. This path is illustrated in Fig. 9.1.

Now consider the complex curvilinear integral

$$\int_{I(\varepsilon,\infty)} \frac{te^{(1-a)t}}{e^t - 1} t^{s-2}\, dt.$$

Since we have to treat t^{s-2} on $I(\varepsilon, \infty)$, we shall choose a complex power t^s for $s \in \mathbf{C}$. Denoting the argument of t by $\arg t$, we can define a single-valued function $\log t$ on $\mathbf{C} - \{z = x + iy \mid y = 0,\ x \ge 0\}$ by

$$\log t = \log|t| + i \arg t \qquad (0 < \arg t < 2\pi).$$

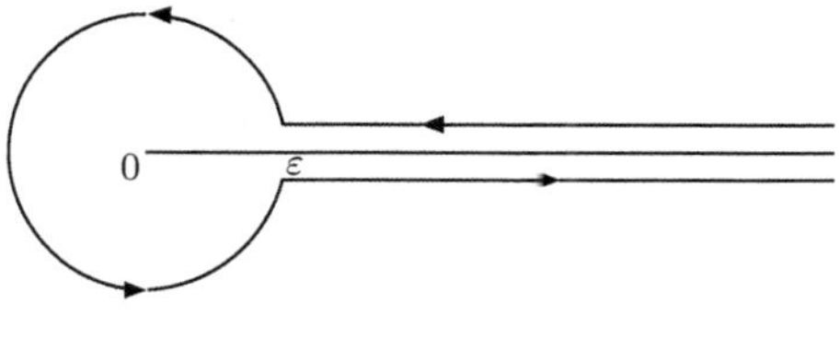

Fig. 9.1 Path of $I(\varepsilon, \infty)$

Using this, a single-valued analytic function t^s on $\mathbf{C}-\{z=x+iy \mid y=0,\ x\geq 0\}$ is defined by

$$t^s = e^{s\log t}.$$

We divide the contour $I(\varepsilon,\infty)$ into three pieces as follows.

C_1: the part of the real axis from ∞ to ε,
$I(\varepsilon)$: the circle of radius ε with center at the origin (oriented counter-clockwise),
C_2: the part of the real axis from ε to ∞.

We have $I(\varepsilon,\infty) = C_1 + I(\varepsilon) + C_2$. For t on C_1, we put $\arg t = 0$, and on C_2 we put $\arg t = 2\pi$. Then the integral on C_1 is given by

$$\begin{aligned}\int_{C_1} \frac{te^{(1-a)t}}{e^t-1}t^{s-2}\,dt &= \int_{\infty}^{\varepsilon} \frac{te^{(1-a)t}}{e^t-1}t^{s-2}\,dt \\ &= -\int_{\varepsilon}^{\infty} \frac{te^{(1-a)t}}{e^t-1}t^{s-2}\,dt\end{aligned}$$

and on C_2 by

$$\begin{aligned}\int_{C_2} \frac{te^{(1-a)t}}{e^t-1}t^{s-2}\,dt &= \int_{\varepsilon}^{\infty} \frac{te^{(1-a)t}}{e^t-1}e^{(s-2)(\log t+2\pi i)}\,dt \\ &= e^{2\pi i(s-2)}\int_{\varepsilon}^{\infty} \frac{te^{(1-a)t}}{e^t-1}t^{s-2}\,dt.\end{aligned}$$

Noting $e^{2\pi i(s-2)} = e^{2\pi i s}$, together we get

$$\int_{I(\varepsilon,\infty)} \frac{te^{(1-a)t}}{e^t-1}t^{s-2}\,dt = \left(e^{2\pi i s}-1\right)\int_{\varepsilon}^{\infty} \frac{te^{(1-a)t}}{e^t-1}t^{s-2}\,dt + \int_{I(\varepsilon)} \frac{e^{(1-a)t}}{e^t-1}t^{s-1}\,dt. \tag{9.5}$$

The circle $I(\varepsilon)$ is parametrized as $t = \varepsilon e^{i\theta}$ $(0\leq\theta\leq 2\pi)$, so on $I(\varepsilon)$ we have $t^{s-1} = \varepsilon^{s-1}e^{i(s-1)\theta}$, and the absolute value of the integral is estimated from above as

$$\begin{aligned}\left|\int_{I(\varepsilon)} \frac{e^{(1-a)t}}{e^t-1}t^{s-1}\,dt\right| &\leq \int_0^{2\pi} \left|\frac{e^{(1-a)\varepsilon e^{i\theta}}}{e^{\varepsilon e^{i\theta}}-1}\varepsilon^{s-1}e^{i(s-1)\theta}i\varepsilon e^{i\theta}\right|\,d\theta \\ &= \varepsilon^{\mathrm{Re}(s)}\int_0^{2\pi}\left|\frac{e^{(1-a)\varepsilon e^{i\theta}}}{e^{\varepsilon e^{i\theta}}-1}e^{is\theta}\right|\,d\theta.\end{aligned} \tag{9.6}$$

We see $\dfrac{\varepsilon}{e^{\varepsilon e^{i\theta}}-1}$ is bounded as a function of ε and θ, so if we take the limit $\varepsilon \to 0$ then, for $\mathrm{Re}(s) > 1$, by (9.6) we get

$$\lim_{\varepsilon\to 0}\int_{I(\varepsilon)} \frac{e^{(1-a)t}}{e^t-1} t^{s-1}\,dt = 0. \tag{9.7}$$

On the other hand, in (9.5), by Cauchy's integral formula, the curvilinear integral on the left-hand side does not depend on the choice of sufficiently small positive number ε ($\varepsilon < 2\pi$), for if we take another $\varepsilon' > 0$, then in the region surrounded by two paths, the integrand is a single-valued holomorphic function. Now we assume that $\mathrm{Re}(s) > 1$. In (9.5), since the left-hand side does not depend on $\varepsilon > 0$, we fix ε on the left-hand side, and on the right-hand side we take a limit as $\varepsilon \to 0$. Then using (9.7), we get

$$\int_{I(\varepsilon,\infty)} \frac{te^{(1-a)t}}{e^t-1} t^{s-2}\,dt = \left(e^{2\pi i s}-1\right)\int_0^\infty \frac{te^{(1-a)t}}{e^t-1} t^{s-2}\,dt. \tag{9.8}$$

Thus we obtain the following curvilinear integral representation of $\zeta(s,a)$.

Proposition 9.1. *Let ε be a positive number with $\varepsilon < 2\pi$. If $\mathrm{Re}(s) > 1$, then the Hurwitz zeta function $\zeta(s,a)$ $(a > 0)$ has the contour integral representation*

$$\zeta(s,a) = \frac{1}{\Gamma(s)(e^{2\pi i s}-1)}\int_{I(\varepsilon,\infty)} \frac{te^{(1-a)t}}{e^t-1} t^{s-2}\,dt.$$

Through this formula, $\zeta(s,a)$ is continued analytically to a meromorphic function on the whole s-plane and has a unique pole at $s = 1$ of order 1 with residue 1.

Proof. The first half is immediately obtained by the integral representation (9.4) and (9.8) of $\zeta(s,a)$. We shall see the latter half below in turn. We prepare the following lemma.

Lemma 9.2. *The integral $\displaystyle\int_{I(\varepsilon,\infty)} \frac{te^{(1-a)t}}{e^t-1} t^{s-2}\,dt$ converges absolutely for arbitrary $s \in \mathbf{C}$ and gives a holomorphic function on the whole s-plane.*

Proof. The curvilinear integral $\displaystyle\int_{C_j} \frac{te^{(1-a)t}}{e^t-1} t^{s-2}\,dt$ $(j = 1, 2)$ converges absolutely for arbitrary $s \in \mathbf{C}$, so it is holomorphic there. Also the curvilinear integral $\displaystyle\int_{I(\varepsilon)} \frac{te^{(1-a)t}}{e^t-1} t^{s-2}\,dt$ converges absolutely for arbitrary s by (9.6) since $\varepsilon > 0$ is fixed, and is differentiable with respect to s, so this is also holomorphic on the whole s-plane. □

As is well known in complex analysis, $\dfrac{1}{\Gamma(s)}$ is a holomorphic function on the whole s-plane, so by Proposition 9.1 and Lemma 9.2, poles of $\zeta(s,a)$ result from zeros of the function $e^{2\pi i s}-1$. Since $\zeta(s,a)$ converges absolutely for $\mathrm{Re}(s)>1$ and is holomorphic there, possible poles are $s=1$ or $s=1-m$ ($m\in\mathbf{N}$). Among these, at $s=1-m$ the function

$$\frac{1}{\Gamma(s)(e^{2\pi i s}-1)}$$

is holomorphic. Hence $\zeta(s,a)$ has a unique pole $s=1$ and its order is 1. The residue is given by

$$\lim_{s\to 1}\frac{(s-1)}{\Gamma(s)(e^{2\pi i s}-1)}\int_{I(\varepsilon,\infty)}\frac{te^{(1-a)t}}{e^t-1}t^{s-2}\,dt=\frac{1}{2\pi i}\int_{I(\varepsilon,\infty)}\frac{te^{(1-a)t}}{e^t-1}t^{-1}\,dt$$
$$=\frac{1}{2\pi i}\int_{I(\varepsilon)}\frac{te^{(1-a)t}}{e^t-1}t^{-1}\,dt=\operatorname*{Res}_{t=0}\left(\frac{e^{(1-a)t}}{e^t-1}\right)=1.$$

The above calculation of the last integral is due to the residue theorem. The only pole of the integrand $\dfrac{e^{(1-a)t}}{e^t-1}$ inside $I(\varepsilon)$ is $t=0$ and it is of order 1, so it is enough to calculate its residue.

We have finished the proof of the latter half of Proposition 9.1. □

Next, we calculate the special values of $\zeta(s,a)$ at non-positive integers $s=1-m$ ($m\in\mathbf{N}$) by using the contour integral representation.

Proposition 9.3. *Let m be a natural number. Then for $a>0$, we have*

$$\zeta(1-m,a)=-\frac{B_m(a)}{m}.$$

Here $B_m(a)$ is the Bernoulli polynomial defined in Sect. 4.3 (p. 55).

Proof. By Proposition 9.1 we have

$$\zeta(1-m,a)=\left(\lim_{s\to 1-m}\frac{1}{\Gamma(s)(e^{2\pi i s}-1)}\right)\int_{I(\varepsilon,\infty)}\frac{te^{(1-a)t}}{e^t-1}t^{-m-1}\,dt.$$

In the last curvilinear integral, the integrand $\dfrac{te^{(1-a)t}}{e^t-1}t^{-m-1}$ becomes a single-valued holomorphic function on $\mathbf{C}\setminus\{0\}$. Notation being as before, in the integral for the contour $I(\varepsilon,\infty)=C_1+I(\varepsilon)+C_2$, the curvilinear integrals on C_1 and C_2 cancel each other. That is,

$$\int_{C_1} \frac{te^{(1-a)t}}{e^t - 1} t^{-m-1} \, dt + \int_{C_2} \frac{te^{(1-a)t}}{e^t - 1} t^{-m-1} \, dt = 0.$$

The calculation of the curvilinear integral on $I(\varepsilon)$ is done by calculation of the residues. By virtue of (5) and (4) of Proposition 4.9 on Bernoulli polynomials, the residue of the integrand $\frac{te^{(1-a)t}}{e^t-1} t^{-m-1}$ at $t = 0$ is given by

$$\operatorname*{Res}_{t=0} \left(\frac{te^{(1-a)t}}{e^t - 1} t^{-m-1} \right) = \frac{B_m(1-a)}{m!} = (-1)^m \frac{B_m(a)}{m!}.$$

On the other hand, the limit is calculated by (9.3) as follows.

$$\begin{aligned} \lim_{s \to 1-m} \frac{1}{\Gamma(s)(e^{2\pi i s} - 1)} &= \lim_{s \to 1-m} \frac{s(s+1) \cdots (s+m-1)}{\Gamma(s+m) \left(e^{2\pi i (s+m-1)} - 1\right)} \\ &= \frac{(-1)(-2) \cdots (1-m)}{2\pi i} \\ &= \frac{(-1)^{m-1}(m-1)!}{2\pi i}. \end{aligned}$$

So we have

$$\zeta(1-m, a) = (-1)^{m-1}(m-1)! \operatorname*{Res}_{t=0} \left(\frac{te^{(1-a)t}}{e^t - 1} t^{-m-1} \right).$$

By this, Proposition 9.3 is proved. □

9.3 The Functional Equation of $\zeta(s, a)$

We fix $t_0 < 0$, and denote by $C(t_0)$ the line on the t-plane defined by $\mathrm{Re}(t) = t_0$ oriented upwards:

$$C(t_0) : t = t_0 + iy \quad (-\infty < y < \infty).$$

We assume $\mathrm{Re}(s) < 0$. In the contour integral representation of the Hurwitz zeta function $\zeta(s, a)$ (Proposition 9.1), we cut out the contour $I(\varepsilon, \infty)$ as in Fig. 9.2, and shift it over the imaginary axis to the line $-C(t_0)$. Here we are taking $0 < a \leq 1$.

When it goes beyond the imaginary axis, it passes the pole $t = 2\pi i n (n \in \mathbf{Z}, n \neq 0)$ of the integrand $\dfrac{te^{(1-a)t}}{e^t - 1} t^{s-2}$, so the residues appear there. The assumptions $\mathrm{Re}(s) < 0$ and $0 < a \leq 1$ are conditions for the integral to converge. Under the condition $0 < a \leq 1$, the curvilinear integral along $C(t_0)$ converges absolutely. Since we can transform as

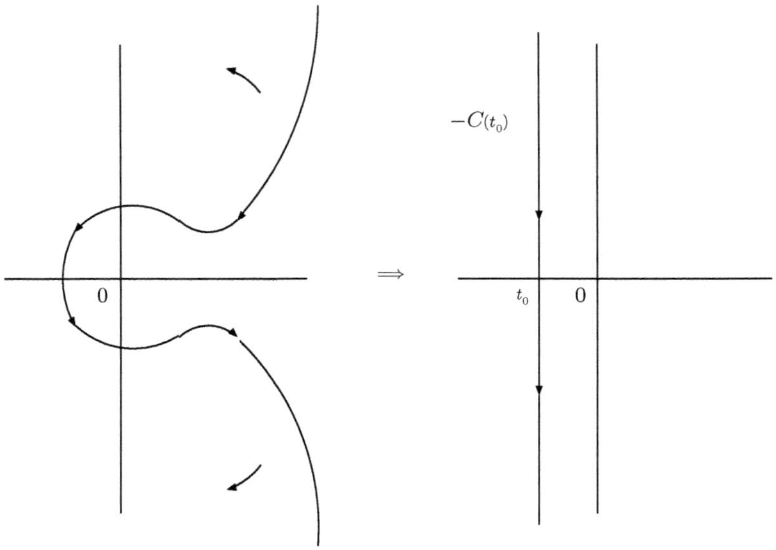

Fig. 9.2 Shift of the contour

$$
\begin{aligned}
\zeta(s,a)&\\
&=\frac{1}{\Gamma(s)(e^{2\pi i s}-1)}\left\{(-2\pi i)\sum_{\substack{n\in\mathbf{Z}\\ n\neq 0}}\frac{e^{(1-a)2\pi i n}}{(2\pi i n)^{1-s}}-\int_{C(t_0)}\frac{te^{(1-a)t}}{e^t-1}t^{s-2}\,dt\right\}\\
&=\frac{1}{\Gamma(s)(e^{2\pi i s}-1)}\left\{(-2\pi i)\sum_{n=1}^{\infty}\frac{e^{-2\pi i n a}e^{\frac{\pi i}{2}(s-1)}+e^{2\pi i n a}e^{\frac{3\pi i}{2}(s-1)}}{(2\pi n)^{1-s}}\right.\\
&\quad\left.-\int_{C(t_0)}\frac{te^{(1-a)t}}{e^t-1}t^{s-2}\,dt\right\},
\end{aligned}
$$

taking the limit $t_0 \longrightarrow -\infty$, we have

$$
\zeta(s,a)=\frac{(-\pi i)(2\pi)^{s-1}}{\Gamma(s)\sin\pi s}\left\{e^{\frac{\pi i s}{2}}\sum_{n=1}^{\infty}\frac{e^{2\pi i n a}}{n^{1-s}}-e^{-\frac{\pi i s}{2}}\sum_{n=1}^{\infty}\frac{e^{-2\pi i n a}}{n^{1-s}}\right\}. \tag{9.9}
$$

Here substituting $1-\{a\}$ for a, we get

$$
\zeta(s,1-\{a\})=\frac{(-\pi i)(2\pi)^{s-1}}{\Gamma(s)\sin\pi s}\left\{e^{\frac{\pi i s}{2}}\sum_{n=1}^{\infty}\frac{e^{-2\pi i n a}}{n^{1-s}}-e^{-\frac{\pi i s}{2}}\sum_{n=1}^{\infty}\frac{e^{2\pi i n a}}{n^{1-s}}\right\}, \tag{9.10}
$$

where the notation $\{a\}$ denotes the fractional part of a: $0 \le \{a\} < 1$, $a - \{a\} \in \mathbf{Z}$. This transformation is valid for $\mathrm{Re}(s) < 0$. By these relations (9.9), (9.10), we get

$$e^{\frac{\pi i s}{2}}\zeta(s,a) + e^{-\frac{\pi i s}{2}}\zeta(s,1-\{a\}) = \frac{(2\pi)^s}{\Gamma(s)}\sum_{n=1}^{\infty}\frac{e^{2\pi i n a}}{n^{1-s}}, \qquad (9.11)$$

and since $\zeta(s,a)$ is analytically continued to a meromorphic function on the whole s-plane, $\sum_{n=1}^{\infty}\frac{e^{2\pi i n a}}{n^{1-s}}$ is also continued analytically to a meromorphic function. Summing up, we get the following theorem.

Theorem 9.4. *We assume that* $0 < a \le 1$. *The Hurwitz zeta function* $\zeta(s,a)$ *is continued analytically to a meromorphic function on the whole* s*-plane. It has a unique pole at* $s = 1$. *It is of order* 1 *and the residue is* 1. *Moreover, the following functional equation is satisfied.*

$$\zeta(s,a) = \frac{(-\pi i)(2\pi)^{s-1}}{\Gamma(s)\sin \pi s}\left\{e^{\frac{\pi i s}{2}}\sum_{n=1}^{\infty}\frac{e^{2\pi i n a}}{n^{1-s}} - e^{-\frac{\pi i s}{2}}\sum_{n=1}^{\infty}\frac{e^{-2\pi i n a}}{n^{1-s}}\right\}.$$

If we put $s = 1 - m$ here and use (4.3) on p. 59, we obtain Proposition 9.3. Furthermore if we put $a = 1$ in (9.11), noting that $\zeta(s,1) = \zeta(s)$, we get

$$\zeta(s) = \frac{(2\pi)^s}{2\Gamma(s)\cos\frac{\pi s}{2}} \times \zeta(1-s) = \frac{(2\pi)^s}{2\Gamma(s)\sin\frac{\pi(s+1)}{2}} \times \zeta(1-s).$$

Then if we use the duplication formula of the gamma function

$$\Gamma(s) = \pi^{-1/2}2^{s-1}\Gamma\left(\frac{s}{2}\right)\Gamma\left(\frac{s+1}{2}\right)$$

and the reflection formula

$$\Gamma(s)\Gamma(1-s) = \frac{\pi}{\sin \pi s}$$

(for these formulas, see [104, Chapter 11]), we get

$$\begin{aligned}\Gamma\left(\frac{s}{2}\right)\zeta(s) &= \pi^{1/2}2^{1-s}\frac{(2\pi)^s}{2\Gamma\left(\frac{s+1}{2}\right)\sin\frac{\pi(s+1)}{2}}\zeta(1-s)\\ &= \pi^{s-1/2}\Gamma\left(\frac{1-s}{2}\right)\zeta(1-s).\end{aligned}$$

Adjusting a little, this gives the usual functional equation of the Riemann zeta function.

Theorem 9.5 (Functional equation of Riemann zeta function). *The Riemann zeta function $\zeta(s)$ can be continued analytically to a meromorphic function on the whole s-plane. If we put*

$$\xi(s) = \pi^{-s/2}\Gamma\left(\frac{s}{2}\right)\zeta(s),$$

it satisfies the functional equation

$$\xi(s) = \xi(1-s).$$

9.4 Special Values of L-Functions and the Functional Equations

In this section, our aim is to define a Dirichlet L-function and to look for its functional equation and special values.

Let χ be a Dirichlet character modulo f. The Dirichlet L-function associated with the character χ is defined by

$$L(s,\chi) = \sum_{n=1}^{\infty} \frac{\chi(n)}{n^s}.$$

The series on the right-hand side converges absolutely for $\mathrm{Re}(s) > 1$ and it is holomorphic there.

First, we consider the special values at positive integers.

The Gaussian sum $g(\chi)$ associated with χ is defined by

$$g(\chi) = \sum_{a=1}^{f} \chi(a)e^{2\pi i a/f}$$

(Definition 8.1, p. 108). If we assume that χ is a primitive character modulo f, then by virtue of Lemma 8.2, we have

$$\chi(n)g(\overline{\chi}) = \sum_{a=1}^{f} \overline{\chi}(a)e^{2\pi i a n/f} \tag{9.12}$$

for arbitrary integer n.

Theorem 9.6. *Let χ be a primitive character modulo f and let k be a natural number. If $\chi(-1) = (-1)^k$, then the special value of the Dirichlet L-function $L(s,\chi)$ at $s = k$ is given by*

$$L(k,\chi) = \frac{(-1)^{k-1}(2\pi i)^k}{2k!f^k} g(\chi) B_{k,\overline{\chi}}.$$

Example 9.7. We give examples first. Let χ be the primitive character modulo 4 and let k be an odd positive integer. Then we have $\overline{\chi} = \chi$ and $g(\chi) = 2i$, so we have

$$L(k,\chi) = \frac{(-1)^{\frac{k+1}{2}}\pi^k}{2^k k!} B_{k,\chi}.$$

We write two of them explicitly below. By the expression in (4.1) in Sect. 4.2, we have $B_{1,\chi} = -\frac{1}{2}$ and $B_{3,\chi} = \frac{3}{2}$, so we get

$$L(1,\chi) = 1 - \frac{1}{3} + \frac{1}{5} - \frac{1}{7} + \cdots = \frac{\pi}{4},$$

$$L(3,\chi) = 1 - \frac{1}{3^3} + \frac{1}{5^3} - \frac{1}{7^3} + \cdots = \frac{\pi^3}{32}.$$

Next, let χ be the primitive character modulo 3. Since we have $B_{1,\chi} = -\frac{1}{3}$, $B_{3,\chi} = \frac{2}{3}$, $g(\chi) = \sqrt{3}i$, we get

$$L(1,\chi) = 1 - \frac{1}{2} + \frac{1}{4} - \frac{1}{5} + \frac{1}{7} - \frac{1}{8} + \cdots = \frac{\pi}{3\sqrt{3}},$$

$$L(3,\chi) = 1 - \frac{1}{2^3} + \frac{1}{4^3} - \frac{1}{5^3} + \frac{1}{7^3} - \frac{1}{8^3} + \cdots = \frac{2^2\pi^3}{3^4\sqrt{3}}.$$

Proof. We give a proof of the above theorem. Using (9.12), we have

$$L(k,\chi)g(\overline{\chi}) = \sum_{a=1}^{f} \overline{\chi}(a) \sum_{n=1}^{\infty} \frac{e^{2\pi i a n/f}}{n^k}.$$

Noting the assumption $\chi(-1) = (-1)^k$, the right-hand side is transformed into

$$\frac{1}{2}\sum_{a=1}^{f} \overline{\chi}(a) \sum_{\substack{n \in \mathbf{Z} \\ n \neq 0}} \frac{e^{2\pi i a n/f}}{n^k}.$$

By virtue of Theorem 4.11 (p. 59) and (4.1) (p. 54), this value can be rewritten as

$$\frac{-(2\pi i)^k}{2k!}\sum_{a=1}^{f}\overline{\chi}(a)B_k(a/f)=\frac{-(2\pi i)^k}{2k!f^{k-1}}B_{k,\overline{\chi}}.$$

Noting $g(\overline{\chi})^{-1}=\chi(-1)g(\chi)/f$ by Lemmas 8.2 and 8.3 in the last section, we can derive the theorem from this. □

We express $L(s,\chi)$ by using Hurwitz zeta functions. Hurwitz zeta functions are very useful to deduce various properties of L-functions. Under the assumption $\mathrm{Re}(s)>1$, we can transform $L(s,\chi)$ as

$$\begin{aligned}L(s,\chi)&=\sum_{a=1}^{f}\sum_{m=0}^{\infty}\frac{\chi(a+fm)}{(a+fm)^s}\\&=\sum_{a=1}^{f}\chi(a)\sum_{m=0}^{\infty}\frac{1}{(a+fm)^s}\\&=f^{-s}\sum_{a=1}^{f}\chi(a)\sum_{m=0}^{\infty}\frac{1}{(m+a/f)^s}\end{aligned}$$

and the next proposition follows.

Proposition 9.8. *Let χ be a primitive Dirichlet character modulo $f>1$. Then in the range* $\mathrm{Re}(s)>1$, *we have*

$$L(s,\chi)=f^{-s}\sum_{a=1}^{f}\chi(a)\zeta\left(s,\frac{a}{f}\right).$$

Through this expression, $L(s,\chi)$ is continued analytically to a holomorphic function on the whole s-plane.

Proof. We have already shown the first half of the assertion. We use Proposition 9.1 for the latter half. By Proposition 9.1, the Hurwitz zeta function $\zeta(s,a/f)$ is continued analytically to a meromorphic function on the whole s-plane, its pole being only situated at $s=1$ and it is of order 1. Hence $L(s,\chi)$ is analytically continued to a meromorphic function on the whole s-plane and it has at most one pole at $s=1$ of order 1. Calculating the residue at $s=1$, we get

$$f^{-1}\sum_{a=1}^{f}\chi(a)=0$$

since $f>1$ and χ is non-trivial, so we see it is holomorphic at $s=1$ too. □

Let us derive the functional equation of $L(s, \chi)$ from the functional equation of Hurwitz zeta functions (Theorem 9.4). We use the following notation $\delta = \delta(\chi)$, defined by

$$\delta = \begin{cases} 0 & \text{if } \chi(-1) = 1, \\ 1 & \text{if } \chi(-1) = -1. \end{cases}$$

So $\chi(-1) = (-1)^{\delta}$, $\delta \in \{0, 1\}$.

Theorem 9.9 (Functional equation). *Let χ be a primitive character modulo f. Then $L(s, \chi)$ satisfies the following functional equation.*

$$\left(\frac{f}{\pi}\right)^{s/2} \Gamma\left(\frac{s+\delta}{2}\right) L(s, \chi) = W(\chi) \left(\frac{f}{\pi}\right)^{(1-s)/2} \Gamma\left(\frac{1-s+\delta}{2}\right) L(1-s, \overline{\chi}),$$

where we put

$$W(\chi) = \frac{g(\chi)}{\sqrt{f} i^{\delta}}.$$

Proof. By using Proposition 9.8 and Theorem 9.4, we continue $L(s, \chi)$ analytically to the region $\mathrm{Re}(s) < 0$, and get

$$\begin{aligned} L(s, \chi) &= f^{-s} \sum_{a=1}^{f} \chi(a) \zeta\left(s, \frac{a}{f}\right) \\ &= \frac{(-\pi i)(2\pi)^{s-1}}{f^s \Gamma(s) \sin \pi s} \left\{ e^{\frac{\pi i s}{2}} \sum_{n=1}^{\infty} \sum_{a=1}^{f} \frac{\chi(a) e^{2\pi i n \frac{a}{f}}}{n^{1-s}} - e^{\frac{-\pi i s}{2}} \sum_{n=1}^{\infty} \sum_{a=1}^{f} \frac{\chi(a) e^{-2\pi i n \frac{a}{f}}}{n^{1-s}} \right\}. \end{aligned}$$

Here, using (9.12), we get

$$L(s, \chi) = \frac{(-\pi i)(2\pi)^{s-1} g(\chi)}{f^s \Gamma(s) \sin \pi s} \left(e^{\frac{\pi i s}{2}} - \chi(-1) e^{\frac{-\pi i s}{2}} \right) L(1-s, \overline{\chi}).$$

Simplifying the expression, we get

$$\Gamma(s) \cos\left(\frac{\pi(s-\delta)}{2}\right) L(s, \chi) = \frac{g(\chi)}{2 i^{\delta}} \left(\frac{2\pi}{f}\right)^{s} L(1-s, \overline{\chi}). \tag{9.13}$$

By using the duplication formula of $\Gamma(s)$ and the relation $\Gamma(s)\Gamma(1-s) = \dfrac{\pi}{\sin \pi s}$, we can rewrite $\Gamma(s)$ as

$$\begin{aligned}\Gamma(s) &= \pi^{-1/2}2^{s-1}\Gamma\left(\frac{s}{2}\right)\Gamma\left(\frac{s+1}{2}\right) \\ &= \pi^{-1/2}2^{s-1}\Gamma\left(\frac{s+\delta}{2}\right)\frac{\pi}{\Gamma\left(\frac{1-s+\delta}{2}\right)\cos\frac{\pi(s-\delta)}{2}}\end{aligned}$$

and then substituting this into (9.13) and simplifying the result, we get the desired functional equation. □

Finally, let us look at special values of $L(s,\chi)$ at negative integers.

Theorem 9.10. *Let χ be a primitive character modulo f. For any positive integer m, we have the following formula:*

$$L(1-m,\chi) = -\frac{B_{m,\chi}}{m}.$$

Proof. By Propositions 9.8 and 9.3, we have

$$\begin{aligned}L(1-m,\chi) &= f^{m-1}\sum_{a=1}^{f}\chi(a)\zeta\left(1-m,\frac{a}{f}\right) \\ &= -f^{m-1}\sum_{a=1}^{f}\chi(a)\frac{B_m(a/f)}{m}.\end{aligned}$$

By (4.1) on p. 54, this is equal to $-\dfrac{B_{m,\chi}}{m}$. □

Remark 9.11. We note that Theorem 9.6 can be derived also from Theorem 9.10 and the functional equation (Theorem 9.9).

Exercise 9.12. Let a be any complex number. Put $f(s) = \zeta(s)\zeta(s-a)$. Find a functional equation of $f(s)$, that is, give a relation between $f(s)$ and $f(b-s)$ for a suitably chosen number b.

Exercise 9.13. Let a be a positive integer. Determine the location of poles of the function $\zeta(s)\zeta(s-a)$ and their residues for each case when a is even or odd.

Exercise 9.14. (1) Let χ be the primitive character with conductor 4. Show that $L(1,\chi) = \frac{\pi}{4}$ by using the integral

$$\int_0^1 \frac{1}{1+x^2}\,dx$$

and Abel's theorem of power series convergence.

(2) Let χ be the primitive character with conductor 3. Show that $L(1,\chi) = \frac{\pi}{3\sqrt{3}}$ by using the integral

$$\int_0^1 \frac{1}{1+x+x^2} dx.$$

Exercise 9.15. (1) Let χ be any Dirichlet character modulo f. Show that for any $x \in \mathbf{C}$ with $Re(s) > 1$, the infinite product

$$\prod_{p:prime, p \nmid f} (1 - \chi(p)p^{-s})^{-1}$$

converges and equal to $L(s, \chi)$.

(2) Show that $L(s, \chi) \neq 0$ for any $s \in \mathbf{C}$ with $Re(s) > 1$.

Chapter 10
Class Number Formula and an Easy Zeta Function of the Space of Quadratic Forms

In this chapter, as an application of quadratic forms and quadratic fields, we give an explicit formula of some simple zeta functions, related to some so-called prehomogeneous vector spaces. We also prove a class number formula of imaginary quadratic fields. Before that, we review the theory of multiplicative structure of ideals of quadratic field without proof.

10.1 Ideal Class Groups of Quadratic Fields

We first review the prime ideal decomposition of the maximal order of quadratic fields. Since there are many good books on prime ideal decompositions of the Dedekind domain, we do not give proofs here. For details, we refer to Lang [67], Weber[1] [101]. Then we supply the theory of ideals of a quadratic order which is not maximal. Since such orders are not Dedekind domains, the theory of ideals is not covered by the above general theory, so we explain the difference.

We fix a quadratic field K of discriminant D_K and denote by $\mathfrak{O}_{max}$ the maximal order of K. For any non-zero ideals $\mathfrak{a}$ and $\mathfrak{b}$ of $\mathfrak{O}_{max}$, we write

$$\mathfrak{a}\mathfrak{b} = \left\{ \sum_{i=1}^{\nu} a_i b_i ; a_i \in \mathfrak{a}, b_i \in \mathfrak{b}, \nu \text{ is an arbitrary natural number} \right\}$$

and call this a product of $\mathfrak{a}$ and $\mathfrak{b}$. It is clear that this is also a non-zero ideal of $\mathfrak{O}_{max}$. We give the following theorem without proof. We denote by χ_K the Dirichlet character associated with K defined in Sect. 6.3.

[1]Heinrich Martin Weber (born on May 5, 1842 in Heidelberg, Germany—died on May 17, 1913 in Strasbourg, Germany (now France)).

T. Arakawa et al., *Bernoulli Numbers and Zeta Functions*, Springer Monographs in Mathematics, DOI 10.1007/978-4-431-54919-2_10, © Springer Japan 2014

Theorem 10.1. *(1) Any (non-zero) ideal of the maximal order $\mathfrak{O}_{max}$ of a quadratic field K is a product of prime ideals. This factorization is unique. For any ideals $\mathfrak{a}$ and $\mathfrak{b}$ of $\mathfrak{O}_{max}$, we have $N(\mathfrak{a}\mathfrak{b}) = N(\mathfrak{a})N(\mathfrak{b})$.*

(2) For any prime $p \in \mathbf{Z}$, the ideal $p\mathfrak{O}_{max}$ is decomposed as follows.

(i) *If $\chi_K(p) = 1$, then $p\mathfrak{O}_{max} = \mathfrak{p}_1\mathfrak{p}_2$, where $\mathfrak{p}_1$ and $\mathfrak{p}_2$ are different prime ideals related by $\mathfrak{p}_2 = \overline{\mathfrak{p}}_1$. Here we put*

$$\overline{\mathfrak{p}}_1 = \{\bar{\alpha};\ \alpha \in \mathfrak{p}_1\}$$

and $x \mapsto \bar{x}$ is the non-trivial automorphism of $K/\mathbf{Q}$.

(ii) *If $\chi_K(p) = -1$ then $p\mathfrak{O}_{max}$ is a prime ideal.*

(iii) *If $\chi_K(p) = 0$, namely if p is a prime factor of D_K, then*

$$p\mathfrak{O}_{max} = \mathfrak{p}^2$$

for a prime ideal $\mathfrak{p}$.

There are no prime ideals of $\mathfrak{O}_{max}$ other than those which appear in (1), (2) *and* (3) *above.*

By this theorem, we have a relation between zeta functions as follows. For a quadratic field K, we define the Dedekind zeta function of K by

$$\zeta_K(s) = \sum_{\mathfrak{a}} \frac{1}{N(\mathfrak{a})^s}.$$

Here the sum runs over all non-zero ideals of $\mathfrak{O}_{max}$ and it converges absolutely for $\mathrm{Re}(s) > 1$. The existence and uniqueness of the prime ideal factorization gives the Euler product decomposition

$$\zeta_K(s) = \prod_{\mathfrak{p}} \frac{1}{1 - N(\mathfrak{p})^{-s}},$$

again for $\mathrm{Re}(s) > 1$. Here the product runs over all prime ideals of $\mathfrak{O}_{max}$. Also the decomposition of primes of $\mathbf{Z}$ in K (Theorem 10.1) gives

$$\zeta_K(s) = \zeta(s)L(s, \chi_K).$$

We shall give a direct proof of this relation later.

We also give the following theorem on character, again without proof.

Proposition 10.2. *Let K and D_K be as before. Let χ be a primitive Dirichlet character modulo D_K and we consider the following property on χ:*

$$\chi(N(\mathfrak{a})) = 1 \textit{ for any ideal } \mathfrak{a} \textit{ of } \mathfrak{O}_{max} \textit{ which is coprime to } D_K.$$

(1) *The Dirichlet character* $\chi_K(n) = \left(\frac{D_K}{n}\right)$ *is a primitive Dirichlet character modulo* $|D_k|$ *and satisfies the above property.*
(2) *Conversely, any non-trivial primitive Dirichlet character modulo* $|D_K|$ *which satisfies the above property is equal to* χ_K.

Now in the rest of this section, we explain a multiplicative structure of ideals including the case of non-maximal orders.

We fix an integer $f \geq 1$ and we consider ideals of an order $\mathfrak{O}_f$ of a quadratic field $K = \mathbf{Q}(\sqrt{m})$. We extend the notion of ideals so that we can consider a group of ideals. For this purpose, we introduce $\mathfrak{O}_f$-modules, called fractional ideals, which are not necessarily contained in $\mathfrak{O}_f$. To define this, we first consider a free $\mathbf{Z}$-submodule L of the quadratic field K. The rank of L as a free $\mathbf{Z}$-module is at most 2. Indeed, if we assume that $rank_{\mathbf{Z}} L \geq 3$, then there are elements ω_1, ω_2, $\omega_3 \in L$ which are the part of a basis over $\mathbf{Z}$. If $c_1\omega_1 + c_2\omega_2 + c_3\omega_3 = 0$ for some $c_i \in \mathbf{Q}$, then there is some non-zero $r \in \mathbf{Z}$ such that $rc_i \in \mathbf{Z}$ for all i, so we have $rc_i = 0$ and hence $c_i = 0$ for $i = 1, 2, 3$. Since $L \subset K$ and K is of dimension two as a vector space over $\mathbf{Q}$, this is a contradiction.

Definition 10.3. A free $\mathbf{Z}$-submodule of rank 2 in the quadratic field $K = \mathbf{Q}(\sqrt{m})$ is called a lattice of K.

For lattices L_1 and L_2, we define their product by

$$L_1 L_2 = \left\{ \sum_{i=1}^{\nu} a_i b_i;\ a_i \in L_1,\ b_i \in L_2,\ \nu \text{ is an arbitrary natural number} \right\}.$$

This product is a lattice. Indeed, for any element $\alpha \in K$, there exists $r \in \mathbf{Z}$ such that $r\alpha \in \mathbf{Z} + \mathbf{Z}\sqrt{m}$ and applying this to a basis of lattices L_1 and L_2, we see that $rL_1, rL_2 \subset \mathbf{Z} + \mathbf{Z}\sqrt{m}$ for some integer r. But since $\mathbf{Z} + \mathbf{Z}\sqrt{m}$ is a ring, we have $r^2 L_1 L_2 \subset \mathbf{Z} + \mathbf{Z}\sqrt{m}$. Hence by virtue of the structure theorem for finitely generated abelian groups, $L_1 L_2$ is a free $\mathbf{Z}$-module of rank 2 at most. It is obvious that its rank is 2, because for any $0 \neq \omega \in L_1$, we have $\omega L_2 \subset L_1 L_2$ and ωL_2 is of rank 2. This multiplication is commutative and associative.

Generalizing slightly the notion of a proper ideal, we call a lattice L of K a fractional ideal of $\mathfrak{O}_f$ if L is an $\mathfrak{O}_f$-module. Then we also have $\mathfrak{O}_f L = L\mathfrak{O}_f = L$, since $1 \in \mathfrak{O}_f$. So $\mathfrak{O}_f$ is an identity element of the semi-group of fractional ideals of $\mathfrak{O}_f$. In this book, if we say just an *ideal* of $\mathfrak{O}_f$, we always understand that it is a subset of $\mathfrak{O}_f$.

Now, in order to introduce a group structure on fractional ideals of an order $\mathfrak{O}_f$, we need inverse elements. In other words, for an ideal $\mathfrak{a}$, we would like to find a lattice L such that $\mathfrak{a}L = \mathfrak{O}_f$. But such lattice does *not* exist in general.

Example 10.4. Take $K = \mathbf{Q}(\sqrt{-2})$ and consider the order $\mathfrak{O}_5 = \mathbf{Z} + 5\sqrt{-2}\mathbf{Z}$ of conductor 5. Put $\mathfrak{a} = 5\mathbf{Z} + 5\sqrt{-2}\mathbf{Z}$. Then this is an ideal of $\mathfrak{O}_5$, but there exists no lattice L such that $\mathfrak{a}L = \mathfrak{O}_5$. The reason is as follows. If such an L exists, then since $5 \in \mathfrak{a}$, we have $5L \subset \mathfrak{O}_5$. So any element α of L is written as $\alpha = a/5 + b\sqrt{-2}$

($a, b \in \mathbf{Z}$), but $(5+5\sqrt{-2})\alpha = (a-10b)+(a+5b)\sqrt{-2}$ and since this should belong to $\mathfrak{O}_5$, we have $a+5b \in 5\mathbf{Z}$, so $a \in 5\mathbf{Z}$. So we get $L \subset \mathbf{Z}+\mathbf{Z}\sqrt{-2}$, therefore we have $\mathfrak{a}L \subset 5(\mathbf{Z}+\mathbf{Z}\sqrt{-2}) \subsetneqq \mathfrak{O}_5$. Notice in this example that $\mathfrak{a}$ is not a proper ideal.

Lemma 10.5. *Let $\mathfrak{a}$ be an $\mathfrak{O}_f$ ideal. There exists the fractional ideal L of $\mathfrak{O}_f$ such that $\mathfrak{a}L = \mathfrak{O}_f$ if and only if $\mathfrak{a}$ is a proper $\mathfrak{O}_f$ ideal. If $\mathfrak{a}$ is proper, L is uniquely determined by $\mathfrak{a}$.*

Proof. First assume that $\mathfrak{a}$ is proper and we construct L such that $\mathfrak{a}L = \mathfrak{O}_f$ explicitly. It is sufficient to give it when $\mathfrak{a}$ is a primitive ideal. Let $\{1,\omega\}$ be the standard basis of the maximal order of K and express the ideal as $\mathfrak{a} = \mathbf{Z}a+\mathbf{Z}(d+f\omega)$ in the standard basis. We denote by $x \mapsto \overline{x}$ the non-trivial automorphism of $K/\mathbf{Q}$, and put $M = \mathbf{Z}+\mathbf{Z}a^{-1}(d+f\overline{\omega}) = a^{-1}\overline{\mathfrak{a}}$. This is obviously an $\mathfrak{O}_f$ module and hence a fractional ideal of $\mathfrak{O}_f$. Taking the product, we have

$$\mathfrak{a}M = \mathbf{Z}a+\mathbf{Z}(d+f\omega)+\mathbf{Z}(d+f\overline{\omega})+\mathbf{Z}a^{-1}(d+f\omega)(d+f\overline{\omega}).$$

Here, because $\mathfrak{a}$ is an ideal, $c := a^{-1}(d+f\omega)(d+f\overline{\omega})$ is an integer (p. 80, Lemma 6.3). Moreover, by the assumption that $\mathfrak{a}$ is a proper ideal, $a, b := 2d + fTr(\omega), c$ are mutually prime (p. 81, Lemma 6.5). But by the above expression we have $a, b, c \in \mathfrak{a}M$, so $1 \in \mathfrak{a}M$. Therefore, by $d+f\omega \in \mathfrak{a}M$, we get $f\omega \in \mathfrak{a}M$, and we see $\mathfrak{O}_f \subset \mathfrak{a}M$. By the above expression, we have $\mathfrak{a}M \subset \mathfrak{O}_f$, so we have $\mathfrak{a}M = \mathfrak{O}_f$ and we proved the first half. Conversely, let $\mathfrak{a}$ be an $\mathfrak{O}_f$ ideal such that there exists L with $\mathfrak{a}L = \mathfrak{O}_f$. Then if $\alpha\mathfrak{a} \subset \mathfrak{a}$ for $\alpha \in K$, then $\alpha\mathfrak{O}_f = \alpha\mathfrak{a}L \subset \mathfrak{a}L = \mathfrak{O}_f$, so $\alpha \in \mathfrak{O}_f$. So $\mathfrak{a}$ is a proper $\mathfrak{O}_f$ ideal. If $\mathfrak{a}M = \mathfrak{a}N = \mathfrak{O}_f$ for fractional $\mathfrak{O}_f$ ideals M, N, then $M = M\mathfrak{O}_f = M(\mathfrak{a}N) = (M\mathfrak{a})N = \mathfrak{O}_fN = N$, so the uniqueness follows. □

We denote the $\mathfrak{O}_f$-module M defined above by $\mathfrak{a}^{-1}$ and call it the inverse ideal of $\mathfrak{a}$.

We call a fractional $\mathfrak{O}_f$ ideal L a proper fractional $\mathfrak{O}_f$-ideal when

$$\{\lambda \in K; \lambda L \subset L\} = \mathfrak{O}_f.$$

For any proper ideal $\mathfrak{a}$ of $\mathfrak{O}_f$, $\mathfrak{a}^{-1}$ is also a proper fractional ideal. For any proper fractional ideal L, we have $rL \subset \mathfrak{O}_f$ for some integer r, and rL is obviously a proper ideal of $\mathfrak{O}_f$, so L has also an inverse M, which is a fractional $\mathfrak{O}_f$ ideal such that $LM = \mathfrak{O}_f$.

Lemma 10.6. *The product L_1L_2 of proper fractional ideals L_1 and L_2 of $\mathfrak{O}_f$ is a proper fractional ideal.*

Proof. It is sufficient to prove when L_1 and L_2 are (integral) ideals. Let $\mathfrak{a}$ and $\mathfrak{b}$ be proper ideals and put $\mathfrak{a}\mathfrak{b} = \mathfrak{c}$. For $\alpha \in K$, assume that $\alpha\mathfrak{c} \subset \mathfrak{c}$. Since $\mathfrak{a}$ is a proper ideal, there exists $\mathfrak{a}^{-1}$ and we have $\mathfrak{a}^{-1}\mathfrak{c} = \mathfrak{a}^{-1}\mathfrak{a}\mathfrak{b} = \mathfrak{b}$. Hence we have $\alpha\mathfrak{b} = \mathfrak{a}^{-1}(\alpha\mathfrak{c}) \subset \mathfrak{a}^{-1}\mathfrak{c} = \mathfrak{b}$ and since $\mathfrak{b}$ is a proper ideal, we have $\alpha \in \mathfrak{O}_f$. □

Summarizing the above, we have

Proposition 10.7. *The set of all the proper fractional ideals of* $\mathfrak{O}_f$ *is a group by the product defined above with the unit element* $\mathfrak{O}_f$.

The group defined in this way is called an ideal group of $\mathfrak{O}_f$. The equivalence of ideals is defined for fractional ideals in the same way as the usual ideals, namely $L \sim L'$ if $L = \alpha L'$ for some $\alpha \in K^\times$, and the quotient group of fractional ideals divided by this equivalence is called an ideal class group (in the wide sense). The class number of $\mathfrak{O}_f$ is the cardinality of the ideal class group of $\mathfrak{O}_f$.

Now we give a certain subset of ideals which, while not including all invertible (i.e. proper) ideals, does include representatives for the whole ideal class group of $\mathfrak{O}_f$.

Definition 10.8. Let m be a natural number. We say that an ideal $\mathfrak{a}$ of $\mathfrak{O}_f$ is coprime to m if

$$\mathfrak{a} + m\mathfrak{O}_f = \mathfrak{O}_f.$$

Lemma 10.9. *An ideal* $\mathfrak{a}$ *of* $\mathfrak{O}_f$ *is coprime to* m *if and only if the norm of* $\mathfrak{a}$ *is coprime to* m. *Any ideal coprime to* f *is a proper ideal.*

Proof. We may assume that $\mathfrak{a} = l\mathfrak{b}$ for a primitive ideal $\mathfrak{b}$. By using the standard basis, we write $\mathfrak{b} = \mathbf{Z}a + \mathbf{Z}(d + f\omega)$. The module $\mathfrak{a} + m\mathfrak{O}_f$ is obviously an ideal of $\mathfrak{O}_f$, so $\mathfrak{a}$ is coprime to m if and only if $1 \in \mathfrak{a} + m\mathfrak{O}_f$, or equivalently there exist $x, y, z, w \in \mathbf{Z}$ such that $l(xa + y(d + f\omega)) + m(z + wf\omega) = 1$. This implies that $lxa + lyd + mz = 1$ and $lyf + mwf = 0$. If there exist such x, y, z, w, then by the second equation, we see that ly is divisible by m and hence by the first equation we see $(m, la) = 1$. Since $N(\mathfrak{a}) = al^2$, we have $(m, N(\mathfrak{a})) = 1$. Conversely if we assume that $(m, la) = 1$, then there exist $x, z \in \mathbf{Z}$ such that $lax + mz = 1$. Since $lax + mz \in \mathfrak{a} + m\mathfrak{O}_f$, this ideal is equal to $\mathfrak{O}_f$ and we have proved the first assertion. Since $f\mathfrak{O}_{max} \subset \mathfrak{O}_f$, for any ideal $\mathfrak{a}$ of $\mathfrak{O}_f$ we have

$$\mathfrak{a} + f\mathfrak{O}_f \subset \mathfrak{a} + f\mathfrak{O}_{max} \subset \mathfrak{O}_f.$$

So if we assume that $\mathfrak{a}$ is coprime to f, we have $\mathfrak{a} + f\mathfrak{O}_{max} = \mathfrak{O}_f$. If $\alpha\mathfrak{a} \subset \mathfrak{a}$, then $\alpha \in \mathfrak{O}_{max}$, so we have

$$\alpha\mathfrak{O}_f = \alpha\mathfrak{a} + f\alpha\mathfrak{O}_{max} \subset \mathfrak{a} + f\mathfrak{O}_{max} = \mathfrak{O}_f.$$

Hence $\alpha \in \mathfrak{O}_f$ and $\mathfrak{a}$ is a proper ideal. □

There exists a proper ideal of $\mathfrak{O}_f$ which is not coprime to f. For example $\mathfrak{a} = 25\mathbf{Z} + 5\sqrt{-2}\mathbf{Z} \subset \mathfrak{O}_5 = \mathbf{Z} + 5\sqrt{-2}\mathbf{Z}$ is not coprime to 5 but is a proper ideal of $\mathfrak{O}_5$.

In order to consider classes of proper $\mathfrak{O}_f$ ideals, it is enough to consider ideals coprime to f by the following lemma.

Lemma 10.10. *Let m be an arbitrary natural number. For any proper ideal $\mathfrak{a}$ of $\mathfrak{O}_f$, there exists a proper ideal of $\mathfrak{O}_f$ which is equivalent to $\mathfrak{a}$ and coprime to m.*

Proof. It is enough to prove it when $\mathfrak{a}$ is a proper primitive ideal. We write a primitive quadratic form corresponding to the class of $\mathfrak{a}$ by

$$\begin{pmatrix} a & b/2 \\ b/2 & c \end{pmatrix}.$$

It is enough to see that we can make a prime to m by changing the quadratic form to an equivalent quadratic form. Let S be the set of all primes which divide m and put

$$S_1 = \{p \in S; p \nmid a\},$$
$$S_2 = \{p \in S; p|a \text{ and } p \nmid c\},$$
$$S_3 = \{p \in S; p|a \text{ and } p|c\}.$$

Put $x = \prod_{p \in S_2} p$ and $y = \prod_{p \in S_1} p$. Then we have $(x, y) = 1$, and if we put $l := ax^2 + bxy + cy^2$, then l is coprime to m. Indeed, if $p \in S_1$ then ax^2 is coprime to p, but $p|(bxy + cy^2)$, so $p \nmid l$, and if $p \in S_2$ then $p|(ax^2 + bxy)$ but $p \nmid cy^2$, so $p \nmid l$. If $p \in S_3$, then since we assumed first that a, b, c are coprime, we have $p \nmid b$, and since $p \nmid xy$, again l is coprime to p. Taking a matrix $A \in SL_2(\mathbf{Z})$ whose first row is (x, y), we have

$$A \begin{pmatrix} a & b/2 \\ b/2 & c \end{pmatrix} {}^t A = \begin{pmatrix} l & * \\ * & * \end{pmatrix},$$

so we are done. □

We denote by $I_0(\mathfrak{O}_f, f)$ the set of ideals of $\mathfrak{O}_f$ which are coprime to f and by $I_0(\mathfrak{O}_{max}, f)$ the set of ideals of $\mathfrak{O}_{max}$ which are coprime to f.

Lemma 10.11. *The mapping $\phi : I_0(\mathfrak{O}_{max}, f) \to I_0(\mathfrak{O}_f, f)$ defined by $\phi(\mathfrak{a}) = \mathfrak{a} \cap \mathfrak{O}_f$ is a bijection. This maps a product to a product and does not change the norm.*

Proof. First we show that this mapping does not change the norm. Let $\mathfrak{a}$ be an ideal of $\mathfrak{O}_{max}$ coprime to f. By the isomorphism theorem of rings, we have $\mathfrak{O}_f/\mathfrak{a} \cap \mathfrak{O}_f \cong (\mathfrak{O}_f + \mathfrak{a})/\mathfrak{a}$. Since we assumed here that $\mathfrak{a}$ is coprime to f, we have $\mathfrak{a} + f\mathfrak{O}_{max} = \mathfrak{O}_{max}$, but since $f\mathfrak{O}_{max} \subset \mathfrak{O}_f$, we also have $\mathfrak{O}_f + \mathfrak{a} = \mathfrak{O}_{max}$. Namely we get

$$|\mathfrak{O}_f/\mathfrak{a} \cap \mathfrak{O}_f| = |\mathfrak{O}_{max}/\mathfrak{a}|,$$

and this means that the norms are equal. In particular, we have seen that the image of the mapping is contained in $I_0(\mathfrak{O}_f, f)$ since the norm is coprime to f.

The surjectivity of the mapping is proved as follows. For $\mathfrak{a}_0 \in I_0(\mathfrak{O}_f, f)$, the $\mathfrak{O}_{max}$ ideal $\mathfrak{a}_0\mathfrak{O}_{max}$ is coprime to f, since $1 \in \mathfrak{O}_f = \mathfrak{a}_0 + f\mathfrak{O}_f \subset \mathfrak{a}_0\mathfrak{O}_{max} + f\mathfrak{O}_{max}$. We have

$$\mathfrak{a}_0 \subset \mathfrak{a}_0\mathfrak{O}_{max} \cap \mathfrak{O}_f = (\mathfrak{a}_0\mathfrak{O}_{max} \cap \mathfrak{O}_f)(\mathfrak{a}_0 + f\mathfrak{O}_f)$$
$$\subset \mathfrak{a}_0 + \mathfrak{a}_0 f\mathfrak{O}_{max} \subset \mathfrak{a}_0 + \mathfrak{a}_0\mathfrak{O}_f = \mathfrak{a}_0.$$

So we have $\phi(\mathfrak{a}_0\mathfrak{O}_{max}) = \mathfrak{a}_0$. Injectivity is proved as follows. If we write $\phi(\mathfrak{a}) = \mathfrak{a}_0$ for a fixed $\mathfrak{a} \in I(f, \mathfrak{O}_{max})$, then as we have shown above, we have $\phi(\mathfrak{a}_0\mathfrak{O}_{max}) = \mathfrak{a}_0$. We also have $N(\mathfrak{a}_0\mathfrak{O}_{max}) = N(\mathfrak{a}_0) = N(\mathfrak{a})$. Since $\mathfrak{a}_0\mathfrak{O}_{max} \subset (\mathfrak{a} \cap \mathfrak{O}_f)\mathfrak{O}_{max} \subset \mathfrak{a}$ and $N(\mathfrak{a}) = N(\mathfrak{a}_0\mathfrak{O}_{max})$, we have $\mathfrak{a} = \mathfrak{a}_0\mathfrak{O}_{max}$. This means that $\mathfrak{a}$ is determined by $\mathfrak{a}_0 = \phi(\mathfrak{a})$. So the mapping is bijective. Now take ideals $\mathfrak{a}, \mathfrak{b} \in I_0(\mathfrak{O}_{max}, f)$. We put $\mathfrak{a}_0 = \phi(\mathfrak{a}) = \mathfrak{a} \cap \mathfrak{O}_f$ and $\mathfrak{b}_0 = \phi(\mathfrak{b}) = \mathfrak{b} \cap \mathfrak{O}_f$. We will show that $\phi(\mathfrak{ab}) = \mathfrak{a}_0\mathfrak{b}_0$. First we show that $\mathfrak{a}_0\mathfrak{b}_0$ is coprime to f. Since $\mathfrak{a}_0$ is coprime to f as shown before, we have $\mathfrak{a}_0 + f\mathfrak{O}_f = \mathfrak{O}_f$, so $\mathfrak{a}_0\mathfrak{b}_0 + f\mathfrak{b}_0 = \mathfrak{b}_0$. Since $\mathfrak{b}_0$ is also coprime to f, we have $\mathfrak{a}_0\mathfrak{b}_0 + f\mathfrak{b}_0 + f\mathfrak{O}_f = \mathfrak{b}_0 + f\mathfrak{O}_f = \mathfrak{O}_f$. Since $f\mathfrak{b}_0 \subset f\mathfrak{O}_f$, we have $\mathfrak{a}_0\mathfrak{b}_0 + f\mathfrak{b} + f\mathfrak{O}_f = \mathfrak{a}_0\mathfrak{b}_0 + f\mathfrak{O}_f$. So we have $\mathfrak{a}_0\mathfrak{b}_0 + f\mathfrak{O}_f = \mathfrak{O}_f$. So $\mathfrak{a}_0\mathfrak{b}_0$ is coprime to f. Now we put $\mathfrak{c}_0 = \mathfrak{a}_0\mathfrak{b}_0$. We will show that $\mathfrak{c}_0 = \mathfrak{ab} \cap \mathfrak{O}_f$. We have already shown that $\mathfrak{a} = \mathfrak{a}_0\mathfrak{O}_{max}$, $\mathfrak{b} = \mathfrak{b}_0\mathfrak{O}_{max}$ and $\mathfrak{c}_0\mathfrak{O}_{max} \cap \mathfrak{O}_f = \mathfrak{c}_0$. We have

$$\mathfrak{a}_0\mathfrak{b}_0\mathfrak{O}_{max} = \mathfrak{a}_0\mathfrak{O}_{max}\mathfrak{b}_0\mathfrak{O}_{max} = \mathfrak{ab},$$

so for $\mathfrak{ab} \in I_0(f, \mathfrak{O}_{max})$, we have $\phi(\mathfrak{ab}) = \mathfrak{a}_0\mathfrak{b}_0$. Therefore a product is mapped to a product. □

So in order to consider the classes of proper $\mathfrak{O}_f$ ideals, we can lift proper $\mathfrak{O}_f$ ideals prime to f in each class to ideals of $\mathfrak{O}_{max}$ and consider these. But this does *not* mean that the classes of $\mathfrak{O}_{max}$ ideals and of $\mathfrak{O}_f$ ideals are the same. In fact, for a principal ideal $\alpha\mathfrak{O}_{max}$, the ideal $\mathfrak{O}_f \cap \alpha\mathfrak{O}_{max}$ is not necessarily a principal ideal of $\mathfrak{O}_f$ and this causes the difference. Also we should note the following point. We denote by $I(\mathfrak{O}_f, f)$ the group of fractional proper $\mathfrak{O}_f$ ideals generated by $I_0(\mathfrak{O}_f, f)$. Then a principal ideal $\alpha\mathfrak{O}_f$ ($\alpha \in K$) is contained in $I(\mathfrak{O}_f, f)$ if and only if we can write $\alpha = \beta/\gamma$ by numbers $\beta, \gamma \in \mathfrak{O}_f$ which are prime to f. Here, we may take γ as a rational integer prime to f since $\alpha = \beta\overline{\gamma}/N(\gamma)$. We denote the set of such principal ideals by $P(\mathfrak{O}_f, f)$. In the case when $\alpha \in K$ is not in $\mathfrak{O}_f$, even if (the numerator and the denominator of) $N(\alpha)$ is prime to f, we cannot conclude that $\alpha\mathfrak{O}_f$ is in $P(\mathfrak{O}_f, f)$. Now, obviously an element $\beta \in \mathfrak{O}_f$ is prime to f if and only if $\beta - a \in f\mathfrak{O}_{max}$ for some integer a such that a is prime to f. So if we denote by $P_{\mathbf{Z}}(\mathfrak{O}_{max}, f)$ the group generated by principal ideals $\beta\mathfrak{O}_{max}$, where $\beta \in \mathfrak{O}_{max}$ and $\beta - a \in f\mathfrak{O}_{max}$ for some integer a prime to f, then $P(\mathfrak{O}_f, f)$ is lifted to this group, and we see that the ideal class group of $\mathfrak{O}_f$ is isomorphic to $I(\mathfrak{O}_{max}, f)/P_{\mathbf{Z}}(\mathfrak{O}_{max}, f)$. If we denote by $P(\mathfrak{O}_{max}, f)$ the group generated by integral principal $\mathfrak{O}_{max}$ ideals prime to f, then by counting $P(\mathfrak{O}_{max}, f)/P_{\mathbf{Z}}(\mathfrak{O}_{max}, f)$, we can calculate the difference between $h(f^2D_K)$ and

$h(D_K)$. This calculation is not so difficult and is explained for example in [101] or [67], but we omit it here. In next section, we use a different method to obtain this difference in the case of imaginary quadratic fields.

10.2 Proof of the Class Number Formula of Imaginary Quadratic Fields

We prove that the class number of an imaginary quadratic field can be written in terms of a generalized Bernoulli number. There are many ways to prove this; here we use properties of theta functions.

Theorem 10.12. *Let K be an imaginary quadratic field, χ_K be the character corresponding with K defined in Sect. 6.3 and w be the half of the number of roots of unity in the maximal order $\mathfrak{O}_{max}$ of K. Then the class number of $\mathfrak{O}_K$ is given by*

$$h(D_K) = -wB_{1,\chi_K}.$$

For the proof, we review the transformation formula of theta functions. Let a, b, c be integers, and we assume that the quadratic form $Q(x,y) = ax^2+bxy+cy^2$ is positive definite. We define Q^{-1} by $Q^{-1}(x,y) = 4(4ac-b^2)^{-1}(cx^2-bxy+ay^2)$. We put

$$\mathfrak{H} = \{\tau \in \mathbf{C};\ \text{the imaginary part of } \tau \text{ is positive}\}$$

(the upper half plane). For $\tau \in \mathfrak{H}$, we write

$$\theta_Q(\tau) = \sum_{x,y\in\mathbf{Z}} e^{\pi i Q(x,y)\tau}.$$

It is easy to see that this converges absolutely and uniformly on compact sets in $\mathfrak{H}$, and is a holomorphic function on $\mathfrak{H}$.

Proposition 10.13. *For $\tau \in \mathfrak{H}$, the following transformation formula holds.*

$$\theta_{Q^{-1}}\left(-\frac{1}{\tau}\right) = \sqrt{4ac-b^2}\,(\tau/2i)\,\theta_Q(\tau).$$

Proof. A general proof can be found for example in [3, p. 25]. Here we give a direct proof restricted to this case. We use the Poisson[2] summation formula which is well

[2]Siméon Denis Poisson (born on June 21, 1781 in Pithiviers, France—died on April 25, 1840 in Sceaux, France).

known in the theory of Fourier[3] transforms. To simplify the notation, for arbitrary complex number $z \in \mathbf{C}$, we write $\mathbf{e}(z) = e^{2\pi i z}$. For $\alpha > 0$, $\tau \in \mathfrak{H}$ and $\xi, \xi^* \in \mathbf{R}$, the following formula is well known:

$$\int_{-\infty}^{\infty} \mathbf{e}(2^{-1}\alpha\tau\xi^2)\mathbf{e}(-\xi\xi^*)d\xi = \frac{\mathbf{e}(-(2\tau\alpha)^{-1}\xi^{*2})}{\sqrt{-i\tau\alpha}}.$$

Here as the square root of $-i\tau\alpha$ (whose real part is positive since $\tau \in \mathfrak{H}$), we take the branch $\sqrt{\alpha} > 0$ for $\tau = i$. When we put

$$f(\xi, \eta) = \mathbf{e}(2^{-1}Q(\xi, \eta)\tau),$$

the Fourier transform of this function with respect to the variables $(\xi, \eta) \in \mathbf{R}^2$ is calculated as follows:

$$\begin{aligned}
&\hat{f}(\xi^*, \eta^*) \\
&= \int_{-\infty}^{\infty}\int_{-\infty}^{\infty} f(\xi, \eta)\mathbf{e}(-\xi\xi^* - \eta\eta^*)d\xi\, d\eta \\
&= \int_{-\infty}^{\infty}\int_{-\infty}^{\infty} \mathbf{e}(2^{-1}\tau(a(\xi + (b/2a)\eta)^2 + (c - b^2/4a)\eta^2))\mathbf{e}(-\xi\xi^* - \eta\eta^*)d\xi\, d\eta \\
&= \int_{-\infty}^{\infty}\int_{-\infty}^{\infty} \mathbf{e}(2^{-1}\tau(a\xi^2 + (c - b^2/4a)\eta^2))\mathbf{e}(-\xi\xi^* - \eta(\eta^* - (b/2a)\xi^*))d\xi\, d\eta \\
&= \frac{1}{\sqrt{-i\tau a}\sqrt{-i\tau(c - b^2/4a)}} \\
&\quad \times \mathbf{e}(-(2\tau a)^{-1}\xi^{*2})\mathbf{e}(-(2\tau(c - b^2/4a))^{-1}(\eta^* - (b/2a)\xi^*)^2) \\
&= \frac{2i}{\tau\sqrt{(4ac - b^2)}} \times \mathbf{e}\left(\frac{-1}{2\tau(ac - b^2/4)}(a\eta^{*2} - b\eta^*\xi^* + c\xi^{*2})\right).
\end{aligned}$$

By this result and the Poisson summation formula (e.g. [102]), we get

$$\sum_{\xi,\eta \in \mathbf{Z}} f(\xi, \eta) = \sum_{\xi^*,\eta^* \in \mathbf{Z}} \hat{f}(\xi^*, \eta^*).$$

Hence we get the transformation formula of theta functions. □

Proof of the Class Number Formula. We show this by calculating the residue of a zeta function in two different ways. Namely, we can derive the formula by the relation $\zeta_K(s) = \zeta(s)L(s, \chi_K)$ and the following calculation:

---

[3]Jean Baptiste Joseph Fourier (born on March 21, 1768 in Auxerre, France—died on May 16, 1830 in Paris, France).

$$\lim_{s\to 1}(s-1)\zeta_K(s) = \frac{\pi h(D_K)}{w\sqrt{|D_K|}}$$

and

$$L(1,\chi_K) = \frac{g(\chi)\pi i}{|D_K|}B_{1,\chi_K} = -\frac{B_{1,\chi_K}\pi}{\sqrt{|D_K|}},$$

$$\lim_{s\to 1}(s-1)\zeta(s) = 1.$$

The latter two equalities are in Theorems 5.4 and 9.6. The point here is the evaluation of $\lim_{s\to 1}(s-1)\zeta_K(s)$. We fix an ideal class of $\mathfrak{O}_{max}$ and denote by C the set of all (integral) ideals of $\mathfrak{O}_{max}$ belonging to this class. We define the following zeta function which is a part of $\zeta_K(s)$:

$$\zeta(s,C) = \sum_{\mathfrak{a}\in C}\frac{1}{N(\mathfrak{a})^s}.$$

Lemma 10.14. *Independently of the choice of C, we have*

$$\lim_{s\to 1}(s-1)\zeta(s,C) = \frac{\pi}{w\sqrt{|D_K|}}.$$

Proof. If we take a fractional ideal L which is equivalent to some ideal in C, then all the other ideals in C are written as $L\alpha(\alpha\in K)$. Since $L\alpha\subset\mathfrak{O}_{max}$, we have $\alpha\in L^{-1}$. We can choose L so that $L^{-1}=\mathfrak{a}\subset\mathfrak{O}_{max}$ by dividing L by some rational number if necessary. Here we can assume that $\mathfrak{a}$ is a primitive ideal without loss of generality. Also we have $L\alpha = L\beta$ if and only if $\alpha=\epsilon\beta$ for some unit ϵ of $\mathfrak{O}_{max}$. Since we assumed that K is imaginary, ϵ is a root of unity in K. Hence we have

$$\zeta(s,C) = \frac{N(\mathfrak{a})^s}{2w}\sum_{\alpha\in\mathfrak{a}}\frac{1}{N(\alpha)^s}.$$

We write $\mathfrak{a}=\mathbf{Z}a+\mathbf{Z}(d+\omega)$ using the standard basis and put $N(d+\omega)=ac$, $b=2d+Tr(\omega)$ as usual. Then for $\alpha = xa+y(d+\omega)$ we have $N(\alpha) = a(ax^2+bxy+cy^2)$. Since $N(\mathfrak{a})=a$, we can cancel a^s in the numerator and the denominator. So we have

$$\zeta(s,C) = \frac{1}{2w}\sum_{x,y\in\mathbf{Z};\ (x,y)\neq(0,0)}(ax^2+bxy+cy^2)^{-s}.$$

We have also $4ac-b^2=|D_K|$. Now our aim is to calculate the residue of this function at $s=1$. We use theta functions. Since $\zeta(s,C)$ converges absolutely uniformly on compact subsets of $\{s \in \mathbf{C}; \mathrm{Re}(s) > 1\}$, exchanging the summation and the integral, and also using the following well-known formula of the gamma function

$$\frac{\Gamma(s)}{n^s} = \int_0^\infty e^{-t}\left(\frac{t}{n}\right)^s \frac{dt}{t} = \int_0^\infty e^{-nt} t^{s-1}\, dt,$$

we get

$$\begin{aligned}\Gamma(s)\zeta(s, C) &= \frac{1}{2w}\int_0^\infty \sum_{(\xi,\eta)\in\mathbf{Z}^2-(0,0)} e^{-(a\xi^2+b\xi\eta+c\eta^2)y} y^{s-1}\, dy \\ &= \frac{\pi^s}{2w}\int_0^\infty (\theta_Q(iy)-1)y^{s-1}\, dy.\end{aligned}$$

Here $\theta_Q(iy) - 1$ is a rapidly decreasing function for $y \to \infty$, so the integral

$$\int_1^\infty (\theta_Q(iy)-1)y^{s-1}\, dy$$

converges for all $s \in \mathbf{C}$ and is a holomorphic function on the whole complex s-plane. On the other hand, if we take $(0, 1)$ as the interval of integration, it might not converge around 0 when s is small. In order to deal with this problem, we use the transformation formula of theta functions. Namely, if we substitute y^{-1} for y in the integral on the interval $(0, 1)$, in the range $\mathrm{Re}(s) > 1$ we have

$$\begin{aligned}&\int_0^1 (\theta_Q(iy)-1)y^{s-1}\, dy \\ &= \int_1^\infty (\theta_Q(iy^{-1})-1)y^{-s-1}\, dy \\ &= \int_1^\infty (\theta_{Q^{-1}}(iy)2y\sqrt{|D_K|^{-1}}-1)y^{-s-1}\, dy \\ &= \int_1^\infty 2(\theta_{Q^{-1}}(iy)-1)y^{-s}\sqrt{|D_K|^{-1}} + 2y^{-s}\sqrt{|D_K|^{-1}} - y^{-s-1}\, dy \\ &= 2\int_1^\infty (\theta_{Q^{-1}}(iy)-1)\frac{y^{-s}}{\sqrt{|D_K|}}\, dy - \frac{2}{\sqrt{|D_K|}(1-s)} - \frac{1}{s}.\end{aligned}$$

The integral in the last expression converges for arbitrary s, so it is holomorphic on the whole complex plane, and the rest has a pole at $s = 0$ and $s = 1$. So the residue of $\zeta(s, C)$ at $s = 1$ is given by

$$\frac{1}{\Gamma(1)}\frac{1}{2w} \times \frac{2\pi}{\sqrt{D_K}} = \frac{\pi}{w\sqrt{|D_K|}}.$$

Hence the lemma is proved.

Therefore we have

$$\lim_{s\to 1}(s-1)\zeta_K(s) = \frac{h(D_K)\pi}{w\sqrt{|D_K|}}.$$

Comparing this with the value of $L(1,\chi_K)$, we get the class number formula. □

Since $\zeta(s)$ and $L(s,\chi)$ have functional equations, $\zeta_K(s)$, which is a product of these two, also has a functional equation and it is easy to calculate its explicit form. But as a matter of fact, if we use the above integral representation, we can obtain the analytic continuation and the functional equation of $\zeta(s,C)$, and through this we can get an alternative proof of the analytic continuation and the functional equation of $\zeta_K(s)$. Namely:

Proposition 10.15. *Let C be an ideal class of the maximal order $\mathfrak{O}_{max}$ of a imaginary quadratic field K, and let the other notation be the same as above. Then $\zeta(s,C)$ can be continued analytically to a meromorphic function on the whole s-plane. If we put $\xi(s,C) = (2\pi)^{-s}|D_K|^{s/2}\Gamma(s)\zeta(s,C)$, then it satisfies the following functional equation.*

$$\xi(1-s,C) = \xi(s,C).$$

In the same way, if we put $\xi_K(s) = (2\pi)^{-s}|D_K|^{s/2}\Gamma(s)\zeta_K(s)$, then we have

$$\xi_K(1-s) = \xi_K(s).$$

Proof. For the proof, we use the integral representation mentioned before. For $\mathrm{Re}(s) > 1$, we have

$$(2w)\pi^{-s}\Gamma(s)\zeta(s,C) = \int_1^\infty (\theta_Q(iy)-1)y^{s-1}\,dy + \frac{2}{\sqrt{|D_K|}}\int_1^\infty (\theta_{Q^{-1}}(iy)-1)y^{-s}\,dy$$
$$-\frac{2}{\sqrt{|D_K|}(1-s)} - \frac{1}{s} \tag{10.1}$$

but the integral on the right-hand side is holomorphic on the whole s-plane. Hence the right-hand side is a meromorphic function on the whole complex s-plane. So the analytic continuation is proved.

The functional equation is shown as follows. Multiplying both sides of the integral representation before by $2^{-s}|D_K|^{s/2}$ and changing $s \to 1-s$, we get

$$\begin{aligned}
&(2w)\xi(1-s,C)\\
&\quad = (2w)|D_K|^{(1-s)/2}(2\pi)^{s-1}\Gamma(1-s)\zeta(1-s,C)\\
&\quad = 2^{s-1}|D_K|^{(1-s)/2}\int_1^\infty (\theta_Q(iy)-1)y^{-s}\,dy
\end{aligned}$$

$$
\begin{aligned}
&+ 2^s|D_K|^{-1/2}|D_K|^{(1-s)/2} \int_1^{\infty} (\theta_{Q^{-1}}(iy) - 1)y^{s-1}\, dy \\
&- \frac{2^s|D_K|^{(1-s)/2}|D_K|^{-1/2}}{s} - \frac{2^{s-1}|D_K|^{(1-s)/2}}{1-s} \\
= {} & 2^s|D_K|^{-s/2}\left(2^{-1}|D_K|^{1/2} \int_1^{\infty} (\theta_Q(iy) - 1)y^{-s}\, dy \right. \\
&\left. + \int_1^{\infty} (\theta_{Q^{-1}}(iy) - 1)y^{s-1}\, dy - \frac{1}{s} - \frac{2^{-1}|D_K|^{1/2}}{1-s}\right).
\end{aligned}
\tag{10.2}
$$

Here the integrals in the parentheses are almost identical to the integral expression (10.1) of $\zeta(s, C)$ if we exchange Q and Q^{-1}. We defined

$$
Q^{-1}(x, y) = \frac{4}{|D_K|}(cx^2 - bxy + ay^2),
$$

and the quadratic forms $ax^2 + bxy + cy^2$ and $cx^2 - bxy + ay^2$ are equivalent since it is the same if we replace $(x, y) \to (-y, x)$. Since we have

$$
(Q^{-1}(x, y))^{-s} = 2^{-2s}|D_K|^s(cx^2 - bxy + ay^2)^{-s}
$$

we see in the same way as before that

$$
(2w)\Gamma(s)\zeta(s, C) = 2^{2s}\pi^s|D_K|^{-s} \int_0^{\infty} (\theta_{Q^{-1}}(iy) - 1)y^{s-1}\, dy.
$$

Just in the same way as in the proof of (10.1), we have

$$
\begin{aligned}
\int_0^{\infty} (\theta_{Q^{-1}}(iy) - 1)y^{s-1} dy &= \int_1^{\infty} (\theta_{Q^{-1}}(iy) - 1)y^{s-1} dy \\
&+ 2^{-1}\sqrt{|D_K|} \int_1^{\infty} (\theta_Q(iy) - 1)y^{-s} dy - \frac{2^{-1}\sqrt{|D_K|}}{1-s} - \frac{1}{s}.
\end{aligned}
$$

(If we note that the discriminant of Q^{-1} is $16|D_K|^{-1}$, we see the symmetry to the case Q.) Therefore comparing this with (10.2), we have

$$
\begin{aligned}
(2w)\xi(1-s, C) &= 2^s|D_K|^{-s/2} \int_0^{\infty} (\theta_{Q^{-1}}(iy) - 1)y^{s-1}\, dy \\
&= (2\pi)^{-s}|D_K|^{s/2}2^{2s}|D_K|^{-s}\pi^s \int_0^{\infty} (\theta_{Q^{-1}}(iy) - 1)y^{s-1}\, dy \\
&= (2w)(2\pi)^{-s}|D_K|^{s/2}\Gamma(s)\zeta(s, C) = (2w)\xi(s, C).
\end{aligned}
$$

Hence the proposition is proved. The functional equation of $\xi_K(s)$ is obvious from this. □

As for the formula for the class number $h(f^2D_K)$ of the ideal class group of a general order $\mathfrak{O}_f$, it is easiest to compare it with the ideal class group of $\mathfrak{O}_{max}$ by a group theoretic method. (For example, see Weber [101, p. 366], Lang [67, p. 95], Zagier [106].) But here we introduce a different method.

We first give the formula for the zeta functions of $\mathfrak{O}_f$ ideals for general $f \geq 1$. Here we do not assume that K is imaginary, since the theory is the same for real quadratic fields.

We fix a natural number f and consider the zeta function of ideals of the order $\mathfrak{O}_f = \mathbf{Z} + \mathbf{Z}f\omega$ of K, which is defined by

$$\zeta(s, f^2D_K) = \sum_{\mathfrak{a}} \frac{1}{N(\mathfrak{a})^s},$$

where the sum runs over arbitrary ideals of $\mathfrak{O}_f$, which are not necessarily proper.

Proposition 10.16. (1) *For a positive integer a, let $r(a)$ be a number of residue classes modulo $2a$ represented by integers b such that $b^2 \equiv f^2D_K \bmod 4a$. Then we have*

$$\zeta(s, f^2D_K) = \zeta(2s)\sum_{a=1}^{\infty} r(a)a^{-s}.$$

(2) *More precisely, we have the next relation.*

$$\zeta(s, f^2D_K) = \zeta(s)L(s, \chi_K)\prod_{\substack{p|f \\ p^m \| f}} \frac{1-p^{(m+1)(1-2s)}-\chi_K(p)p^{-s}(1-p^{m(1-2s)})}{1-p^{1-2s}}.$$

Here we denote by m the highest power of p dividing f (hence m depends on p).

Proof. A primitive ideal of $\mathfrak{O}_f$ is written as $\mathfrak{a} = \mathbf{Z}a + \mathbf{Z}(d + f\omega)$ $(a = N(\mathfrak{a}))$ and there exists an integer c such that $N(d + f\omega) = ac$, so if we put $b = 2d + fTr(\omega)$ then $b^2 \equiv f^2D_K \bmod 4a$ (p. 79, Lemma 6.2 and p. 80, Lemma 6.3). For positive integers a, a', we have $\mathbf{Z}a + \mathbf{Z}(d + f\omega) = \mathbf{Z}a' + \mathbf{Z}(d' + f\omega)$ if and only if $a = a'$ and $d \equiv d' \bmod a$. So if we fix a, two primitive ideals which have the above bases are equal if and only if $b \equiv b' \bmod 2a$ for $b = 2d + fTr(\omega)$ and $b' = 2d' + fTr(\omega)$. Besides, if $b^2 \equiv f^2D_K \bmod 4a$ for some integer b, then since $Tr(\omega)^2 - D_K = 4N(\omega)$, we have $b^2 \equiv (fTr(\omega))^2 \bmod 4$ and there exists an integer d such that $b = 2d + fTr(\omega)$. So we can define a primitive ideal $\mathbf{Z}a + (d + f\omega)\mathbf{Z}$. Taking ideals which are not primitive into account, (1) is proved. Next we show (2). For this purpose, we should just calculate $r(a)$. Here we introduce a local version of $r(a)$. For a positive integer a and a fixed prime p, we denote by e the maximum non-negative integer such that $p^e|a$. If p is odd, we define $r_p(a)$ to be the number of integers b mod p^e such that $b^2 \equiv f^2 D_K$ mod p^e. If $p = 2$, we define $r_p(a) = r_2(a)$ to be the number of integers b mod 2^{e+1} such that $b^2 \equiv f^2 D_K$ mod 2^{e+2}. Obviously we have $r_p(a) = r_p(p^e)$ By the Chinese remainder theorem, we have $r(a) = \prod_{p|a} r_p(a)$. So we have $\sum_{a=1}^{\infty} r(a)a^{-s} = \prod_p \sum_{e=0}^{\infty} r_p(p^e)p^{-es}$ and it is enough to calculate the above local Dirichlet series for each prime p. First we assume that p is an odd prime. Let $f = p^m f_0$, where f_0 is a positive integer prime to p. We consider the congruence equation $b^2 \equiv p^{2m} f_0^2 D_K$ mod p^e with unknown b. If $e \leq 2m$ then $b^2 \equiv 0$ mod p^e. Hence if $e = 2e_0$ here, then $b \equiv 0$ mod p^{e_0} so $r_p(p^e) = p^e/p^{e_0} = p^{e_0}$. If $e = 2e_0 + 1$, then $b \equiv 0$ mod p^{e_0+1}, and $r_p(p^e) = p^e/p^{e_0+1} = p^{e_0}$. So we have $r_p(p^e) = p^{e_0}$ in both cases. Now we assume that $2m < e$. Then we have $b = p^m b_0$ and $b_0^2 \equiv D_K f_0^2$ mod p^{e-2m}. If $p|D_K$ then p divides D_K precisely once, so if $e - 2m \geq 2$, there is no such b_0. Hence it can have solutions only when $e - 2m = 1$, and in this case, the congruence implies nothing but $p|b_0$, so $p^{m+1}|b$ is the condition. But we are counting b mod p^{2m+1}, so we have $r_p(p^{2m+1}) = p^m$, and we have $r_p(p^e) = 0$ for $e > 2m + 1$. If $p \nmid D_K$, then the congruence equation $b_0^2 \equiv D_K f_0^2$ mod p^{e-2m} has solutions only when $\chi_K(p) = 1$ and the number of solutions mod p^{e-2m} is two in that case. Since we must count these as residues mod p^{e-m}, the number is $2p^m$. After all, if $p \nmid D_K$ and $e > 2m$, we have $r_p(p^e) = p^m(1+\chi_K(p))$, independently of e. Summing up, we have the following formula for $r_p(p^e)$.

$$r_p(p^e) = \begin{cases} p^{[e/2]} & 0 \leq e \leq 2m, \\ (1 + \chi_K(p))p^m & 2m + 1 \leq e \text{ and } p \nmid D_K, \\ p^m & e = 2m + 1 \text{ and } p|D_K, \\ 0 & 2m + 2 \leq e \text{ and } p|D_K. \end{cases} \tag{10.3}$$

Here $[e/2]$ denotes the biggest integer which does not exceed $e/2$, that is, we put $[e/2] = e_0$ if $e = 2e_0$ or $2e_0 + 1$ for some integer e_0. Also when $p = 2$, we obtain the same formula (10.3) by similar consideration. For example, assume that $0 \leq e \leq 2m$. If $e = 2e_0$ is even, then $b^2 \equiv 2^{2m} f_0^2 D_K$ mod 2^{e+2} means that $b = 2^{e_0} b_0$ and $b_0^2 \equiv f_0^2 D_K$ mod 4. This last relation imposes the condition that b_0 is even or odd. So the number of solutions b mod 2^{e+1} is $2^{e_0+1}/2 = 2^{e_0}$. If $e = 2e_0 + 1$ is odd, then $2e_0 + 2 \leq 2m$, so $b = 2^{e_0+1} b_0$. Again the relation $b_0^2 \equiv f_0^2 D_K$ mod 2 tells that b_0 is odd or even. So the number of solutions b mod 2^{2e_0+2} is $2^{2e_0+2-(e_0+1)}/2 = 2^{e_0}$. The case $e \geq 2m + 1$ is calculated similarly. Here we note that if D_K is odd, then $\chi_K(2) = 1$ if $D_K \equiv 1$ mod 8 and $\chi_K(2) = -1$ if $D_K \equiv 5$ mod 8, and that the congruence equation $x^2 \equiv y$ mod 2^{l+2} with $l \geq 1$ for a fixed y and unknown x has no solution when $y \equiv 5$ mod 8 but two solutions x mod 2^{l+1} when $y \equiv 1$ mod 8. We should also note that, by definition, if D_K is even then $D_K \equiv 0$ mod 4, and if $D_K \equiv 4$ mod 8 then $D_K/4 \equiv 3$ mod 4. The rest is the same and is omitted here. By using (10.3), we see

$$\sum_{e=0}^{\infty} r_p(p^e)p^{-es} = \frac{(1+p^{-s})-(1+p^{s-1})p^{(1-2s)(m+1)}}{1-p^{1-2s}}$$
$$+\begin{cases} p^{m(1-2s)-s} & \text{if } p|D_K, \\ \dfrac{(1+\chi_K(p))p^{m(1-2s)-s}}{1-p^{-s}} & \text{if } p\nmid D_K. \end{cases}$$

The last term written for each case $p|D_K$ or $p \nmid D_K$ separately has a unified expression

$$\frac{(1+\chi_K(p))p^{m(1-2s)-s}}{1-\chi_K(p)p^{-s}}.$$

Reducing to a common denominator, the right-hand side becomes

$$\frac{(1+p^{-s})(1-p^{(m+1)(1-2s)}-\chi_K(p)p^{-s}(1-p^{m(1-2s)}))}{(1-\chi_K(p)p^{-s})(1-p^{1-2s})}.$$

Hence multiplying this by the Euler p-factor of $\zeta(2s)$, we get

$$(1-p^{-2s})^{-1}\sum_{e=0}^{\infty} r(p^e)p^{-es} = \frac{1-p^{(m+1)(1-2s)}-\chi_K(p)p^{-s}(1-p^{m(1-2s)})}{(1-p^{-s})(1-\chi_K(p)p^{-s})(1-p^{1-2s})}.$$

□

Next we put

$$\zeta(s,\mathfrak{O}_f) = \sum_{\mathfrak{a}} \frac{1}{N(\mathfrak{a})^s}.$$

But this time, we define the above sum so that $\mathfrak{a}$ runs only over proper ideals of $\mathfrak{O}_f$.

Here we introduce the Möbius[4] function $\mu(m)$. This is a function of natural numbers to $\{0, \pm 1\}$, defined as follows:

(1) $\mu(1) = 1$.
(2) When the prime factor decomposition is given by $m = p_1 \cdots p_r$ with different r primes, we define $\mu(m) = (-1)^r$.
(3) When m has a square factor, that is, if m is divisible by a square of a prime, we define $\mu(m) = 0$.

---

[4] August Ferdinand Möbius (born on November 17, 1790 in Schulpforta, Saxony (now Germany)—died on September 26, 1868 in Leipzig, Germany).

If m_1 and m_2 are coprime natural numbers, then we have $\mu(m_1)\mu(m_2) = \mu(m_1m_2)$.

The next formula is known as the Möbius inversion formula.

Lemma 10.17. *Let $F(n)$ and $G(n)$ be functions defined for all natural numbers n. We assume that, for an arbitrary natural number n, the following equality holds.*

$$F(n) = \sum_{d|n} G(d).$$

Then for an arbitrary natural number n, we have

$$G(n) = \sum_{d|n} \mu\left(\frac{n}{d}\right) F(d).$$

The proof is easy and is omitted here (see [83]).

Proposition 10.18. (1) *Using the Möbius function μ, we have*

$$\zeta(s, f^2 D_K) = f^{-s} \sum_{e|f} e^s \zeta(s, \mathfrak{O}_e),$$

$$\zeta(s, \mathfrak{O}_f) = f^{-s} \sum_{d|f} \mu(f/d) d^s \zeta(s, d^2 D_K).$$

(2) *The next relation holds.*

$$\zeta(s, \mathfrak{O}_f) = \zeta(s) L(s, \chi_K)$$
$$\times \prod_{\substack{p|f \\ p^m \| f}} \frac{(1-p^{-s})(1-\chi_K(p)p^{-s}) - p^{m-1-2sm}(1-p^{1-s})(\chi_K(p)-p^{1-s})}{1-p^{1-2s}}.$$

Proof. The first equality of (1) comes from the facts that any ideal of $\mathfrak{O}_f$ is a proper ideal of $\mathfrak{O}_e$ for some $e|f$ and that if the ideal is regarded as an ideal of $\mathfrak{O}_e$, then the norm becomes f/e times the original one. The second equality follows from the Möbius inversion formula, applying it to $F(f) = f^s \zeta(s, f^2 D_K)$ and $G(e) = e^s \zeta(s, \mathfrak{O}_e)$. In order to show (2), we calculate the right-hand side of the second equality of (1) by using Proposition 10.16. Since the Möbius function is multiplicative, it is enough to calculate the Euler p-factors. Here when $f = \prod_{i=1}^t p_i^{m_i}$, by the definition of the Möbius function, we can assume that d runs only over the numbers $\prod p_i^{m_i - j_i}$ with $j_i = 0, 1$. So if we write

$$f(m, p^{-s}) = \frac{1 - p^{(m+1)(1-2s)} - \chi_K(p)p^{-s}(1 - p^{m(1-2s)})}{1 - p^{1-2s}}.$$

then we have

$$\zeta(s,\mathfrak{O}_f)=\zeta(s)L(s,\chi_K)\prod_{p|f}(f(m,p^{-s})-p^{-s}f(m-1,p^{-s}))\,.$$

Calculating the latter product, we get the result. □

Remark 10.19. The explicit expression of the zeta function in Propositions 10.16 and 10.18 was given in [54, 106]. (See these papers for applications.)

Now, when K is an imaginary quadratic field, we obtain the class number of $\mathfrak{O}_f$ from the above formula for $\zeta(s,\mathfrak{O}_f)$. As before, we denote by C an ideal class of proper ideals of $\mathfrak{O}_f$ and write

$$\zeta(s,C,\mathfrak{O}_f)=\sum_{\mathfrak{a}\in C}\frac{1}{N(\mathfrak{a})^s}.$$

Let w_f be half the number of roots of unity in $\mathfrak{O}_f$. For a primitive quadratic form $Q(x,y)=ax^2+bxy+cy^2$ with discriminant f^2D_K corresponding to C, we have

$$(2w_f)\zeta(s,C,\mathfrak{O}_f)=\sum_{(x,y)\in\mathbf{Z}^2-(0,0)}(ax^2+bxy+cy^2)^{-s}.$$

This is proved in the same way as in the case of the maximal order. If we use the transformation formula of $\theta_Q(\tau)$, the residue at $s=1$ of this Dirichlet series is obtained in the same way as before, independently of C, and given by

$$\lim_{s\to1}(s-1)\zeta(s,C,\mathfrak{O}_f)=\frac{\pi}{w_f f\sqrt{|D_K|}}.$$

By this, we have

$$\lim_{s\to1}(s-1)\zeta(s,\mathfrak{O}_f)=h(f^2D_K)\frac{\pi}{w_f f\sqrt{|D_K|}}.$$

On the other hand, using the Euler product expression in the formula for $\zeta(s,\mathfrak{O}_f)$ in Proposition 10.18, we have

$$\begin{aligned}\lim_{s\to1}(s-1)\zeta(s,\mathfrak{O}_f)&=(\lim_{s\to1}(s-1)\zeta_K(s))\times\prod_{p|f}(1-\chi_K(p)p^{-1})\\&=\frac{\pi h(D_K)}{w\sqrt{|D_K|}}\prod_{p|f}(1-\chi_K(p)p^{-1}).\end{aligned}$$

Hence we get the class number formula

$$h(f^2 D_K) = \frac{h(D_K)}{w/w_f} \times f \prod_{p|f} (1 - \chi_K(p)p^{-1})$$

as was given in Chap. 6, Theorem 6.12 (2).

The class number formula for a real quadratic field K (or for indefinite quadratic forms) is fairly complicated compared with this, and it is remote from our standpoint from Bernoulli numbers, so we omit it here.

10.3 Some L-Functions Associated with Quadratic Forms

We introduce here one of the simplest example of zeta functions associated with so-called prehomogeneous vector spaces. The content of this section is based on [47].

Let p be a prime. We assume that $p \equiv 3 \bmod 4$. We define a Dirichlet character ψ modulo p by $\psi(a) = \left(\frac{a}{p}\right)$. From this, we define a function $\tilde{\psi}$ on the set L^* of all 2×2 half-integral symmetric matrices as follows. First, for

$$T = \begin{pmatrix} a & b/2 \\ b/2 & c \end{pmatrix} \in L^*,$$

if $T \bmod p = 0$ or $\det(T) \not\equiv 0 \bmod p$, we define $\tilde{\psi}(T) = 0$. If $T \bmod p \neq 0$ and $\det(T) \equiv 0 \bmod p$, then there exists $g \in GL_2(\mathbf{Z}/p\mathbf{Z})$ such that

$${}^t g T g \equiv \begin{pmatrix} e & 0 \\ 0 & 0 \end{pmatrix} \bmod p.$$

Indeed since the rank of T mod p is 1, there exists g such that the second column of Tg mod p is 0, and so is the second column of ${}^t gTg$, but since this is symmetric, the $(2, 1)$ component is also 0. We define $\psi(T) = \psi(e)$ for such T. This is well defined since if we assume that

$$\begin{pmatrix} x & y \\ z & w \end{pmatrix} \begin{pmatrix} e & 0 \\ 0 & 0 \end{pmatrix} \begin{pmatrix} x & z \\ y & w \end{pmatrix} = \begin{pmatrix} e' & 0 \\ 0 & 0 \end{pmatrix}$$

for $e, e' \not\equiv 0 \bmod p$ and $\begin{pmatrix} x & y \\ z & w \end{pmatrix} \in GL_2(\mathbf{Z}/p\mathbf{Z})$, then we have $e' = x^2 e$ (and $z = 0$, $x \in (\mathbf{Z}/p\mathbf{Z})^\times$).

We denote by L_+^* the set of all positive definite half-integral symmetric matrices in L^*. Recall that T_1 and $T_2 \in L^*$ are said to be equivalent if we have ${}^t g T_1 g = T_2$ for some $g \in SL_2(\mathbf{Z})$. We define an L-function of L_+^* with *character* $\tilde{\psi}$ by

$$L(s, L_+^*, \psi) = \sum_{T \in L_+^*/\sim} \frac{\tilde{\psi}(T)}{\epsilon(T)\det(T)^s}.$$

Here $L_+^*/\sim$ denotes a complete set of representatives of L_+^* up to the equivalence and $\epsilon(T)$ the order of the finite group $\{g \in SL_2(\mathbf{Z});\ {}^t gTg = T\}$. (This L-function was defined first by K. Hashimoto in relation to dimension formulas of automorphic forms. This is also an example of L-functions of prehomogeneous vector spaces. As we see later, $\tilde{\psi}$ can be regarded as a character of some ideal class group, so the term "L-function with character $\tilde{\psi}$" is not strange.)

Let us try to calculate an example. If $T = \begin{pmatrix} a & b/2 \\ b/2 & c \end{pmatrix} \in L^*$ and $\operatorname{rank}(T \bmod p) = 1$, then $\tilde{\psi}(T) = \psi(a)$ if $p \nmid a$ and $\tilde{\psi}(T) = \psi(c)$ if $p|a$. Indeed if $p|a$, then by the condition on the rank, we have $p|b$ and $c \in (\mathbf{Z}/p\mathbf{Z})^\times$, and if $p \nmid a$, then we have

$$\begin{pmatrix} 1 & 0 \\ -b/2a & 1 \end{pmatrix} \begin{pmatrix} a & b/2 \\ b/2 & c \end{pmatrix} \begin{pmatrix} 1 & -b/2a \\ 0 & 1 \end{pmatrix} = \begin{pmatrix} a & 0 \\ 0 & c - \frac{b^2}{4a} \end{pmatrix} \equiv \begin{pmatrix} a & 0 \\ 0 & 0 \end{pmatrix} \bmod p.$$

In the case $p = 3$, for $\det(2T) = 3, 12, 15, 24, 27, \ldots$, we have

$$\tilde{\psi}\begin{pmatrix} 1 & 1/2 \\ 1/2 & 1 \end{pmatrix} = 1, \quad \tilde{\psi}\begin{pmatrix} 1 & 0 \\ 0 & 3 \end{pmatrix} = 1, \quad \tilde{\psi}\begin{pmatrix} 2 & 1 \\ 1 & 2 \end{pmatrix} = -1, \quad \tilde{\psi}\begin{pmatrix} 1 & 0 \\ 0 & 6 \end{pmatrix} = 1,$$
$$\tilde{\psi}\begin{pmatrix} 2 & 0 \\ 0 & 3 \end{pmatrix} = -1, \quad \tilde{\psi}\begin{pmatrix} 1 & 1/2 \\ 1/2 & 7 \end{pmatrix} = 1, \quad \tilde{\psi}\begin{pmatrix} 3 & 3/2 \\ 3/2 & 3 \end{pmatrix} = 0, \ldots$$

We also have $\epsilon(T) = 6$ for $T = \begin{pmatrix} a & a/2 \\ a/2 & a \end{pmatrix}$ $(a = 1, 2, 3 \ldots)$ and $\epsilon(T) = 2$ for the remaining T with $\operatorname{rank}(T \bmod 3) = 1$. From this we get

$$\begin{aligned} L(s, L_+^*, \psi) &= \frac{2^{2s}}{6 \cdot 3^s} + \frac{1}{\cdot 3^s}\left(\frac{1}{2} - \frac{1}{6}\right) + \frac{1}{6^s}\left(\frac{1}{2} - \frac{1}{2}\right) + \frac{2^{2s}}{2 \cdot 3^{3s}} + \cdots \\ &= \frac{2^{2s-1}}{3^{s+1}} + \frac{1}{3^{s+1}} + \frac{2^{2s-1}}{3^{3s}} + \cdots. \end{aligned}$$

Theorem 10.20. *This L-function is essentially the Riemann zeta function. More precisely we have*

$$L(s, L_+^*, \psi) = -\frac{2^{2s-1} B_{1,\psi}}{p^s} \zeta(2s - 1).$$

In particular, for an arbitrary natural number m, we have

$$L(1-m, L_+^*, \psi) = \frac{p^{m-1}}{2^{2m} \cdot m} B_{2m} B_{1,\psi}.$$

Before proving this, we prove a lemma. For an imaginary quadratic field K, its discriminant D_K and a positive integer f, we put $d = f^2 D_K$ and

$$\mathcal{P}(d) = \left\{ T = \begin{pmatrix} a & b/2 \\ b/2 & c \end{pmatrix} \in L_+^{*,prim};\ 4ac - b^2 = -d \right\}.$$

Here $L_+^{*,prim}$ is the subset of L_+^* consisting of primitive ones. We also put $\mathcal{S}(d) = \mathcal{P}(d)/\sim$.

Lemma 10.21. *We assume that $d(<0)$ is divisible by p.*

(1) *If $\mathbf{Q}(\sqrt{d}) \neq \mathbf{Q}(\sqrt{-p})$, then $\sum_{T \in \mathcal{S}(d)} \tilde{\psi}(T) = 0$.*

(2) *If $\mathbf{Q}(\sqrt{d}) = \mathbf{Q}(\sqrt{-p})$, then $\tilde{\psi}(T) = 1$ for any $T \in \mathcal{P}(d)$.*

Proof. We first see that $\tilde{\psi}(T)$ is the value of ψ at the norm of the ideal corresponding to T if $\operatorname{rank}(T \bmod p) = 1$. In this case, by definition, there exist integers x, y coprime with each other such that $\tilde{\psi}(T) = \psi(e)$ for $e = ax^2 + bxy + cy^2$. Since $4ae = (2ax + by)^2 + (4ac - b^2)y^2$ and since we assumed that $4ac - b^2$ is divisible by p, we have $\psi(ae) = \psi(2ax + by)^2$. By assumption, we have $p \nmid e$, and replacing by an equivalent one if necessary, we may assume that $p \nmid a$. Hence we have $\psi(2ax + by) \neq 0$ and $\psi(2ax + by)^2 = 1$, so we have $\psi(a) = \psi(e)$. This is the value at the norm. Therefore, since we have $\chi_K = \psi$ if $K = \mathbf{Q}(\sqrt{d}) = \mathbf{Q}(\sqrt{-p})$, we get $\tilde{\psi}(T) = 1$ by Proposition 10.2 and Lemma 10.11. If $K = \mathbf{Q}(\sqrt{d}) \neq \mathbf{Q}(\sqrt{-p})$, then $\chi_K \neq \psi$. So $\tilde{\psi}(T)$ takes -1 on some elements of $\mathcal{S}(d)$ again by Proposition 10.2. That is, $\tilde{\psi}$ is a character of the group $\mathcal{S}(d)$ which is not trivial, so the sum over the group becomes 0. □

Proof of Theorem. First, it is obvious that $\tilde{\psi}(eT) = \psi(e)\tilde{\psi}(T)$ for any integer e prime to p and any $T \in L_+^*$, and since $\epsilon(eT) = \epsilon(T)$, we reduce the L-function to the sum over primitive quadratic forms. Namely if we put

$$L^{prim}(s, L_+^*, \psi) = \sum_{T \in L_+^{*,prim}/\sim} \frac{\tilde{\psi}(T)}{\epsilon(T)\det(T)^s},$$

then we have

$$
\begin{aligned}
L(s, L_+^*, \psi) &= \sum_{e=1}^{\infty} \sum_{T \in L_+^{*,prim}/\sim} \frac{\psi(e)\tilde{\psi}(T)}{\epsilon(eT)\det(eT)^s} \\
&= \sum_{e=1}^{\infty} \frac{\psi(e)}{e^{2s}} L^{prim}(s, L_+^*, \psi) \\
&= L(2s, \psi) L^{prim}(s, L_+^*, \psi).
\end{aligned}
$$

Here $L(2s, \psi)$ is the Dirichlet L-function. Next, if we apply the previous lemma to $L_+^{*,prim}$, only the part for T with $4\det(T) = -d = f^2 p$ (f are positive integers) remains alive. Since $\epsilon(T)$ has a common value for all the elements in $\mathcal{P}(d)$, we denote this by $\epsilon(d)$. Then we have

$$
L^{prim}(s, L_+^*, \psi) = 2^{2s} \sum_{f=1}^{\infty} \frac{|\mathcal{S}(d)|}{p^s f^{2s} \epsilon(d)}.
$$

By virtue of the relation between primitive quadratic forms and proper primitive ideals of an order of a quadratic field, $|\mathcal{S}(d)|$ is the class number $h(f^2 D_K)$ of the order $\mathfrak{O}_f$ of the imaginary quadratic field $K = \mathbf{Q}(\sqrt{-p})$. By the class number formula, we have

$$
L^{prim}(s, L_+^*, \psi) = 2^{2s} \frac{h(D_K)}{|\mathfrak{O}_{max}^\times| p^s} \sum_{f=1}^{\infty} f^{1-2s} \prod_{\substack{q|f \\ q:prime}} \left(1 - \frac{1}{q}\psi(q)\right).
$$

Here by Theorem 10.12, we have

$$
\frac{h(D_K)}{|\mathfrak{O}_{max}^\times| p^s} = -\frac{B_{1,\psi}}{2p^s}.
$$

The above sum is calculated as

$$
\begin{aligned}
\sum_{f=1}^{\infty} f^{1-2s} \prod_{q|f} \left(1 - \frac{1}{q}\psi(q)\right) &= \sum_{f=1}^{\infty} f^{1-2s} \prod_{\substack{q|f \\ q:prime}} \left(1 + \frac{1}{q}\psi(q)\mu(q)\right) \\
&= \sum_{f=1}^{\infty} f^{1-2s} \sum_{m|f} \frac{1}{m}\psi(m)\mu(m).
\end{aligned}
$$

If we put $f = mn$ here, the above sum becomes

$$\sum_{m,n=1}^{\infty} n^{1-2s} m^{-2s} \psi(m)\mu(m) = \zeta(2s-1) \prod_{q:prime} (1-\psi(q)q^{-2s})$$
$$= \zeta(2s-1)L(2s,\psi)^{-1}.$$

Combining all the above considerations together, we get the expression of $L(s, L_+^*, \psi)$ in the theorem. As for special values, it is an immediate consequence of the fact that the values of the Riemann zeta function at integers not more than 0 are given by $\zeta(1-m) = -B_m/m$ (p. 72, Theorem 5.4). □

The above relatively easy considerations gave us a chance to start thinking that zeta functions of prehomogeneous vector spaces consisting of symmetric matrices would be easy objects. As things turned out, we found out that zeta functions for the vector space of symmetric matrices of any degree are described by previously known standard objects such as the Mellin transform of Eisenstein series of half-integral weight and the Riemann zeta functions, and they have a decisive application for dimension formulas of automorphic forms. See [46,48,49] for these developments.

Exercise 10.22. (1) Let D_K be the discriminant of a quadratic field K and p an odd prime such that $\chi_K(p) = \left(\frac{D_K}{p}\right) = 1$. Show that there exists $b \in \mathbf{Z}$ such that $b^2 \equiv D_K \bmod 4p$. Show also that for such b, the module $\mathfrak{p} = \mathbf{Z}p + \mathbf{Z}\frac{b+\sqrt{D_K}}{2}$ is an ideal of $\mathfrak{O}_{max}$ whose norm is p.

(2) Assume $D_K \equiv 1 \bmod 8$ and put $\mathfrak{p} = \mathbf{Z}2 + \mathbf{Z}\frac{1+\sqrt{D_K}}{2}$. Show that this is an ideal of $\mathfrak{O}_{max}$ and $N(\mathfrak{p}) = 2$. Also when $D_K \equiv 4 \bmod 8$ (i.e. $D_K \equiv 12 \bmod 16$), give a prime ideal $\mathfrak{p}$ with $N(\mathfrak{p}) = 2$.

(3) Show that if $\mathfrak{a}$ is an ideal of $\mathfrak{O}_{max}$ and $N(\mathfrak{a})$ is prime to D_K, then $\chi_K(N(\mathfrak{a})) = 1$.

Exercise 10.23. (1) Let K be a quadratic field and $\mathfrak{O}_f$ is the order of conductor f. Let $\mathfrak{a}$ and $\mathfrak{b}$ are proper $\mathfrak{O}_f$ ideals which are prime to f. Show that if $\mathfrak{a} = \mathfrak{b}\alpha$ for some $\alpha \in K^\times$, then $\alpha = \beta/\gamma$ for some $\beta \in \mathfrak{O}_f$ prime to f and $\gamma \in \mathbf{Z}$ prime to f.

(2) We put $K = \mathbf{Q}(\sqrt{3})$, $\mathfrak{O}_2 = \mathbf{Z}+\mathbf{Z}(2\sqrt{3})$ and $\alpha = 4+\sqrt{3}$. Show that, although $N(\alpha) = 13$ is prime to 2, there is no pair $\beta, \gamma \in \mathfrak{O}_2$ prime to 2 such that $\alpha = \beta/\gamma$.

Exercise 10.24. Let $\mathfrak{a}$ be a proper $\mathfrak{O}_f$ ideal. We define the norm $N(\mathfrak{a})$ of $\mathfrak{a}$ by the cardinality of the set $\mathfrak{O}_f/\mathfrak{a}$.

(1) If $\mathfrak{a} = \mathbf{Z}a + \mathbf{Z}\frac{b+\sqrt{D}}{2}$ with $a > 0$, then show that $N(\mathfrak{a}) = a$.
(2) Assume that $\mathfrak{a} = \mathbf{Z}\omega_1 + \mathbf{Z}\omega_2$. Then show that $N(\mathfrak{a}) = |(\omega_1\overline{\omega_2} - \omega_2\overline{\omega_1})/\sqrt{D}|$.
(3) Let $\alpha \in \mathfrak{O}_f$. Then show that $N(\alpha\mathfrak{a}) = |N(\alpha)|N(\mathfrak{a})$.

Exercise 10.25. Let $\mathfrak{a} = \mathbf{Z}a + \mathbf{Z}\frac{b+\sqrt{D}}{2}$ be a proper primitive $\mathfrak{O}_f$ ideal and put

$$\overline{\mathfrak{a}} = \mathbf{Z}a + \frac{-b+\sqrt{D}}{2}\mathbf{Z}.$$

(1) Show that $\bar{\mathfrak{a}}$ is also a proper primitive $\mathfrak{O}_f$ ideal and that

$$\mathfrak{a}\bar{\mathfrak{a}} = a\mathfrak{O}_f.$$

(2) Denote by $Q(x, y)$ a representative of quadratic form corresponding to the narrow ideal class containing $\mathfrak{a}$. Then show that the following conditions are equivalent.

(i) $Q(x, y)$ belongs to an ambig class.
(ii) $\mathfrak{a}^2 = \alpha\mathfrak{O}_f$ with $N(\alpha) > 0$.

Exercise 10.26. Let $\mathfrak{a} = \mathbf{Z}a + \mathbf{Z}\frac{b+\sqrt{D}}{2}$ be a proper primitive $\mathfrak{O}_f$ ideal, where $D = f^2 D_K$ for some fundamental discriminant D_K. Assume that $b \equiv 0 \bmod a$. Show that $\mathfrak{a}^2$ is a principal ideal $a\mathfrak{O}_f$.

Exercise 10.27 (Product of "einig" pair of ideals). Denote by D_K the fundamental discriminant of a quadratic field K. Fix a discriminant $D = f^2 D_K$ and consider two proper primitive $\mathfrak{O}_f$ ideals

$$\mathfrak{a}_i = \mathbf{Z}a_i + \mathbf{Z}\frac{b_i + \sqrt{D}}{2},$$

where $a_i > 0$ $(i = 1, 2)$. As is shown in Chap. 6, there exist integers c_1 and c_2 such that $D = b_1^2 - 4a_1c_1 = b_2^2 - 4a_2c_2$, and we have $gcd(a_1, b_1, c_1) = gcd(a_2, b_2, c_2) = 1$. In this exercise, we will obtain the free basis of the product

$$\mathfrak{a}_1\mathfrak{a}_2 = \mathbf{Z}a_1a_2 + \mathbf{Z}a_1\frac{b_2 + \sqrt{D}}{2} + \mathbf{Z}a_2\frac{b_1 + \sqrt{D}}{2} + \mathbf{Z}\frac{b_1b_2 + D + (b_1 + b_2)\sqrt{D}}{4}.$$

We assume that $gcd(a_1, a_2, (b_1 + b_2)/2) = 1$. (Here we note that $b_1 \equiv b_2 \equiv 2$ since $b_1^2 \equiv D \equiv b_2^2 \bmod 4$.) Such a pair of ideals is called *einig* by Dirichlet (see [29]). We fix integers x, y, z such that $xa_1 + ya_2 + z(b_1 + b_2)/2 = 1$. Define a rational number b_0 by the relation

$$xa_1\frac{b_2 + \sqrt{D}}{2} + ya_2\frac{b_1 + \sqrt{D}}{2} + z\frac{b_1b_2 + D + (b_1 + b_2)\sqrt{D}}{4} = \frac{b_0 + \sqrt{D}}{2}.$$

(1) Show that b_0 is an integer.
(2) We put

$$\alpha = x\frac{b_1 - b_2}{2} + c_1z,$$

$$\beta = y\frac{b_2 - b_1}{2} + c_2z,$$

$$\gamma = -(c_2x + c_1y)$$

Then, show the following equalities.

$$a_2\left(\frac{b_1+\sqrt{D}}{2}\right)=a_2\left(\frac{b_0+\sqrt{D}}{2}\right)+a_1a_2\alpha,$$

$$a_1\left(\frac{b_2+\sqrt{D}}{2}\right)=a_1\left(\frac{b_0+\sqrt{D}}{2}\right)+a_1a_2\beta,$$

$$\frac{b_1b_2+D+(b_1+b_2)\sqrt{D}}{4}=\left(\frac{b_1+b_2}{2}\right)\frac{b_0+\sqrt{D}}{2}+a_1a_2\gamma.$$

(3) Show that

$$\mathfrak{a}_1\mathfrak{a}_2=\mathbf{Z}a_1a_2+\mathbf{Z}\frac{b_0+\sqrt{D}}{2}$$

and $N(\mathfrak{a}_1\mathfrak{a}_2)=a_1a_2$.

(4) If we put $c_0=\dfrac{b_0^2-D}{4a_1a_2}$, then this should be an integer. Show that

$$c_0=(c_1z+b_1x)(c_2z+b_2y)+(a_1x+a_2y)(c_2x+c_1y)-\left(\frac{b_1+b_2}{2}\right)^2xy$$
$$=\alpha\beta-\gamma.$$

(Note, e.g. that $b_1^2-b_2^2=(D+4a_1c_1)-(D+4a_2c_2)=4(a_1c_1-a_2c_2)$.)

(5) For variables x_1, x_2, y_1, y_2, put

$$X=(x_1+y_1\alpha)(x_2+y_2\beta)-c_0y_1y_2=x_1x_2+\alpha x_2y_1+\beta x_1y_2+\gamma y_1y_2,$$

$$Y=a_1x_1y_2+a_2x_2y_1+\frac{b_1+b_2}{2}y_1y_2.$$

Then show that

$$(a_1x_1^2+b_1x_1y_1+c_1y_1^2)(a_2x_2^2+b_2x_2y_2+c_2y_2^2)=a_1a_2X^2+b_0XY+c_0Y^2.$$

Exercise 10.28 (General case). Fix a discriminant $D=f^2D_K$ and take two proper primitive $\mathfrak{O}_f$ ideals $\mathfrak{a}_1$ and $\mathfrak{a}_2$ as in the last exercise. This time, we assume that

$$gcd\left(a_1,a_2,\frac{b_1+b_2}{2}\right)=l,$$

where l might be greater than 1. Take integers x, y, z such that

$$xa_1+ya_2+z\frac{b_1+b_2}{2}=l.$$

We write $a_1 = la'_1$, $a_2 = la'_2$ and $\frac{b_1+b_2}{2} = lB$.

(1) Put

$$b_0 = xa_1b_2 + ya_2b_1 + z\frac{b_1b_2 + D}{2}$$

Show directly that b_0 is an integer and divisible by l.

(2) Put

$$\alpha = x\frac{b_1 - b_2}{2} + c_1z,$$
$$\beta = y\frac{b_2 - b_1}{2} + c_2z,$$
$$\gamma = -(xc_2 + yc_1)$$

Show the following equalities.

$$a_2\left(\frac{b_1 + \sqrt{D}}{2}\right) = a'_2\left(\frac{b_0 + l\sqrt{D}}{2}\right) + la'_1a'_2\alpha,$$
$$a_1\left(\frac{b_2 + \sqrt{D}}{2}\right) = a'_1\left(\frac{b_0 + l\sqrt{D}}{2}\right) + la'_1a'_2\beta,$$
$$\left(\frac{b_1 + \sqrt{D}}{2}\right)\left(\frac{b_2 + \sqrt{D}}{2}\right) = B\left(\frac{b_0 + l\sqrt{D}}{2}\right) + la'_1a'_2\gamma.$$

(3) We put $m = gcd(l, \alpha, \beta, \gamma)$. Show that $m = 1$ by the following steps.

(i)

$$\frac{b_1 - b_2}{2} = a'_1\alpha - a'_2\beta \in m\mathbf{Z}.$$

Hence $c_1z \in m\mathbf{Z}$.

(ii)

$$B(c_1z) - a'_2\gamma = c_1 + \frac{b_2 - b_1}{2}Bx.$$

Hence $c_1 \in m\mathbf{Z}$.

(iii) We have $\frac{b_1+b_2}{2} = lB \in m\mathbf{Z}$ and hence $b_1 \in m\mathbf{Z}$. Also we have $a_1 = la'_1 \in m\mathbf{Z}$.

(iv) $m = 1$ since $gcd(a_1, b_1, c_1) = 1$.

(4) Show that

$$\mathfrak{a}_1\mathfrak{a}_2 = \mathbf{Z}\, la_1'a_2' + \mathbf{Z}\frac{b_0 + l\sqrt{D}}{2} = l\left(\mathbf{Z}a_1'a_2' + \mathbf{Z}\frac{b_0/l + \sqrt{D}}{2}\right),$$

where b_0/l is an integer.

(5) Show that

$$N(\mathfrak{a}_1\mathfrak{a}_2) = N(\mathfrak{a}_1)N(\mathfrak{a}_2).$$

(6) Find integers a_3, b_3, c_3, X and Y similarly as in the last exercise such that

$$(a_1x_1^2 + b_1x_1y_2 + c_1y_1^2)(a_2x_2^2 + b_2x_2y_2 + c_2y_2^2) = a_3X^2 + b_3XY + c_3Y^2.$$

(Hint: Write down the product of general elements of $\mathfrak{a}_1$ and $\mathfrak{a}_2$ by the basis of $\mathfrak{a}_1\mathfrak{a}_2$ and take the norms. For example, put $a_3 = a_1'a_2'$, $b_3 = b_0/l$, $c_3 = \dfrac{b_3^2 - D}{4a_1'a_2'}$, and $X = lx_1x_2 + \alpha x_2y_1 + \beta x_1y_2 + \gamma y_1y_2$, $Y = a_1'x_2y_1 + a_1'x_1y_2 + By_1y_2$.)

Exercise 10.29. (1) We fix $D = f^2D_K$. Show that for a natural number m and a proper $\mathfrak{O}_f$ ideal $\mathfrak{a}$, there exists a proper $\mathfrak{O}_f$ ideal $\mathfrak{b}$ which is equivalent to $\mathfrak{a}$ and whose norm $N(\mathfrak{b})$ is prime to m.

(2) For two proper $\mathfrak{O}_f$ ideals $\mathfrak{a}_1$, $\mathfrak{a}_2$, show that $N(\mathfrak{a}_1\mathfrak{a}_2) = N(\mathfrak{a}_1)N(\mathfrak{a}_2)$ by using Exercise 10.27 but not using Exercise 10.28.

Exercise 10.30. (1) For $K = \mathbf{Q}(\sqrt{5})$, let $\mathfrak{O}_f$ be the order of K of conductor f. By using Proposition 10.18, count the number of proper $\mathfrak{O}_f$ ideals of norm 4 and 16 for $f = 1, 2, 3$. Also, give them all explicitly.

(2) For $K = \mathbf{Q}(\sqrt{11})$ and the order $\mathfrak{O}_3$ of K of conductor 3, give all the proper ideals of norm 9 explicitly.

Exercise 10.31. For $K = \mathbf{Q}(\sqrt{7})$, we take the order $\mathfrak{O}_3$ of conductor 3. Put $L = 3\mathbf{Z} + \sqrt{7}\mathbf{Z}$. Show that L is a proper $\mathfrak{O}_3$ lattice, though it is not an integral ideal of $\mathfrak{O}_3$ (i.e. not contained in $\mathfrak{O}_3$).

Exercise 10.32. For $K = \mathbf{Q}(\sqrt{82})$, let $\mathfrak{O}_{max} = \mathbf{Z} + \mathbf{Z}\sqrt{82}$ be the maximal order and $\mathfrak{O}_2 = \mathbf{Z} + 2\sqrt{82}\mathbf{Z}$ the order of conductor 2.

(1) Show that $\mathfrak{a} = 2\mathbf{Z} + \sqrt{82}\mathbf{Z}$ is an ideal of $\mathfrak{O}_{max}$.

(2) Show that $\mathfrak{a}$ is not prime to 2 and that $\mathfrak{a} \cap \mathfrak{O}_2$ is not a proper ideal of $\mathfrak{O}_2$.

Exercise 10.33. Assume that $\mathfrak{a}$ is a proper $\mathfrak{O}_f$ ideal. We write $\mathfrak{a}$ by the standard basis as $\mathfrak{a} = l(\mathbf{Z}a + \mathbf{Z}(d + f\omega))$ with $N(d + f\omega) = ac$. We assume that $\mathfrak{a} + f\mathfrak{O}_{max} = \mathfrak{O}_f$. Prove that $\mathfrak{a} + f\mathfrak{O}_f = \mathfrak{O}_f$ by the following steps.

(1) Two numbers l and f are coprime.
(Hint: If there exists a prime p such that $p|l$ and $p|f$, then $\mathfrak{a} + f\mathfrak{O}_{max} \subset p\mathfrak{O}_{max}$, which is a contradiction.)

(2) Two numbers a and f are coprime.
(Hint: We have $ac = N(d + f\omega) = d^2 + dfTr(\omega) + f^2N(\omega)$. If there is a prime p such that $p|a$ and $p|f$, then $p|d$, and $\mathfrak{a} \subset p\mathfrak{O}_{max}$, which is again a contradiction.)

(3) We have $\mathfrak{a} + f\mathfrak{O}_f = \mathfrak{O}_f$.
(Hint: By (1) and (2), two numbers al and f are coprime and we have $x, y \in \mathbf{Z}$ such that $xal + yf = 1$.)

Chapter 11
p-adic Measure and Kummer's Congruence

In modern number theory, the p-adic method or p-adic way of thinking plays an important role. As an example, there are objects called p-adic L-functions which correspond to the Dirichlet L-functions, and in fact the natural setup to understand the Kummer congruence described in Sect. 3.2 is in the context of the p-adic L-functions. To be precise, a modified version (by a suitable "Euler factor") of Kummer's congruence guarantees the existence of the p-adic L-function.

To discuss this aspect fully is beyond the scope of this book, but in this chapter we explain the p-adic integral expression of the Bernoulli number and prove Kummer's congruence using it. Interested readers are advised to read books such as Iwasawa [51], Washington [100], Lang [66].

We assume the basics of p-adic numbers. For this we refer readers to Serre [83, Ch. 1] or Gouvea [37]. The results in this chapter are not used in other chapters.

11.1 Measure on the Ring of p-adic Integers and the Ring of Formal Power Series

In this section we review the general correspondence between measures on the ring of p-adic integers $\mathbf{Z}_p$ and the ring of formal power series. We use this setup in the next section to define the Bernoulli measure on $\mathbf{Z}_p$ and to express Bernoulli numbers as integrals. This expression turns out to be very useful in proving Kummer's congruence relation.

Let $\overline{\mathbf{Q}}_p$ be the algebraic closure of the field $\mathbf{Q}_p$ of p-adic numbers. The p-adic absolute value $|\ \ |$ of $\mathbf{Q}_p$ (normalized by $|p| = 1/p$) is extended uniquely to $\overline{\mathbf{Q}}_p$. We use the same notation $|\ \ |$ for this extension. Then $\overline{\mathbf{Q}}_p$ is not complete with respect to this absolute value, and the completion is denoted by $\mathbf{C}_p$. The absolute value $|\ \ |$ also extends naturally to $\mathbf{C}_p$. Let $\mathcal{O}_p$ be the ring of integers of $\mathbf{C}_p$:

$$\mathcal{O}_p = \{x \in \mathbf{C}_p \mid |x| \leq 1\}.$$

T. Arakawa et al., *Bernoulli Numbers and Zeta Functions*, Springer Monographs in Mathematics, DOI 10.1007/978-4-431-54919-2_11, © Springer Japan 2014

Remark 11.1. Like the complex number field $\mathbf{C}$, the field $\mathbf{C}_p$ is complete and algebraically closed. To do analysis in the p-adic setting, we need this big field.

First we review the general theory of measures on $\mathbf{Z}_p$.

Denote the $\mathbf{Z}$-module $\mathbf{Z}/p^n\mathbf{Z}$ by X_n and the canonical map from X_{n+1} to X_n by π_{n+1}, so $\pi_{n+1} : X_{n+1} \to X_n$ is defined by

$$x \bmod p^{n+1}\mathbf{Z} \longmapsto x \bmod p^n\mathbf{Z}.$$

The system of pairs (X_n, π_n) gives a projective system and we have the projective limit $\varprojlim X_n$:

$$\varprojlim X_n = \left\{(x_n) \in \prod_{n\geq 1} X_n \mid \pi_{n+1}(x_{n+1}) = x_n\right\}.$$

The ring of p-adic integers $\mathbf{Z}_p$ is identified with this projective limit $\varprojlim X_n$.

Definition 11.2 (Measure on $\mathbf{Z}_p$). A set of functions $\mu = \{\mu_n\}_{n=1}^{\infty}$ is called an $\mathcal{O}_p$-valued measure on $\mathbf{Z}_p$ if the following two conditions are satisfied:

(i) Each μ_n is an $\mathcal{O}_p$-valued function on X_n, $\mu_n : X_n \longrightarrow \mathcal{O}_p$.
(ii) For any $n \in \mathbf{N}$ and $x \in X_n$, the distribution property

$$\mu_n(x) = \sum_{\substack{y\in X_{n+1}\\ \pi_{n+1}(y)=x}} \mu_{n+1}(y)$$

holds.

The set of $\mathcal{O}_p$-valued measures on $\mathbf{Z}_p$ is denoted by $\mathcal{M}(\mathbf{Z}_p, \mathcal{O}_p)$. This has an $\mathcal{O}_p$-module structure. Further, the norm of $\mu = \{\mu_n\} \in \mathcal{M}(\mathbf{Z}_p, \mathcal{O}_p)$ is defined as

$$\|\mu\| = \sup_{n\in\mathbf{N},\, x\in X_n} |\mu_n(x)|.$$

Also, the $\mathcal{O}_p$-module of continuous $\mathcal{O}_p$-valued functions on $\mathbf{Z}_p$ is denoted by $C(\mathbf{Z}_p, \mathcal{O}_p)$, and the norm $\|\varphi\|$ of an element $\varphi \in C(\mathbf{Z}_p, \mathcal{O}_p)$ is defined by

$$\|\varphi\| = \sup_{x\in\mathbf{Z}_p} |\varphi(x)|.$$

For $\varphi \in C(\mathbf{Z}_p, \mathcal{O}_p)$ and $\mu = \{\mu_n\} \in \mathcal{M}(\mathbf{Z}_p, \mathcal{O}_p)$, the integral on $\mathbf{Z}_p$ is defined by

$$\int_{\mathbf{Z}_p} \varphi(x)d\mu(x) = \lim_{n\to\infty} \sum_{r=0}^{p^n-1} \varphi(r)\mu_n(r).$$

(We use the abbreviated notation $\mu_n(r)$ for $\mu_n(r \bmod p^n)$. A similar abbreviation will be used in the following.) The convergence of the limit on the right-hand side is guaranteed by the following estimate: when $n < m$, we have

$$\begin{aligned}
&\Big|\sum_{r=0}^{p^n-1} \varphi(r)\mu_n(r) - \sum_{l=0}^{p^m-1} \varphi(l)\mu_m(l)\Big| \\
&= \Big|\sum_{r=0}^{p^n-1}\Big(\varphi(r)\mu_n(r) - \sum_{q=0}^{p^{m-n}-1} \varphi(r+p^nq)\mu_m(r+p^nq)\Big)\Big| \\
&= \Big|\sum_{r=0}^{p^n-1}\Bigg(\sum_{q=0}^{p^{m-n}-1} (\varphi(r)-\varphi(r+p^nq))\,\mu_m(r+p^nq)\Bigg)\Big| \\
&\le \max_{r,q} |\varphi(r)-\varphi(r+p^nq)|\,\|\mu\|.
\end{aligned}$$

For each natural number k, the binomial polynomial

$$\binom{t}{k} = \frac{t(t-1)\cdots(t-k+1)}{k!}$$

in t is a continuous function on $\mathbf{Z}_p$.

To $\mu = \{\mu_n\} \in \mathcal{M}(\mathbf{Z}_p, \mathcal{O}_p)$ we associate $f \in \mathcal{O}_p[[X]]$ in the following manner. Set $\Lambda = \mathcal{O}_p[[X]]$, $\Lambda_n = ((1+X)^{p^n}-1)\Lambda$ and consider the projective system $\{(\Lambda/\Lambda_n, \varpi_n)\}$ by the natural map $\varpi_n : \Lambda/\Lambda_n \longrightarrow \Lambda/\Lambda_{n-1}$. Define $f_n(X) \in \Lambda/\Lambda_n$ by

$$f_n(X) = \sum_{r=0}^{p^n-1} \mu_n(r)(1+X)^r = \sum_{r=0}^{p^n-1}\sum_{k=0}^{r} \mu_n(r)\binom{r}{k}X^k = \sum_{k=0}^{p^n-1} c_{n,k}X^k.$$

Here we understand that the equalities are mod Λ_n and put

$$c_{n,k} = \sum_{r=0}^{p^n-1} \mu_n(r)\binom{r}{k}.$$

Since we have

$$(\varpi_n f_n)(X) = \varpi_n\left(\sum_{r=0}^{p^n-1} \mu_n(r)(1+X)^r\right)$$

$$= \varpi_n \left(\sum_{r'=0}^{p^{n-1}-1} \sum_{l=0}^{p-1} \mu_n(r' + p^{n-1}l)(1+X)^{r'}(1+X)^{p^{n-1}l} \right)$$

$$= \sum_{r'=0}^{p^{n-1}-1} \mu_{n-1}(r')(1+X)^{r'}$$

$$= f_{n-1}(X),$$

the system (f_n) is an element in the projective limit $\varprojlim \Lambda/\Lambda_n$. Now we have the isomorphism

$$\Lambda \cong \varprojlim \Lambda/\Lambda_n, \quad \Lambda \ni g \longmapsto (g_n) \in \varprojlim \Lambda/\Lambda_n,$$

where, for $g \in \Lambda$, the system (g_n) is given by $g_n = g \bmod \Lambda_n$. Through this isomorphism, the above $\{f_n\}$ corresponds to $f \in \Lambda$ by

$$f(X) = \sum_{m=0}^{\infty} c_m X^m,$$

where

$$c_m = \lim_{n\to\infty} \sum_{r=0}^{p^n-1} \mu_n(r) \binom{r}{m}$$

$$= \int_{\mathbf{Z}_p} \binom{x}{m} d\mu(x).$$

We therefore have obtained a map from $\mathcal{M}(\mathbf{Z}_p, \mathcal{O}_p)$ to $\mathcal{O}_p[[X]]$. An important fact is that this map gives a natural *isomorphism* between $\mathcal{M}(\mathbf{Z}_p, \mathcal{O}_p)$ and the ring of formal power series $\mathcal{O}_p[[X]]$, often referred to as the Iwasawa isomorphism. The way to associate a measure to an element in $\mathcal{O}_p[[X]]$ is described as follows.

For $f = \sum_{m=0}^{\infty} c_m X^m \in \mathcal{O}_p[[X]]$, define $\mu = \{\mu_n\}$ by

$$\mu_n(r) = \frac{1}{p^n} \sum_{\zeta^{p^n}=1} \zeta^{-r} f(\zeta - 1) \qquad (r \in X_n), \tag{11.1}$$

the sum running over all p^n-th roots ζ of 1. Since $|\zeta - 1| < 1$, $f(\zeta - 1)$ converges. For each $m \geq 0$, we have

$$\frac{1}{p^n} \sum_{\zeta^{p^n}=1} \zeta^{-r}(\zeta - 1)^m = \frac{1}{p^n} \sum_{\zeta^{p^n}=1} \sum_{j=0}^{m} \zeta^{-r} \binom{m}{j} (-1)^{m-j} \zeta^j$$

$$= \sum_{\substack{0 \leq j \leq m \\ j \equiv r \bmod p^n}} \binom{m}{j} (-1)^{m-j}.$$

So this is contained in $\mathcal{O}_p$. In particular, if $p^n > r > m$, then this is zero. When ζ is a primitive p^ν-th root of 1 ($\nu \geq 1$), the equality

$$|\zeta - 1|^{\varphi(p^\nu)} = |p| \qquad (\varphi \text{ is the Euler function})$$

holds and hence

$$|(\zeta - 1)^m| = |p^{m/\varphi(p^\nu)}|.$$

From this, we conclude that p^e divides the quantity

$$\sum_{\substack{0 \leq j \leq m \\ j \equiv r \bmod p^n}} \binom{m}{j} (-1)^{m-j}$$

for $e = m/\phi(p^n) - n$. Therefore,

$$\mu_n(r) = \sum_{m=0}^{\infty} c_m \left(\frac{1}{p^n} \sum_{\zeta^{p^n}=1} \zeta^{-r} (\zeta - 1)^m \right)$$

is convergent and the value is in $\mathcal{O}_p$. To check the distribution property (ii) of the measure, we need to calculate the following value:

$$\sum_{y \in X_{n+1},\ \pi_{n+1}(y)=x} \mu_{n+1}(y) = \sum_{a \bmod p} \mu_{n+1}(x + p^n a)$$

$$= \frac{1}{p^{n+1}} \sum_{\zeta^{p^{n+1}}=1} \left(\sum_{a \bmod p} \zeta^{-(x+p^n a)} \right) f(\zeta - 1).$$

Using the identity

$$\sum_{a \bmod p} \zeta^{-p^n a} = \begin{cases} 0 & \text{if } \zeta^{p^n} \neq 1, \\ p & \text{if } \zeta^{p^n} = 1 \end{cases}$$

for a p^{n+1}-th root ζ of 1, we have

$$\sum_{\substack{a \bmod p \\ \zeta^{p^{n+1}}=1}} \zeta^{-x-p^n a} = p \sum_{\zeta^{p^n}=1} \zeta^{-x},$$

so we have

$$\sum_{\substack{y \in X_{n+1} \\ \pi_{n+1}(y)=x}} \mu_{n+1}(y) = \mu_n(x)$$

which is to be proved. If we define the formal power series $\tilde{f} \in \mathcal{O}_p[[X]]$ corresponding to this measure defined as before, then the coefficients c'_k of X^k of this series are given by

$$c'_k = \lim_{n\to\infty} \sum_{r=0}^{p^n-1} \mu_n(r)\binom{r}{k}$$

$$= \lim_{n\to\infty} \sum_{m=0}^{\infty} c_m \sum_{r=0}^{p^n-1} \binom{r}{k} \sum_{\substack{0\le j\le m \\ j\equiv r \bmod p^n}} \binom{m}{j}(-1)^{m-j}.$$

We fix k. To calculate the coefficient of c_m in the expression of c'_k in the right-hand side above, we fix m. We have $\binom{r}{k} = 0$ for $k > r$ so we may assume that $k \le r$. Taking n big enough, we assume that $m < p^n$. Then, if $j \equiv r \bmod p^n$ for some j with $0 \le j \le m$, we have $j = r$ since we also have $0 \le r \le p^n - 1$ by definition. So we may assume that $k \le r = j \le m$. So the coefficient of c_m is given by

$$\sum_{r=k}^{m} \binom{r}{k}\binom{m}{r}(-1)^{m-r} = \sum_{i=0}^{m-k} \binom{m-k}{i}\binom{m}{k}(-1)^{m-k-i} = \begin{cases} 1 & \text{if } m = k, \\ 0 & \text{if } m \ne k. \end{cases}$$

Hence we have $c'_k = c_k$. So we have $\tilde{f} = f$ and two mappings are inverse with each other and we see that the set $\mathcal{M}(\mathbf{Z}_p, \mathcal{O}_p)$ of $\mathcal{O}_p$-valued measures and the space of formal power series $\mathcal{O}_p[[X]]$ are bijective.

More precisely, we can introduce a product for both spaces and show that these are isomorphic as $\mathcal{O}_p$ algebras, as given in the following theorem whose complete proof is omitted (see e.g. Lang [66, Ch.4]).

For two measures $\mu,\ \nu \in \mathcal{M}(\mathbf{Z}_p, \mathcal{O}_p)$, we define an $\mathcal{O}_p$-valued function $(\mu * \nu)_n$ on X_n by

$$(\mu * \nu)_n(x) = \sum_{y=0}^{p^n-1} \mu_n(y)\nu_n(x-y) \qquad (x \in X_n). \tag{11.2}$$

Then $\mu * \nu = \{(\mu * \nu)_n\}$ becomes an element of $\mathcal{M}(\mathbf{Z}_p, \mathcal{O}_p)$. We call this a convolution product of μ and ν. The set $\mathcal{M}(\mathbf{Z}_p, \mathcal{O}_p)$ becomes an $\mathcal{O}_p$ algebra by this product $\mu * \nu$.

Theorem 11.3 (Iwasawa isomorphism). *Between the space $\mathcal{M}(\mathbf{Z}_p, \mathcal{O}_p)$ of $\mathcal{O}_p$-valued measures and the ring of formal power series $\mathcal{O}_p[[X]]$, there is an $\mathcal{O}_p$ algebra isomorphism $P : \mathcal{M}(\mathbf{Z}_p, \mathcal{O}_p) \longrightarrow \mathcal{O}_p[[X]]$ given by*

$$\mathcal{M}(\mathbf{Z}_p, \mathcal{O}_p) \ni \mu = \{\mu_n\} \overset{P}{\longmapsto} f(X) = \sum_{m=0}^{\infty} c_m X^m \in \mathcal{O}_p[[X]].$$

Here, c_m is determined by μ:

$$c_m = \int_{\mathbf{Z}_p} \binom{x}{m} d\mu(x),$$

and conversely μ_n is determined by f:

$$\mu_n(x) = \frac{1}{p^n} \sum_{\zeta^{p^n}=1} \zeta^{-x} f(\zeta - 1).$$

For convenience of the description below, we recall Mahler's[1] theorem giving the necessary and sufficient condition for an $\mathcal{O}_p$-valued function on $\mathbf{Z}_p$ to be continuous.

Theorem 11.4. *The function $\varphi : \mathbf{Z}_p \longrightarrow \mathcal{O}_p$ is continuous if and only if it can be written as*

$$\varphi(x) = \sum_{n=0}^{\infty} a_n \binom{x}{n}, \qquad a_n \in \mathcal{O}_p, \quad |a_n| \longrightarrow 0.$$

If this is the case, the coefficients a_n are uniquely determined by φ and given by

$$a_n = \sum_{k=0}^{n} (-1)^{n-k} \binom{n}{k} \varphi(k).$$

We omit the proof (cf. Lang [66, §4.1]).

If we use Theorem 11.4, we can understand a part of Theorem 11.3 more intuitively as follows. Fix $x_0 \in \mathbf{Z}$. Denote by φ the characteristic polynomial of $x_0 + p^n\mathbf{Z}_p$. Then by the definition of the p-adic measure, we see easily that

$$\int_{\mathbf{Z}_p} \varphi(x) d\mu(x) = \mu_n(x_0).$$

So if we replace $\varphi(x)$ by the expansion $\varphi(x) = \sum_{m=0}^{\infty} a_m \binom{x}{m}$ in Theorem 11.4, we have

$$\mu_n(x_0) = \sum_{m=0}^{\infty} a_m c_m = \sum_{m=0}^{\infty} c_m \sum_{k=0}^{m} (-1)^{m-k} \binom{m}{k} \varphi(k).$$

[1]Kurt Mahler (born on July 26, 1903 in Krefeld, Prussian Rhineland—died on February 25, 1988 in Canberra, Australia).

Now, for any ζ with $\zeta^{p^n} = 1$, we have

$$\sum_{m=0}^{\infty} c_m(\zeta - 1)^m = \sum_{m=0}^{\infty} c_m \sum_{k=0}^{m} \binom{m}{k} (-1)^{m-k} \zeta^k.$$

Since $\varphi(k) = 1$ if $k \equiv x_0 \bmod p^n$ and $\varphi(k) = 0$ otherwise, we have

$$\frac{1}{p^n} \sum_{\zeta^{p^n}=1} \zeta^{-x_0} \sum_{m=0}^{\infty} c_m(\zeta - 1)^m = \sum_{m=0}^{\infty} c_m \sum_{k=0}^{m} \binom{m}{k} (-1)^{m-k} \varphi(k).$$

So we get the expression of $\mu(x)$ by f in Theorem 11.3.

We describe here several useful properties of the correspondence P in Theorem 11.3 between measures and formal power series. Let the maximal ideal of $\mathcal{O}_p$ be

$$\mathcal{P} = \{z \in \mathcal{O}_p \mid |z| < 1\}.$$

For $z \in \mathcal{P}$, define the function $(1+z)^x$ in x by

$$(1+z)^x := \sum_{n=0}^{\infty} \binom{x}{n} z^n.$$

By Mahler's theorem, $(1+z)^x$ is a continuous function of $x \in \mathbf{Z}_p$. When x is a non-negative integer, this definition of $(1+z)^x$ coincides with the usual binomial expansion. We have the relation

$$(1+z)^x (1+z)^{x'} = (1+z)^{x+x'} \qquad (x,\ x' \in \mathbf{Z}_p). \tag{11.3}$$

This is obvious for $x,\ x' \in \mathbf{N}$, and the general case for $x,\ x' \in \mathbf{Z}_p$ follows from the fact that the set $\mathbf{N}$ of natural numbers is dense in $\mathbf{Z}_p$.

In the following, we list several properties of measures and corresponding power series, which will be used later.

Property (1). Let $z \in \mathcal{P}$. If μ corresponds to f (i.e. $P\mu = f$), then

$$f(z) = \int_{\mathbf{Z}_p} (1+z)^x \, d\mu(x).$$

In particular, by putting $z = 0$,

$$f(0) = \int_{\mathbf{Z}_p} d\mu(x).$$

Proof. Writing $f(X) = \sum_{n=0}^{\infty} c_n X^n$, we have by Theorem 11.3

$$\int_{\mathbf{Z}_p} (1+z)^x \, d\mu(x) = \int_{\mathbf{Z}_p} \sum_{n=0}^{\infty} \binom{x}{n} z^n \, d\mu(x)$$
$$= \sum_{n=0}^{\infty} z^n \int_{\mathbf{Z}_p} \binom{x}{n} d\mu(x)$$
$$= \sum_{n=0}^{\infty} c_n z^n = f(z).$$

□

We call the map λ from $C(\mathbf{Z}_p, \mathcal{O}_p)$ to $\mathcal{O}_p$ a bounded linear functional on $C(\mathbf{Z}_p, \mathcal{O}_p)$ if the following conditions (i), (ii) are satisfied:

(i) For any $\varphi,\ \varphi' \in C(\mathbf{Z}_p, \mathcal{O}_p)$ and any $a,\ b \in \mathcal{O}_p$,

$$\lambda(a\varphi + b\varphi') = a\lambda(\varphi) + b\lambda(\varphi').$$

(ii) There exists a positive constant $M > 0$ such that for any $\varphi \in C(\mathbf{Z}_p, \mathcal{O}_p)$,

$$|\lambda(\varphi)| \leq M \|\varphi\|.$$

The norm of λ is defined by

$$\|\lambda\| = \sup_{\substack{\varphi \in C(\mathbf{Z}_p, \mathcal{O}_p) \\ \varphi \neq 0}} \frac{|\lambda(\varphi)|}{\|\varphi\|}.$$

Let λ be a bounded linear functional on $C(\mathbf{Z}_p, \mathcal{O}_p)$. For $x \in X_n = \mathbf{Z}/p^n\mathbf{Z}$, write the characteristic function of $x + p^n\mathbf{Z}_p$ as $\varphi_{x,n}$. If we put

$$\mu_n(x) = \lambda(\varphi_{x,n}),$$

then $\mu = \{\mu_n\}$ is an $\mathcal{O}_p$-valued measure on $\mathbf{Z}_p$ (i.e. $\mu \in \mathcal{M}(\mathbf{Z}_p, \mathcal{O}_p)$). Conversely, given $\mu = \{\mu_n\} \in \mathcal{M}(\mathbf{Z}_p, \mathcal{O}_p)$, if we put

$$\lambda(\varphi) = \int_{\mathbf{Z}_p} \varphi(x) \, d\mu(x),$$

then λ is a bounded linear functional on $C(\mathbf{Z}_p, \mathcal{O}_p)$. This correspondence between λ and μ is easily seen to be one to one.

Moreover, for $h \in C(\mathbf{Z}_p, \mathcal{O}_p)$ and $\mu \in \mathcal{M}(\mathbf{Z}_p, \mathcal{O}_p)$, the map

$$\varphi \longmapsto \int_{\mathbf{Z}_p} \varphi(x)h(x)\,d\mu(x), \qquad (\varphi \in C(\mathbf{Z}_p, \mathcal{O}_p))$$

is a bounded linear functional on $C(\mathbf{Z}_p, \mathcal{O}_p)$. Let $h\mu$ be the corresponding measure. It is an interesting problem to compute the formal power series corresponding to the measure $h\mu$ when μ corresponds to $f = P\mu \in \mathcal{O}_p[[X]]$. Properties (2) and (3) below give examples of this correspondence.

For $f \in \mathcal{O}_p[[X]]$, put

$$(\mathbb{U}f)(X) = f(X) - \frac{1}{p}\sum_{\zeta^p=1} f(\zeta(1+X)-1). \tag{11.4}$$

Since

$$\frac{1}{p}\sum_{\zeta^p=1} (\zeta(1+X)-1))^l \in \mathbf{Z}_p[X]$$

for non-negative integers l, we have $\mathbb{U}f \in \mathcal{O}_p[[X]]$.

Property (2). Let $f \in \mathcal{O}_p[[X]]$ and μ_f be the corresponding measure. Also, let ψ be the characteristic function of $\mathbf{Z}_p^\times$. Then the formal power series corresponding to the measure $\psi\mu_f$ is $\mathbb{U}f$, i.e., $\psi\mu_f = \mu_{\mathbb{U}f}$. More precisely, we have for any $\varphi \in C(\mathbf{Z}_p, \mathcal{O}_p)$

$$\int_{\mathbf{Z}_p} \varphi(x)\psi(x)\,d\mu_f(x) = \int_{\mathbf{Z}_p} \varphi(x)\,d\mu_{\mathbb{U}f}(x).$$

This can also be written as

$$\int_{\mathbf{Z}_p^\times} \varphi(x)\,d\mu_f(x) = \int_{\mathbf{Z}_p} \varphi(x)\,d\mu_{\mathbb{U}f}(x).$$

Proof. Write the power series corresponding to the measure $\psi\mu_f$ as g. When $z \in \mathcal{P}$, by Property (1) we have

$$g(z) = \int_{\mathbf{Z}_p} (1+z)^x \psi(x)\,d\mu_f(x).$$

Let ζ be a pth root of 1. Regarding ψ also as a function on $\mathbf{Z}/p\mathbf{Z}$ via $\psi(a \bmod p) = \psi(a + p\mathbf{Z}_p)$, and putting

$$\hat{\psi}(\zeta) = \frac{1}{p} \sum_{a \in \mathbf{Z}/p\mathbf{Z}} \psi(a)\zeta^{-a},$$

(Fourier transform on $\mathbf{Z}/p\mathbf{Z}$) we have

$$\psi(a) = \sum_{\zeta^p=1} \hat{\psi}(\zeta)\zeta^a$$

by a simple calculation (inverse Fourier transform). Since

$$\hat{\psi}(\zeta) = \begin{cases} -\frac{1}{p} & \text{if } \zeta \neq 1, \\ \frac{p-1}{p} & \text{if } \zeta = 1, \end{cases}$$

by the definition of ψ, we obtain

$$\begin{aligned} g(z) &= \int_{\mathbf{Z}_p} (1+z)^x \psi(x)\, d\mu_f(x) \\ &= \int_{\mathbf{Z}_p} (1+z)^x \sum_{\zeta^p=1} \hat{\psi}(\zeta)\zeta^x\, d\mu_f(x) \\ &= \sum_{\zeta^p=1} \hat{\psi}(\zeta) \int_{\mathbf{Z}_p} (1+z)^x \zeta^x\, d\mu_f(x) \\ &= \sum_{\zeta^p=1} \hat{\psi}(\zeta) \int_{\mathbf{Z}_p} \Big(1 + (\zeta(1+z)-1)\Big)^x d\mu_f(x) \\ &= \sum_{\zeta^p=1} \hat{\psi}(\zeta) f(\zeta(1-z)-1) = f(z) - \frac{1}{p} \sum_{\zeta^p=1} f(\zeta(1+z)-1). \end{aligned}$$

This shows $g = \mathbb{U}f$. (Here we define the power ζ^x for $x \in \mathbf{Z}_p$ by

$$\zeta^x = (1+\zeta-1)^x = \sum_{n=0}^{\infty} \binom{x}{n} (\zeta-1)^n.$$

If we choose $a \in \mathbf{Z}$ so that $x - a \in p\mathbf{Z}_p$, we have $\zeta^x = \zeta^a$.) □

Define the differential operator D on the ring of formal power series $\mathcal{O}_p[[X]]$ by

$$D = (1+X)D_X, \qquad \text{where} \quad D_X = \frac{d}{dX}.$$

Property (3). For $f \in \mathcal{O}_p[[X]]$, the power series corresponding to the measure $x\mu_f$ is Df. Hence the power series corresponding to the measure $x^k\mu_f$ (k natural number) is $D^k f$ and the equalities

$$\int_{\mathbf{Z}_p} x^k \, d\mu_f(x) = \int_{\mathbf{Z}_p} d\mu_{D^k f}(x) = (D^k f)(0)$$

hold.

Proof. It is enough to show this when $k = 1$. Let $g \in \mathcal{O}_p[[X]]$ be the power series corresponding to the measure $x\mu_f$. By Property (1), we have for $z \in \mathcal{P}$

$$g(z) = \int_{\mathbf{Z}_p} x(1+z)^x \, d\mu_f(x).$$

Put $f(X) = \sum_{n=0}^{\infty} a_n X^n$, $g(X) = \sum_{n=0}^{\infty} b_n X^n$. Using

$$X\binom{X}{n} = (n+1)\binom{X}{n+1} + n\binom{X}{n}$$

and Theorem 11.3, we have

$$\begin{aligned} b_n &= \int_{\mathbf{Z}_p} \binom{x}{n} d\mu_g(x) = \int_{\mathbf{Z}_p} \binom{x}{n} x \, d\mu_f(x) \\ &= (n+1)\int_{\mathbf{Z}_p} \binom{x}{n+1} d\mu_f(x) + n\int_{\mathbf{Z}_p} \binom{x}{n} d\mu_f(x) \\ &= (n+1)a_{n+1} + na_n. \end{aligned}$$

On the other hand, Df is computed as

$$\begin{aligned} (Df)(X) &= ((1+X)D_X f)(X) \\ &= (1+X)(a_1 + 2a_2X + \cdots + na_nX^{n-1} + \cdots) \\ &= \sum_{n=0}^{\infty} ((n+1)a_{n+1} + na_n)X^n. \end{aligned}$$

This gives $g = Df$. □

In general, for a power series $f(X)$, we define a new power series $f^*(Z)$ in Z by setting $X = e^Z - 1$:

$$f^*(Z) = f(e^Z - 1). \tag{11.5}$$

For example, when

$$f(X) = (1+X)^a = \sum_{n=0}^{\infty} \binom{a}{n} X^n,$$

we have

$$f^*(Z) = e^{aZ} = \sum_{n=0}^{\infty} \frac{a^n Z^n}{n!}.$$

Note the identity

$$(D_Z^k f^*)(0) = (D^k f)(0) \tag{11.6}$$

since

$$D_Z f^*(Z) = (1+X) D_X f(X) = Df(X).$$

The next property is the basis of the fact that the isomorphism P in Theorem 11.3 is an $\mathcal{O}_p$ algebra isomorphism.

Property (4). Let the measures $\mu,\ \nu$ correspond respectively to the power series $f,\ g \in \mathcal{O}_p[[X]]$ (i.e., $\mu = \mu_f,\ \nu = \mu_g$). Then the power series corresponding to the convolution $\mu * \nu$ is fg:

$$\mu_f * \mu_g = \mu_{fg}.$$

Proof. By Eq. (11.1), we have

$$\mu_n(r) = \frac{1}{p^n} \sum_{\zeta^{p^n}=1} \zeta^{-r} f(\zeta - 1),$$

$$\nu_n(k-r) = \frac{1}{p^n} \sum_{\zeta^{p^n}=1} \zeta^{-k+r} g(\zeta - 1).$$

Substituting this into the right-hand side of (11.2), we obtain

$$(\mu * \nu)_n(k) = \sum_{r=0}^{p^n-1} \frac{1}{p^n} \sum_{\zeta^{p^n}=1} \zeta^{-r} f(\zeta - 1) \frac{1}{p^n} \sum_{\xi^{p^n}=1} \xi^{-k+r} g(\xi - 1)$$

$$= \frac{1}{p^n} \sum_{\zeta} \sum_{\xi} f(\zeta - 1) g(\xi - 1) \xi^{-k} \cdot \frac{1}{p^n} \sum_{r=0}^{p^n-1} (\xi/\zeta)^r$$

$$= \frac{1}{p^n} \sum_{\zeta} f(\zeta - 1) g(\zeta - 1) \zeta^{-k}$$
$$= \mu_{fg,n}(k).$$

Here ζ and ξ run through all p^n-th roots of 1. From this, Property (4) follows. □

11.2 Bernoulli Measure

We define a specific measure called the Bernoulli measure. Recall that the first Bernoulli polynomial is by definition equal to

$$B_1(x) = x - \frac{1}{2}.$$

In the following, p denotes an *odd* prime. For each natural number n and $x \in X_n = \mathbf{Z}/p^n\mathbf{Z}$, set

$$E_n(x) = B_1\left(\left\{\frac{x}{p^n}\right\}\right),$$

where in the right-hand side, we regard x as an integer representing $x \bmod p^n$, and for $w \in \mathbf{R}$, $\{w\}$ is the real number satisfying $0 \leq \{w\} < 1$ and $w - \{w\} \in \mathbf{Z}$ (the fractional part of w). Then $E = \{E_n\}$ is a measure on $\mathbf{Z}_p$ but is not $\mathcal{O}_p$-valued. We modify this as follows in order to have an $\mathcal{O}_p$-valued measure. Take an invertible element c in $\mathbf{Z}_p$ (i.e. $c \in \mathbf{Z}_p^\times$), and for $x \in X_n = \mathbf{Z}/p^n\mathbf{Z}$, let

$$E_{c,n}(x) = E_n(x) - cE_n(c^{-1}x).$$

We understand $c^{-1}x$ as an element in $X_n = \mathbf{Z}/p^n\mathbf{Z}$. It is easy to see that $E_c = \{E_{c,n}\}$ is an $\mathcal{O}_p$-valued measure. We call this the Bernoulli measure.

Proposition 11.5. (1) *The formal power series corresponding to the Bernoulli measure E_c is given by*

$$f_c(X) = \frac{1}{X} - \frac{c}{(1+X)^c - 1}.$$

(2) *Let k be a natural number. For $c \in \mathbf{Z}_p^\times$ with $c^k \neq 1$, we have*

$$\frac{B_k}{k} = \frac{(-1)^k}{1-c^k} \int_{\mathbf{Z}_p} x^{k-1}\, dE_c.$$

In particular, if $p - 1 \nmid k$, then $B_k/k \in \mathbf{Z}_{(p)}$.

Proof. (1) Since $c \in \mathbf{Z}_p^\times$, we see $f_c \in \mathbf{Z}_p[[X]]$, the first two terms of $f_c(X)$ being

$$f_c(X) = \frac{c-1}{2} + \frac{1-c^2}{12}X + \cdots .$$

Let $\mu = \{\mu_n\}$ be the measure on $\mathbf{Z}_p$ corresponding to f_c by Theorem 11.3. For $r \in X_n = \mathbf{Z}/p^n\mathbf{Z}$ we have

$$\begin{aligned}\mu_n(r) &= \frac{1}{p^n}\sum_{\zeta^{p^n}=1} \zeta^{-r} f_c(\zeta - 1)\\ &= \frac{1}{p^n} f_c(0) + \frac{1}{p^n}\sum_{\zeta^{p^n}=1,\,\zeta\neq 1} \zeta^{-r}\left(\frac{1}{\zeta-1} - \frac{c}{\zeta^c - 1}\right).\end{aligned}$$

Now we use Lemma 8.5 on p. 110. For $\zeta^{p^n} = 1,\ \zeta \neq 1$ and $f = p^n$, the lemma gives

$$\frac{1}{\zeta^c - 1} = \frac{1}{f}\sum_{j=1}^{f-1} j\zeta^{cj}$$

since $(c, p) = 1$. By this, if we choose l so that $cl \equiv k \bmod p^n,\ 0 \le l < p^n$, we obtain

$$\begin{aligned}\frac{1}{f}\sum_{\zeta^{p^n}=1,\,\zeta\neq 1} \zeta^{-k}\frac{c}{\zeta^c - 1} &= \frac{c}{f^2}\sum_{\zeta^{p^n}=1} \zeta^{-k}\sum_{j=1}^{f-1} j\zeta^{cj} - \frac{c(f-1)}{2f}\\ &= \frac{cl}{f} - \frac{c(f-1)}{2f}\\ &= c\left\{\frac{c^{-1}k}{p^n}\right\} - \frac{c}{2} + \frac{c}{2f}\end{aligned}$$

and by substituting this into the formula for $\mu_n(r)$ above and noting that $f_c(0) = (c-1)/2$, we have

$$\begin{aligned}\mu_n(r) &= \frac{c-1}{2f} + \left(\left\{\frac{r}{p^n}\right\} - \frac{1}{2} + \frac{1}{2f} - c\left\{\frac{c^{-1}r}{p^n}\right\} + \frac{c}{2} - \frac{c}{2f}\right)\\ &= \left(\left\{\frac{r}{p^n}\right\} - \frac{1}{2}\right) - c\left(\left\{\frac{c^{-1}r}{p^n}\right\} - \frac{1}{2}\right).\end{aligned}$$

By the definition of the Bernoulli measure, we conclude $\mu_n(r) = E_{c,n}(r)$, i.e., $\mu = E_c$ and the power series corresponding to E_c is f_c.

The proof of (2) goes as follows. By Property (3) and Eq. (11.5) we have

$$\int_{\mathbf{Z}_p} x^{k-1}\, dE_c = (D^{k-1} f_c)(0) = (D_Z^{k-1} f_c^*)(0).$$

Here by definition (11.5), we have

$$\begin{aligned} f_c^*(Z) = f_c(e^Z - 1) &= \frac{1}{e^Z - 1} - \frac{c}{e^{cZ} - 1} \\ &= \sum_{n=1}^{\infty} (1 - c^n)(-1)^n B_n \frac{Z^{n-1}}{n!}, \end{aligned}$$

so we have

$$(D_Z^{k-1} f_c^*)(0) = (1 - c^k)(-1)^k \frac{B_k}{k}$$

and thus

$$\int_{\mathbf{Z}_p} x^{k-1}\, dE_c = (1 - c^k)(-1)^k \frac{B_k}{k}.$$

This gives (2). □

11.3 Kummer's Congruence Revisited

The "right" formulation of Kummer's congruence is the following.

Theorem 11.6. *Suppose p is an odd prime.*

(1) *Assume that m is a positive even integer such that $p - 1 \nmid m$. Then $B_m/m \in \mathbf{Z}_{(p)}$.*
(2) *Let a be a positive integer, and m and n positive even integers satisfying $m \equiv n \bmod (p-1)p^{a-1}$ and $m \not\equiv 0 \bmod (p-1)$. Then we have*

$$(1 - p^{m-1})\frac{B_m}{m} \equiv (1 - p^{n-1})\frac{B_n}{n} \mod p^a.$$

To prove this, we need the following integral expression of the Bernoulli number, a refined version of Proposition 11.5 (2).

Proposition 11.7. *Let k be a positive even integer and take $c \in \mathbf{Z}_p^\times$. Then we have*

$$(1 - c^k)(1 - p^{k-1})\frac{B_k}{k} = \int_{\mathbf{Z}_p^\times} x^{k-1}\, dE_c.$$

Proof. The power series that corresponds to the Bernoulli measure E_c is f_c in Proposition 11.5. As in (11.4), define from f_c a new power series g by

$$g(X) = \mathbb{U} f_c(X) = f_c(X) - \frac{1}{p} \sum_{\zeta^p=1} f_c(\zeta(1+X)-1).$$

We have $g \in \mathcal{O}_p[[X]]$ and so we let $\mu = \mu_g$ be the measure on $\mathcal{O}_p$ obtained from g. By Property (2) on p. 192 we have

$$\int_{\mathbf{Z}_p^\times} x^{k-1}\, dE_c = \int_{\mathbf{Z}_p} x^{k-1}\, d\mu.$$

Further, using Property (3) on p. 194 and (11.6) one sees

$$\int_{\mathbf{Z}_p} x^{k-1}\, d\mu = (D^{k-1} g)(0) = (D_Z^{k-1} g^*)(0).$$

We compute the value $(D_Z^{k-1} g^*)(0)$. First,

$$g^*(Z) = \frac{1}{e^Z-1} - \frac{c}{e^{cZ}-1} - \frac{1}{p} \sum_{\zeta^p=1} \left(\frac{1}{\zeta e^Z - 1} - \frac{c}{\zeta^c e^{cZ} - 1} \right).$$

Here, since

$$\frac{1}{p} \sum_{\zeta^p=1} \frac{1}{\zeta X - 1} = \frac{1}{X^p - 1},$$

we get

$$\begin{aligned} g^*(Z) &= \frac{1}{e^Z-1} - \frac{c}{e^{cZ}-1} - \left(\frac{1}{e^{pZ}-1} - \frac{c}{e^{cpZ}-1} \right) \\ &= \sum_{k=0}^{\infty} (1-c^k)(-1)^k \frac{B_k}{k!} Z^{k-1} - \sum_{k=0}^{\infty} (1-c^k)(-1)^k \frac{B_k}{k!} (pZ)^{k-1} \\ &= \sum_{k=1}^{\infty} (1-c^k)(1-p^{k-1})(-1)^k \frac{B_k}{k!} Z^{k-1}. \end{aligned}$$

Hence if k is even we have

$$(D_Z^{k-1} g^*)(0) = (1-c^k)(1-p^{k-1}) \frac{B_k}{k},$$

and the proposition is established. □

Proof of Theorem 11.6. The first assertion is already given in Theorem 3.2, but we give here an alternative proof for that too. Since we assumed $m \not\equiv 0 \bmod p-1$, we can take $c \in \mathbf{Z}$ such that $(c, p) = 1$ and $c^m \not\equiv 1 \bmod p$. For instance one may take a primitive root mod p. From the proposition above, we have

$$(1-c^n)(1-p^{n-1})\frac{B_n}{n} = \int_{\mathbf{Z}_p^\times} x^{n-1}\, dE_c$$

and

$$(1-c^m)(1-p^{m-1})\frac{B_m}{m} = \int_{\mathbf{Z}_p^\times} x^{m-1}\, dE_c.$$

The assumption $m \equiv n \bmod (p-1)p^{a-1}$ gives $c^{n-m} \equiv 1 \bmod p^a$, and since we assumed $(1-c^m, p) = 1$, we have also $(1-c^n, p) = 1$. Since E_c is an $\mathcal{O}_p$ measure, the above integral values are in $\mathcal{O}_p$ and we see that B_n/n and $B_m/m \in \mathbf{Z}_{(p)}$. Since $x^{m-1} \equiv x^{n-1} \bmod p^a$ if $x \in \mathbf{Z}_p^\times$, and since E_c is an $\mathcal{O}_p$-valued measure, we have

$$(1-c^n)\left((1-p^{n-1})\frac{B_n}{n} - (1-p^{m-1})\frac{B_m}{m}\right) \in p^a\mathcal{O}_p.$$

The left-hand side being contained in $\mathbf{Z}_p$, we conclude

$$(1-p^{n-1})\frac{B_n}{n} - (1-p^{m-1})\frac{B_m}{m} \in p^a\mathbf{Z}_p.$$

This proves the theorem. □

Theorem 3.2 is a corollary of Theorem 11.6. Indeed, if $a < m \leq n$, then by Theorem 11.6, we have

$$\begin{aligned}
&(1-p^{m-1})\frac{B_m}{m} - (1-p^{n-1})\frac{B_n}{n} \\
&= (1-p^{m-1})\left(\frac{B_m}{m} - \frac{B_n}{n}\right) + \frac{B_n}{n}(p^{n-m}-1)p^{m-1} \\
&\equiv 0 \bmod p^a.
\end{aligned}$$

Since $p-1 \nmid n$, we have $B_n/n \in \mathbf{Z}_{(p)}$ by Theorem 11.6. Since $a \leq m-1$, we have $p^{m-1}B_n/n \in p^a\mathbf{Z}_{(p)}$. Hence we have

$$\frac{B_m}{m} - \frac{B_n}{n} \equiv 0 \bmod p^a.$$

Exercise 11.8. Give an example of an odd prime p and integers $2 \leq a = m < n$ such that the congruence in Theorem 3.2 does not hold. Check that for the same choice of a, n, m and p, the congruence of Theorem 11.6 surely holds.
Hint: For example, put $p = 5$, $a = m = 2$ and $n = 22$ and use the following values:

$$B_2 = \frac{1}{6}, \qquad B_{22} = \frac{854513}{138}.$$

Exercise 11.9. Show that the Bernoulli number B_n is given by the limit (p-adic limit in $\mathbf{Q}_p$)

$$\lim_{m\to\infty} \frac{1}{p^m} \sum_{i=0}^{p^m-1} i^n.$$

(For a function $f : \mathbf{Z}_p \to \mathbf{Q}_p$ with a suitable condition, the limit

$$\lim_{m\to\infty} \frac{1}{p^m} \sum_{i=0}^{p^m-1} f(i)$$

is sometimes referred to as the Volkenborn integral of f over $\mathbf{Z}_p$. See [94, 95] for details.)

Chapter 12
Hurwitz Numbers

12.1 Hurwitz Numbers

In this section, we briefly introduce Hurwitz's generalization of Bernoulli numbers, known as the Hurwitz numbers.

As we have seen in Proposition 1.17, Bernoulli numbers appear as the coefficients of expansions of trigonometric functions. Hurwitz replaced the trigonometric function by a "lemniscate function", which, in modern terms, is an elliptic function having complex multiplication by the ring of Gaussian integers $\mathbf{Z}[\sqrt{-1}]$, and considered its expansion. It was Gauss who first studied the lemniscate function or more general elliptic functions, but most of his work on them did not appear during his lifetime. In his *Disquisitiones* [35], Gauss mentioned at the beginning of Chap. 7 that his theory of cyclotomy developed there could also be applied to the division of the lemniscate. Inspired by this comment, Abel[1] and Jacobi started their investigation on elliptic functions and laid the foundations of the theory.

Let us define Hurwitz numbers. For this, put

$$\varpi = 2\int_0^1 \frac{dx}{\sqrt{1-x^4}} \quad (= 2.622057\ldots).$$

This is one-half of the arc-length of the lemniscate defined by

$$r^2 = \cos(2\theta),$$

and is analogous to the similar quantity for the case of the circle:

$$\pi = 2\int_0^1 \frac{dx}{\sqrt{1-x^2}}.$$

[1]Niels Henrik Abel (born on August 5, 1802 in Frindoe, Norway—died on April 6, 1829 in Froland, Norway).

T. Arakawa et al., *Bernoulli Numbers and Zeta Functions*, Springer Monographs in Mathematics, DOI 10.1007/978-4-431-54919-2_12, © Springer Japan 2014

Let $\wp(z)$ be the Weierstrass $\wp$-function with periods ϖ and $\varpi\sqrt{-1}$ defined by

$$\wp(z) = \frac{1}{z^2} + \sum_{\substack{\lambda \in \mathbf{Z}\varpi + \mathbf{Z}\varpi\sqrt{-1} \\ \lambda \neq 0}} \left(\frac{1}{(z-\lambda)^2} - \frac{1}{\lambda^2} \right). \tag{12.1}$$

The function $\wp(z)$ is a doubly periodic meromorphic function on the complex plane with periods ϖ and $\varpi\sqrt{-1}$, having double poles at the lattice points $\mathbf{Z}\varpi + \mathbf{Z}\varpi\sqrt{-1}$ and holomorphic elsewhere (for the basic properties of the elliptic function, see [104]). Owing to the fact that the lattice $\mathbf{Z}\varpi + \mathbf{Z}\varpi\sqrt{-1}$ is preserved by multiplication by $\sqrt{-1}$, the function $z \mapsto \wp(\lambda z)$ for any $\lambda \in \mathbf{Z}[\sqrt{-1}]$ is expressible as a rational function of $\wp(z)$ and $\wp'(z)$. By virtue of this property, $\wp(z)$ is said to have complex multiplication by $\mathbf{Z}[\sqrt{-1}]$. In particular, it is easily seen from the definition that $\wp(z)$ is an even function and

$$\wp(\sqrt{-1}z) = -\wp(z).$$

Also, $\wp(z)$ satisfies the differential equations

$$\wp'(z)^2 = 4\wp(z)^3 - 4\wp(z)$$

and

$$\wp''(z) = 6\wp(z)^2 - 2.$$

Write the Laurent expansion of $\wp(z)$ at $z = 0$ as

$$\wp(z) = \frac{1}{z^2} + \sum_{n=2}^{\infty} \frac{2^n H_n}{n} \frac{z^{n-2}}{(n-2)!} \tag{12.2}$$

and use this to define the Hurwitz numbers H_n. From the property $\wp(\sqrt{-1}z) = -\wp(z)$, we see that $H_n = 0$ unless n is a multiple of 4. The analogous expansion for the Bernoulli numbers is

$$\frac{1}{\sin^2(x)} = \frac{1}{x^2} + \sum_{n=2}^{\infty} \frac{(-1)^{\frac{n}{2}-1} 2^n B_n}{n} \frac{x^{n-2}}{(n-2)!},$$

which can be obtained from the expansion of cot given in Proposition 1.17 by differentiation, since $\cot'(x) = -\frac{1}{\sin^2(x)}$. In a similar manner to the proof of Proposition 1.15, namely by comparing the coefficients of

$$\wp''(z) = 6\wp(z)^2 - 2, \tag{12.3}$$

Table 12.1 Hurwitz numbers

n	4	8	12	16	20	24	28	32
H_n	$\frac{1}{10}$	$\frac{3}{10}$	$\frac{567}{130}$	$\frac{43659}{170}$	$\frac{392931}{10}$	$\frac{1724574159}{130}$	$\frac{2498907956391}{290}$	$\frac{1671769422825579}{170}$

we have

$$H_4 = \frac{1}{10},$$

and the recursion

$$(2n-3)(4n-1)(4n+1)H_{4n} = 3\sum_{i=1}^{n-1}(4i-1)(4n-4i-1)\binom{4n}{4i}H_{4i}\,H_{4(n-i)}$$

for $n \geq 2$. In particular, we see from this that H_{4n} is a positive rational number. Several values are in Table 12.1.

The analogy between Hurwitz and Bernoulli numbers is more apparent when we look at the special values of zeta functions. Namely, first we rewrite the formula (4.4) as

$$\sum_{\substack{r\in\mathbf{Z}\\ r\neq 0}} \frac{1}{r^{2n}} = (-1)^{n-1}\frac{(2\pi)^{2n}}{(2n)!}B_{2n} \quad (n \geq 1).$$

Note the sum on the left runs over all non-zero integers. If we replace the rational integers here by Gaussian integers, we have the formula

$$\sum_{\substack{\lambda\in\mathbf{Z}+\mathbf{Z}\sqrt{-1}\\ \lambda\neq 0}} \frac{1}{\lambda^{4n}} = \frac{(2\varpi)^{4n}}{(4n)!}H_{4n}.$$

To deduce this, expand the right-hand side of the definition (12.1) of $\wp(z)$ by using $\frac{1}{(z-\lambda)^2} = \frac{1}{\lambda^2} + \frac{2z}{\lambda^3} + \frac{3z^2}{\lambda^4} + \cdots$ and compare with (12.2).

Hurwitz proved a Clausen–von Staudt type theorem for H_n. To state this, we need the following classical theorem of Fermat and Euler.

Theorem 12.1. *Any prime number congruent to* 1 *modulo* 4 *can be written as a sum of two squares essentially in the unique way.*

Various proofs of this theorem are known since Euler. See [50] or [77] for some of the standard proofs. For Zagier's "One-sentence proof", which simplifies a proof by Heath-Brown, see [108].

Suppose now that p is a prime number congruent to 1 modulo 4 and write

$$p = a^2 + b^2.$$

We may assume a is odd and b is even. We change the sign of a if necessary in order to have

$$a \equiv b + 1 \pmod 4.$$

(Note that the right-hand side is independent of the choice of the sign of b since b is even.) These conditions uniquely determine a and so we denote it by a_p to make the dependence on p explicit. For instance, we have

$$5 = (-1)^2 + 2^2,\ 13 = 3^2 + 2^2,\ 17 = 1^2 + 4^2,$$

so

$$a_5 = -1,\ a_{13} = 3,\ a_{17} = 1.$$

Hurwitz's theorem is then stated as:

Theorem 12.2. *For each $n \geq 1$, we have*

$$H_{4n} = \frac{1}{2} + \sum_{\substack{p \equiv 1 \bmod 4 \\ p-1 \mid 4n}} \frac{(2a_p)^{\frac{4n}{p-1}}}{p} + G_n$$

with an integer G_n. In the sum, p runs over prime numbers congruent to 1 *modulo* 4 *such that $p - 1$ divides* $4n$.

As examples of the theorem, we have

$$\begin{aligned}
H_4 &= \frac{1}{10} = \frac{1}{2} + \frac{(-2)}{5},\\
H_8 &= \frac{3}{10} = \frac{1}{2} + \frac{(-2)^2}{5} - 1,\\
H_{12} &= \frac{567}{130} = \frac{1}{2} + \frac{(-2)^3}{5} + \frac{6}{13} + 5,\\
H_{16} &= \frac{43659}{170} = \frac{1}{2} + \frac{(-2)^4}{5} + \frac{2}{17} + 253,\\
H_{20} &= \frac{392931}{10} = \frac{1}{2} + \frac{(-2)^5}{5} + 39299,\\
H_{24} &= \frac{1724574159}{130} = \frac{1}{2} + \frac{(-2)^6}{5} + \frac{6^2}{13} + 13265939.
\end{aligned}$$

Hurwitz also proved that every G_n for $n > 1$ is odd and determined its class modulo 8. The proof [44] of Theorem 12.2 is not as elementary as the original Clausen–von Staudt theorem. It uses the complex multiplication of the $\wp$-function as well as the notion of Hurwitz integral series (cf. Chap. 7). See [27, 58] for a modern interpretation of the theorem and Kummer-type congruence for H_n.

12.2 A Short Biography of Hurwitz

Adolf Hurwitz was born on March 26, 1859 in Hildesheim, Germany.[2] His teacher at the Gymnasium was Schubert,[3] who is famous for his enumerative geometry ("Schubert calculus"). Impressed by Hurwitz's genius, Schubert came every Sunday as a house tutor and persuaded Hurwitz's father, who was not at all affluent, to send his son to the university to study mathematics. Already during his time at the Gymnasium, while he was still 17, Hurwitz wrote a joint paper with Schubert on enumerative geometry. In the spring of 1877, he entered Munich Technical University, where Klein[4] was. From the fall of the same year until the spring of 1879, he attended the lectures of Kummer, Weierstrass, and Kronecker at Berlin University. Returning once again to Munich, and then moving to Leipzig together with Klein, he obtained his doctoral degree there with research on modular functions. He became a "Privatdozent" at Göttingen University in 1882 and then, on the recommendation of Lindemann,[5] a professor at Königsberg University in 1884. Hilbert[6] and Minkowski[7] were among the students there. The exchanges among the three of them in those early days and the influence which this had on Hilbert are vividly described in [78]. In 1892, he moved to the ETH in Zürich as a successor of Frobenius,[8] a position he was to hold until his death. At the same time he was offered a professorship at Göttingen as the successor of Schwarz, who moved to Berlin, but he had already accepted the position at Zürich (the director at Zürich at that time, Bleuler, went especially to Königsberg to make the contract).

Hurwitz's health was not good. He got typhoid fever twice and suffered from frequent migraines. And in 1905, one of his kidneys had to be removed. In the midst of all this, he continued to do research and published in a large number of areas.

---

[2]The description in this section is based on Freudenthal [34], Hilbert [41], and Pólya [76].

[3]Hermann Cäsar Hannibal Schubert (born on May 22, 1848 in Potsdam, Germany—died on July 20, 1911 in Hamburg, Germany).

[4]Felix Christian Klein (born on April 25, 1849 in Düsseldorf, Prussia (now Germany)—died on June 22, 1925 in Göttingen, Germany).

[5]Carl Louis Ferdinand von Lindemann (born on April 12, 1852 in Hannover, Hanover (now Germany)—died on March 6, 1939 in Munich, Germany) who proved in 1882 that π is a transcendental number.

[6]David Hilbert (born on January 23, 1862 in Königsberg, Prussia (now Kaliningrad, Russia)—died on February 14, 1943 in Göttingen, Germany).

[7]Hermann Minkowski (born on June 22, 1864 in Alexotas, Russian Empire (now Kaunas, Lithuania)—died on January 12, 1909 in Göttingen, Germany).

[8]Ferdinand Georg Frobenius (born on October 26, 1849 in Berlin-Charlottenburg, Prussia (now Germany)—died on August 3, 1917 in Berlin, Germany).

His articles have been collected and comprise two volumes. As well as the Hurwitz zeta functions and Hurwitz numbers treated in this book, there are many other topics which carry his name, such as the Hurwitz class number relations, Hurwitz's genus formula, the Hurwitz estimates for the size of automorphism groups of Riemann surfaces and his investigations on the quaternions and other algebras. And according to Hilbert, continued fractions were also a favorite topic of Hurwitz. (He wrote approximately five articles on the subject.)

Hurwitz seems to have been an exceptionally modest and courteous man. According to Pólya,[9] who was the main editor of his collected works, he never failed to tip his hat in greeting even to service people whom he met on the street. Moreover, he was so skilled at the piano that in his youth he hesitated between a career as a mathematician or as a pianist. He died on the 18th of November, 1919, at the age of 60.

Exercise 12.3. Deduce the recursion of H_n given in the text from (12.3).

Exercise 12.4. Compute the residues $2H_n \mod 16$ for first several n, and find the pattern. (For a proof, see [44].)

[9]George Pólya (born on December 13, 1887 in Budapest, Hungary—died on September 7, 1985 in Palo Alto, USA).

Chapter 13
The Barnes Multiple Zeta Function

In this chapter, we introduce Barnes' multiple zeta function, which is a natural generalization of the Hurwitz zeta function, give an analytic continuation, and then express their special values at negative integers by using Bernoulli polynomials. Furthermore, for double zeta functions, we take up the problem of finding a functional equation. This problem has several interesting points. For example, a kind of Lerch[1] type zeta function appears in the functional equation, and also the proof has the flavor of the theory of automorphic forms.

The motivation to take up Barnes' multiple zeta function here is that this is deeply related with the very important and interesting conjecture about constructing class fields over real quadratic fields, or totally real algebraic number fields (Stark–Shintani[2] conjecture, see [86, 87, 89] for example). Although we do not treat it in this book, the value at $s = 0$ of the first-order derivative of Barnes' multiple zeta-function is expressed by special values of the multiple gamma function. Shintani saw that for the construction of class fields of real quadratic fields, special values of the double gamma function play an important role, and he proposed a detailed conjecture that certain special values of the double gamma functions generate the class fields, and solved part of his conjecture (cf. [85–87]). For the construction of class fields over imaginary quadratic fields, modular forms such as the j-invariant play an extremely important role, but in the case of real quadratic fields, any viewpoint from modular forms is not considered in the above papers. The Lerch type function treated in this chapter was introduced with the intention of giving more or less such a viewpoint to the construction problem of class fields over real quadratic fields. For details, see [6, 7].

[1]Mathias Lerch (born on February 20, 1860 in Milinov, Bohemia (now Czech Republic)—died on August 3, 1922 in Susice, Czechoslovakia (now Czech Republic)).

[2]Takuro Shintani (born on February 4, 1943 in Kita-kyushu, Japan—died on November 14, 1980 in Tokyo, Japan).

T. Arakawa et al., *Bernoulli Numbers and Zeta Functions*, Springer Monographs in Mathematics, DOI 10.1007/978-4-431-54919-2_13,

13.1 Special Values of Multiple Zeta Functions and Bernoulli Polynomials

Let all $a,\ w_1,\ w_2,\ \dots, w_r$ be positive real numbers. We put $\mathbf{w} = (w_1, \dots, w_r)$ and define a zeta function $\zeta(s, \mathbf{w}, a)$ by

$$\zeta(s, \mathbf{w}, a) = \sum_{m_1=0}^{\infty} \cdots \sum_{m_r=0}^{\infty} (a + m_1 w_1 + \cdots + m_r w_r)^{-s}.$$

This zeta function converges absolutely for $\mathrm{Re}(s) > r$. This kind of zeta function was systematically studied by Barnes[3] [11, 12].

For a while, we assume $\mathrm{Re}(s) > r$. We see that $\zeta(s, \mathbf{w}, a)$ has an integral representation:

$$\begin{aligned}
\zeta(s, \mathbf{w}, a) &= \frac{1}{\Gamma(s)} \sum_{m_1=0}^{\infty} \cdots \sum_{m_r=0}^{\infty} \int_0^{\infty} t^{s-1} e^{-(a+m_1 w_1+\cdots+m_r w_r)t}\, dt \\
&= \frac{1}{\Gamma(s)} \int_0^{\infty} t^{s-1} \sum_{m_1=0}^{\infty} \cdots \sum_{m_r=0}^{\infty} e^{-(a+m_1 w_1+\cdots+m_r w_r)t}\, dt \\
&= \frac{1}{\Gamma(s)} \int_0^{\infty} t^{s-1} \frac{e^{-at}}{(1-e^{-w_1 t}) \cdots (1-e^{-w_r t})}\, dt \\
&= \frac{1}{\Gamma(s)} \int_0^{\infty} t^{s-1} \frac{e^{(w_1+\cdots+w_r-a)t}}{(e^{w_1 t}-1) \cdots (e^{w_r t}-1)}\, dt.
\end{aligned}$$

Since the series $\zeta(s, \mathbf{w}, a)$ converges absolutely for $\mathrm{Re}(s) > r$, by virtue of Lebesgue's dominant convergence theorem for the Lebesgue[4] integral, the above calculation is justified for the same range of s. When $r = 1$, this is essentially the Hurwitz zeta function. Namely, for $w_1 > 0$, we have

$$\zeta(s, w_1, a) = w_1^{-s} \zeta\left(s, \frac{a}{w_1}\right).$$

Now, a multiple zeta-function $\zeta(s, \mathbf{w}, a)$ as well as the Hurwitz zeta function, can be represented by a contour integral. Indeed, taking $\varepsilon > 0$ small enough and denoting

---

[3]Ernest Barnes (born on April 1, 1874 in Birmingham, England—died on November 29, 1953 in Sussex, England).

[4]Henri Léon Lebesgue (born on June 28, 1875 in Beauvais (Oise), France—died on July 26, 1941 in Paris, France).

by $I(\varepsilon, \infty)$ the contour in Sect. 9.2, we have

$$\zeta(s, \mathbf{w}, a) = \frac{1}{\Gamma(s)(e^{2\pi i s} - 1)} \int_{I(\varepsilon,\infty)} t^{s-1} \frac{e^{(w_1+\cdots+w_r-a)t}}{(e^{w_1 t} - 1)\cdots(e^{w_r t} - 1)} dt. \quad (13.1)$$

An advantage of this representation is that $\zeta(s, \mathbf{w}, a)$ is continued analytically to a meromorphic function on the whole s-plane by this, because the contour integral on the right-hand side converges absolutely for arbitrary $s \in \mathbf{C}$ and it defines a holomorphic function on the whole s-plane.

A point which possibly becomes a pole of $\zeta(s, \mathbf{w}, a)$ is among zeros

$$s = r,\ r-1,\ \ldots,\ 0,\ -1,\ -2,\ \ldots$$

of $e^{2\pi i s} - 1$ which are not in the range of absolute convergence. But $\Gamma(s)$ has poles of order 1 at integer points less than or equal to 0, so points $s = 1 - m$ $(m \in \mathbf{N})$ cannot appear as poles and $\zeta(s, \mathbf{w}, a)$ is holomorphic there.

Let us look for special values of $\zeta(s, \mathbf{w}, a)$ at integer points $s = 1 - m$ $(m \in \mathbf{N})$. For any integer $k \geq 0$, and for any fixed a and $\mathbf{w}$, we define constants $C_k(\mathbf{w}, a)$ by

$$\frac{e^{at} t^r}{\prod_{i=1}^{r}(e^{w_i t} - 1)} = \sum_{k=0}^{\infty} C_k(\mathbf{w}, a) t^k,$$

Proposition 13.1. *We take $m \in \mathbf{N}$. If we write $a > 0$ as*

$$a = a_1 w_1 + a_2 w_2 + \cdots + a_r w_r \qquad (a_1, a_2, \ldots, a_r \in \mathbf{R}),$$

we have

$$\begin{aligned} &\zeta(1-m, \mathbf{w}, a) \\ &= (-1)^r (m-1)! \sum_{\substack{m_1+\cdots-m_r=m+r-1 \\ m_1\geq 0, \ldots, m_r \geq 0}} \left(\prod_{j=1}^{r} \frac{B_{m_j}(a_j)}{m_j!} \right) w_1^{m_1-1} \cdots w_r^{m_r-1} \\ &= (-1)^r (m-1)! C_{m+r-1}(\mathbf{w}, a). \end{aligned}$$

Proof. The proof is based on the same method as the one used for looking at special values of the Hurwitz zeta functions. In the same way as in the proof of Proposition 9.3, we have

$$\zeta(1-m, \mathbf{w}, a) = \left(\lim_{s\to 1-m} \frac{1}{\Gamma(s)(e^{2\pi i s} - 1)} \right) \int_{I(\varepsilon)} t^{-m} \frac{e^{(w_1+\cdots+w_r-a)t}}{(e^{w_1 t} - 1)\cdots(e^{w_r t} - 1)} dt$$

$$= \frac{(-1)^{m-1}(m-1)!}{2\pi i} \int_{I(\varepsilon)} t^{-m} \prod_{j=1}^{r} \frac{e^{(1-a_j)w_j t}}{e^{w_j t}-1}\, dt.$$

Now by using the generating function of Bernoulli polynomials (p. 56, (4.2)) and Proposition 4.9 (4), the Laurent expansion of the integrand is obtained as follows:

$$t^{-m} \prod_{j=1}^{r} \frac{e^{(1-a_j)w_j t}}{e^{w_j t}-1} = t^{-m} \sum_{m_1=0}^{\infty} \cdots \sum_{m_r=0}^{\infty} \prod_{j=1}^{r} \left(B_{m_j}(1-a_j) \frac{(w_j t)^{m_j-1}}{m_j!} \right)$$

$$= \sum_{m_1=0}^{\infty} \cdots \sum_{m_r=0}^{\infty} \left(\prod_{j=1}^{r} \frac{(-1)^{m_j} B_{m_j}(a_j) w_j^{m_j-1}}{m_j!} \right) t^{m_1+\cdots+m_r-r-m}.$$

The residue at $t = 0$ of this integrand is given by

$$\sum_{\substack{m_1+\cdots+m_r=m+r-1 \\ m_1\geq 0, \ldots, m_r \geq 0}} \left(\prod_{j=1}^{r} (-1)^{m_j} \frac{B_{m_j}(a_j) w_j^{m_j-1}}{m_j!} \right).$$

Using the generating function of Bernoulli polynomials in Proposition 4.9, we have

$$\sum_{m_1,\ldots,m_r=0}^{\infty} \frac{B_{m_1}(a_1)\cdots B_{m_r}(a_r)}{m_1!\cdots m_r!} w_1^{m_1} \cdots w_r^{m_r} t^{m_1+\cdots+m_r}$$

$$= \prod_{i=1}^{r} \frac{(w_i t) e^{a_i w_i t}}{e^{w_i t}-1} = (w_1 \cdots w_r) \frac{t^r e^{at}}{\prod_{i=1}^{r}(e^{w_i t}-1)},$$

so we get the final equality of the proposition. □

13.2 The Double Zeta Functions and Dirichlet Series

Now we set $\mathbf{w} = (w_1, w_2)$. It will be a very interesting problem to study what kind of zeta functions appear if we cut out the contour integral representation (13.1) of the Barnes double zeta function $\zeta(s, \mathbf{w}, a)$, in the same way as we did when we obtained the functional equation of the Hurwitz zeta functions (Theorem 9.4).

In order to answer this problem, we introduce a new Dirichlet series. For each real algebraic irrational number [5] α, let us consider the following Dirichlet series:

[5] A real algebraic number which is not a rational number.

$$\xi(s,\alpha) = \sum_{n=1}^{\infty} \frac{\cot \pi n\alpha}{n^s}.$$

Since $\cot(x)$ is unbounded, we need to show that this converges when $\mathrm{Re}(s)$ is big enough.

By virtue of Roth's theorem on Diophantine approximation in transcendental number theory, the following things are known. Assume that α is an algebraic irrational number. When we take any positive number ε, there exists a positive constant $C(\varepsilon)$ such that

$$\left|\alpha - \frac{m}{n}\right| > C(\varepsilon)n^{-2-\varepsilon} \tag{13.2}$$

for arbitrary integers m, n ($n > 0$). Now for a real number x, we use the notation $\langle\langle x\rangle\rangle$ for the distance to the nearest integer from x. We have $0 \le \langle\langle x\rangle\rangle \le 1/2$. Using this notation, for any integer n, (13.2) is expressed as

$$\langle\langle n\alpha\rangle\rangle > C(\varepsilon)n^{-1-\varepsilon}.$$

When $0 < x < \pi/2$, the inequality $\sin x > \frac{2}{\pi}x$ holds, so

$$\begin{aligned} |\cot \pi n\alpha| &\le \frac{1}{\sin \pi\langle\langle n\alpha\rangle\rangle} < \frac{1}{2\langle\langle n\alpha\rangle\rangle} \\ &< \frac{1}{2C(\varepsilon)} n^{1+\varepsilon}. \end{aligned}$$

By this estimate, if we put $\sigma = \mathrm{Re}(s)$, we can take

$$\frac{1}{2C(\varepsilon)} \sum_{n=1}^{\infty} \frac{n^{1+\varepsilon}}{n^{\sigma}}$$

as a majorant series of $\xi(s,\alpha)$, and it converges for $\sigma > 2 + \varepsilon$. Since we can take arbitrary ε, $\xi(s,\alpha)$ converges for[6] $\mathrm{Re}(s) > 2$, and moreover it converges uniformly on compact sets, so it is a holomorphic function of s in that range.

We denote by $SL_2(\mathbf{Z})$ the modular group which appeared before (p. 76). If we let an element $V = \begin{pmatrix} a & b \\ c & d \end{pmatrix}$ of $SL_2(\mathbf{Z})$ act on a real algebraic irrational number α by

[6] In fact, if we use Roth's theorem more efficiently, we can show absolute convergence for $\mathrm{Re}(s) > 1$ (see [6]).

$$V\alpha = \frac{a\alpha + b}{c\alpha + d},$$

then $V\alpha$ is again a real algebraic irrational number.

Let ω and a be positive numbers and consider the following double zeta function:

$$\zeta_2(s,\omega,a) = \sum_{m,n=0}^{\infty} (m + n\omega + a)^{-s}.$$

In the former notation, we have $\zeta_2(s,\omega,a) = \zeta(s,(1,\omega),a)$. We put

$$\gamma(s) = \frac{(2\pi)^s}{\Gamma(s)\sin(\pi s/2)}.$$

The zeta function $\zeta_2(s,\omega,a)$ converges absolutely for $\mathrm{Re}(s) > 2$, is holomorphic there, and is continued analytically to a meromorphic function on the whole s-plane, and its poles are $s = 1,\ 2$ and of order 1.

Using the contour integral representation (13.1), the principal part of the Laurent expansion at $s = 1$ is given by

$$\zeta_2(s,\omega,a) = \left(\frac{1+\omega}{2\omega} - \frac{a}{\omega}\right)\frac{1}{s-1} + \cdots. \tag{13.3}$$

Proposition 13.2. *Let* $\alpha > 0$ *be a real algebraic irrational number. We put* $V = \begin{pmatrix} n & -1 \\ 1 & 0 \end{pmatrix}$ $(n \in \mathbf{Z})$. *Then for the range* $\mathrm{Re}(s) > 2$, *we get the following transformation formula:*

$$\alpha^{s-1}\xi(s,V\alpha) - \xi(s,\alpha) = (1-\alpha^{s-1})\cot(\pi s/2)\zeta(s) - \gamma(s)\zeta_2(1-s,\alpha,1). \tag{13.4}$$

Proof. We start from the contour integral representation of the double zeta function $\zeta_2(s,\alpha,1)$ given by

$$\zeta_2(s,\alpha,1) = \frac{1}{\Gamma(s)(e^{2\pi i s}-1)}\int_{I(\varepsilon,\infty)} t^{s-1}\frac{e^{\alpha t}}{(e^t-1)(e^{\alpha t}-1)}\,dt. \tag{13.5}$$

This expression is obtained if we put $r = 2,\ w_1 = 1,\ w_2 = \alpha,\ a = 1$ in (13.1). Since the right-hand side is continued analytically to a meromorphic function on the whole s-plane, we consider the range $\mathrm{Re}(s) < -1$. We take $t_0 < 0$ and let $C(t_0)$ be the vertical line $\mathrm{Re}(t) = t_0$, oriented upwards:

$$C(t_0) : t = t_0 + iy \quad (-\infty < y < \infty).$$

In the same way as in the proof of Theorem 9.4, shifting the contour $I(\varepsilon, \infty)$ to the left to $C(t_0)$ and using the residue theorem, we can write

$$\zeta_2(s,\alpha,1) = \frac{1}{\Gamma(s)(e^{2\pi i s}-1)}\left\{ -\int_{C(t_0)} t^{s-1}\frac{e^{\alpha t}}{(e^t-1)(e^{\alpha t}-1)}\,dt \right.$$

$$\left. -2\pi i \sum_{n\in\mathbf{Z}-\{0\}} \left(\frac{(2\pi i n)^{s-1}}{1-e^{-2\pi i n\alpha}} + \frac{(2\pi i n/\alpha)^{s-1}}{e^{2\pi i n/\alpha}-1} \times \frac{1}{\alpha} \right) \right\},$$

and then taking the limit $t_0 \to -\infty$ and noting

$$\frac{1}{1-e^{-2\pi i n\alpha}} = \frac{1}{2i}\cot \pi n\alpha + \frac{1}{2} \quad \text{and} \quad \frac{1}{e^{2\pi i n/\alpha}-1} = \frac{1}{2i}\cot\frac{\pi n}{\alpha} - \frac{1}{2},$$

we have

$$\zeta_2(s,\alpha,1)$$

$$= \frac{-2\pi i(2\pi)^{s-1}}{\Gamma(s)(e^{2\pi i s}-1)}\left\{ -\frac{e^{\pi i s/2}(1+e^{\pi i s})}{2}\left(\xi(1-s,\alpha) + \alpha^{-s}\xi\left(1-s,\frac{1}{\alpha}\right)\right)\right.$$

$$\left. + \frac{e^{\pi i s/2}(1-e^{\pi i s})}{2i}(1-\alpha^{-s})\zeta(1-s)\right\}.$$

The transformation up to here is valid for $\mathrm{Re}(s) < -1$. Substituting $1-s$ for s and using $\Gamma(s)\Gamma(1-s) = \pi/\sin \pi s$, in the range $\mathrm{Re}(s) > 2$ we have

$$\zeta_2(1-s,\alpha,1)$$

$$= (2\pi)^{-s}\Gamma(s)\sin \pi s \left\{ \frac{e^{-\pi i s/2}(1-e^{-\pi i s})}{1-e^{-2\pi i s}}\left(\xi(s,\alpha) + \alpha^{s-1}\xi\left(s,\frac{1}{\alpha}\right)\right)\right.$$

$$\left. + \frac{ie^{-\pi i s/2}(1+e^{-\pi i s})}{1-e^{-2\pi i s}}(1-\alpha^{s-1})\zeta(s)\right\}$$

$$= \frac{1}{\gamma(s)}\left\{\xi(s,\alpha) - \alpha^{s-1}\xi\left(s,-\frac{1}{\alpha}\right) + (1-\alpha^{s-1})\cot(\pi s/2)\zeta(s)\right\},$$

and simplifying this, we have

$$\alpha^{s-1}\xi\left(s,-\frac{1}{\alpha}\right) - \xi(s,\alpha) = (1-\alpha^{s-1})\cot(\pi s/2)\zeta(s) - \gamma(s)\zeta_2(1-s,\alpha,1).$$

Therefore, noting $\xi(\alpha+n,s) = \xi(\alpha,s)$, we obtain the assertion of the theorem. □

The transformation formula (13.4) can be extended to the following form. For $x \in \mathbf{R}$, we denote by $\{x\}$ the real number which satisfies $x - \{x\} \in \mathbf{Z}$, $0 \le \{x\} < 1$ (fractional part of x).

Theorem 13.3. *For a real algebraic irrational number* α *and* $V = \begin{pmatrix} a & b \\ c & d \end{pmatrix} \in SL_2(\mathbf{Z})$ *which satisfies* $c > 0$ *and* $c\alpha + d > 0$*, the following transformation formula holds in the range* $\mathrm{Re}(s) > 2$.

$$\begin{aligned}&\eta^{s-1}\xi(s, V\alpha) - \xi(s,\alpha)\\ &\quad = (1-\eta^{s-1})\cot(\pi s/2)\zeta(s) - \gamma(s) \sum_{j \bmod c} \zeta_2(1-s,\eta,x_j+y_j\eta).\end{aligned}$$

Here we put $\eta = c\alpha + d$ *and* j *runs over integers which represent residue classes modulo c, and for each* j mod c*, we put* $x_j = 1 - \{jd/c\}$, $y_j = \{j/c\}$.

Proof. See [6] and [7].

If we put $V = \begin{pmatrix} n & -1 \\ 1 & 0 \end{pmatrix}$ in this theorem, we get the transformation formula in Proposition 13.2. □

From now on, we always assume that α is a real quadratic irrational number. Then there exists $V = \begin{pmatrix} a & b \\ c & d \end{pmatrix} \in SL_2(\mathbf{Z})$ which satisfies the conditions

$$V\alpha = \alpha, \quad c\alpha + d > 0. \tag{13.6}$$

This is easily understood if we think as follows. Let F be the real quadratic field $\mathbf{Q}(\alpha)$. Let $L = \mathbf{Z}\alpha + \mathbf{Z}$ be a lattice generated by α and 1 and $\mathcal{O}(L)$ be an order associated to L, that is, $\mathcal{O}(L) = \{u \in F \mid uL \subset L\}$. We denote by $E(L)$ the group of all totally positive units of $\mathcal{O}(L)$. We take an element η of $E(L)$ such that $\eta \ne 1$. Since $\eta \in L$ and $\eta\alpha \in L$, if we define a 2×2 matrix V with coefficients in $\mathbf{Z}$ by

$$\eta\begin{pmatrix}\alpha\\1\end{pmatrix} = V\begin{pmatrix}\alpha\\1\end{pmatrix}, \qquad V = \begin{pmatrix} a & b \\ c & d \end{pmatrix},$$

then we have $V \in SL_2(\mathbf{Z})$ and $\eta = c\alpha + d > 0$. In order to make $c > 0$, it is sufficient to take η so that $\eta > 1$ if $\alpha > \overline{\alpha}$ ($\overline{\alpha}$ is the conjugate of α), and to take η so that $0 < \eta < 1$ if $\alpha < \overline{\alpha}$.

Applying Theorem 13.3 to this situation, we get the following theorem.

Theorem 13.4. *When a real quadratic irrational number* α *is given, we take a totally positive unit* η *of* F *such that*

$$\eta\begin{pmatrix}\alpha\\1\end{pmatrix}=V\begin{pmatrix}\alpha\\1\end{pmatrix},\qquad V=\begin{pmatrix}a&b\\c&d\end{pmatrix}\in SL_2(\mathbf{Z})$$

with $c>0$. *Then in the range* $\mathrm{Re}(s)>2$, *we have*

$$\xi(s,\alpha)=-\cot(\pi s/2)\zeta(s)+\frac{\gamma(s)}{1-\eta^{s-1}}\sum_{j \bmod c}\zeta_2(1-s,\eta,x_j+y_j\eta). \qquad (13.7)$$

Through this expression, $\xi(s,\alpha)$ *is analytically continued to a meromorphic function on the whole* s*-plane, and the poles are* $s=1$ *and* $s=1+\frac{2\pi i n}{\log\eta}$ $(n\in\mathbf{Z},\ n\neq 0)$ *and they are of order* 1. *Moreover, its values at* $s=2m-1$ $(m\in\mathbf{Z},\ m\geq 2)$ *are expressed using Bernoulli polynomials:*[7]

$$\xi(2m-1,\alpha)=\frac{(-1)^{m-1}(2\pi)^{2m-1}}{1-\eta^{2m-2}}\sum_{k=0}^{2m}\sum_{j \bmod c}\frac{B_k(x_j)B_{2m-k}(y_j)}{k!(2m-k)!}\eta^{2m-k-1}.$$

Proof. The expression (13.7) is immediately obtained by using $V\alpha=\alpha$ in Theorem 13.3. Poles can possibly appear only at $s=2$, $s=1$, $s=1+\frac{2\pi i n}{\log\eta}$ $(n\in\mathbf{Z},\ n\neq 0)$, and $s=0$. For $s=-2,-4,-6,\ldots$ we have $\zeta(s)=0$, so we note that these are not poles. As for $s=2$, as we remarked in the footnote on p. 213, $\xi(s,\alpha)$ converges absolutely for $\mathrm{Re}(s)>1$ and is holomorphic there, so this cannot be a pole. Now we consider $s=0$. Taking the result that the principal part of $\zeta_2(s,\omega,a)$ at the pole $s=1$ is given by (13.3) and the fact $\zeta(0)=-\frac{1}{2}$ into consideration, the residue of the function $\xi(s,\alpha)$ at $s=0$ (possible pole of order at most 1) is given by

$$\frac{1}{\pi}-\frac{2}{\pi(1-\eta^{-1})}\sum_{j \bmod c}\left(\frac{1+\eta}{2\eta}-\frac{x_j+y_j\eta}{\eta}\right).$$

We see easily that this value is zero. So $\xi(s,\alpha)$ is holomorphic at $s=0$.

Lastly, the values at $s=2m-1$ $(m\in\mathbf{Z},\ m\geq 2)$ are, by virtue of (13.7), given by

$$\xi(2m-1,\alpha)=\frac{(-1)^{m-1}(2\pi)^{2m-1}}{(1-\eta^{2m-2})(2m-2)!}\sum_{j \bmod c}\zeta_2(2-2m,\eta,x_j+y_j\eta)$$

and by Proposition 13.1, we get the expression using Bernoulli polynomials. □

[7] When α is a unit of a real quadratic field, this kind of special value has been calculated by Berndt [14].

13.3 $\xi(s, \alpha)$ and Continued Fractions

As explained in detail in *Zetafunktionen und quadratische Körper* by D. Zagier [107, §13], any quadratic irrational number β can be expanded into a continued fraction[8] of the following form.

$$\beta = b_1 - \cfrac{1}{b_2 - \cfrac{1}{b_3 - \cfrac{1}{b_4 - \ddots}}} \qquad (b_j \in \mathbf{Z},\ b_2, b_3, \ldots \geq 2).$$

It is important that all the coefficients $b_j (j \geq 2)$ except for the first one are integers bigger than or equal to 2. Here the sequence $\{b_1, b_2, \ldots\}$ is periodic, namely, there exist some natural numbers ν and r such that for j not less than ν, we have

$$b_{r+j} = b_j \qquad (j \geq \nu).$$

We call the smallest such natural number r the period of β. We write this continued fraction expansion as

$$\beta = [[b_1, b_2, \ldots, b_{\nu-1}, \overline{b_\nu, \ldots, b_{\nu+r-1}}]]. \tag{13.8}$$

The sequence $\{b_\nu, \ldots, b_{\nu+r-1}\}$ is called a fundamental period. Moreover when $\nu = 1$, that is, if it is periodic from the first term, we say that the continued fraction expansion of β is purely periodic.

For a real quadratic irrational number β, We denote by $\overline{\beta}$ the conjugate number of β. Now we assume that the continued fraction expansion of a real quadratic irrational number ω is purely periodic and we write

$$\omega = [[\overline{n_1, n_2, \ldots, n_r}]] \qquad (n_j \geq 2).$$

Then we have

$$\frac{1}{\overline{\omega}} = [[\overline{n_r, n_{r-1}, \ldots, n_1}]]$$

[107, p. 136, (17)]. By this expression, we have

$$\omega > 1 \quad \text{and} \quad 0 < \overline{\omega} < 1. \tag{13.9}$$

---

[8] The continued fractions treated here are different from the usual continued fractions of real quadratic irrational numbers.

If a real quadratic irrational number ω satisfies condition (13.9), we say that ω is reduced.[9] Conversely, if ω satisfies this condition, we can show that the continued fraction expansion of ω is periodic, so ω is reduced if and only if the continued fraction expansion of ω is purely periodic.

Let F be a real quadratic field and regard F as a subfield of the real numbers $\mathbf{R}$. Moreover we define a mapping of F into $\mathbf{R} \times \mathbf{R}$ by

$$F(\theta) = (\theta, \overline{\theta}) \in \mathbf{R} \times \mathbf{R},$$

and identify θ with the point $(\theta, \overline{\theta})$ of $\mathbf{R} \times \mathbf{R}$. Now we assume that $\omega \in F - \mathbf{Q}$ is reduced and that the continued fraction expansion of ω is given by

$$\omega = [[\, \overline{n_1, n_2, \ldots, n_r} \,]].$$

We define an integer sequence $\{n_j\}_{j \in \mathbf{Z}}$ by extending the continued fraction expansion $\{n_1, n_2, \ldots, n_r\}$ by periodicity so that $n_{j+r} = n_j$ for any $j \in \mathbf{Z}$. For a positive integer i we define a reduced number ω_i by

$$\omega_i = [[\, \overline{n_i, n_{i+1}, \ldots, n_{i+r-1}} \,]]. \tag{13.10}$$

Of course we have $\omega_1 = \omega_{r+1} = \omega$. For $n \in \mathbf{Z}$, we put

$$V(n) = \begin{pmatrix} n & -1 \\ 1 & 0 \end{pmatrix} \in SL_2(\mathbf{Z}).$$

From the continued fraction expansion (13.10), we get

$$\omega_i = n_i - \frac{1}{\omega_{i+1}} = V(n_i)\omega_{i+1},$$

so we have

$$\omega = V(n_1)V(n_2)\cdots V(n_r)\omega.$$

We define now a lattice L of F by

$$L = \mathbf{Z}\omega + \mathbf{Z}$$

and take a generator ε of the group $E(L)$ of totally positive units of $\mathcal{O}(L)$ so that $\varepsilon > 1$. We define V by the relation

[9] This is different from the usual definition of reduced for the setting of ordinary continued fractions. Usually a real quadratic irrational number α is called reduced when $\alpha > 1$ and $0 > \overline{\alpha} > -1$.

$$\varepsilon \begin{pmatrix} \omega \\ 1 \end{pmatrix} = V \begin{pmatrix} \omega \\ 1 \end{pmatrix}, \qquad V = \begin{pmatrix} a & b \\ c & d \end{pmatrix} \in SL_2(\mathbf{Z}).$$

If we think of the stabilizer subgroup $\Gamma_\omega = \{M \in SL_2(\mathbf{Z}) \mid M\omega = \omega\}$ of ω, we see Γ_ω is generated by V and $\pm I_2$ (I_2 is the identity matrix). Because $c > 0$ and r is the period of ω, we have

$$V = V(n_1)V(n_2)\cdots V(n_r).$$

For any $V_0 \in SL_2(\mathbf{Z})$ and a real algebraic irrational number α, we use the notation of an automorphy factor

$$J(V_0, \alpha) = c\alpha + d. \tag{13.11}$$

As in the case when α is in the complex upper half plane, this has a property $J(V_1V_2, \alpha) = J(V_1, V_2\alpha)J(V_2, \alpha)$ for any V_1, $V_2 \in SL_2(\mathbf{Z})$. Then using this property of the automorphy factor (13.11), we have

$$\begin{aligned} \varepsilon = J(V, \omega) &= J(V(n_1), \omega_2)J(V(n_2), \omega_3) \cdots J(V(n_r), \omega) \\ &= \omega_2\omega_3 \cdots \omega_r\omega. \end{aligned}$$

So we have

$$\varepsilon = \omega_1\omega_2 \cdots \omega_r. \tag{13.12}$$

Now, we put $\xi_0 = 1$, $\xi_1 = \omega = \omega_{r+1}$, $\xi_2 = \omega_{r+1}\omega_r, \ldots, \xi_r = \omega_{r+1}\omega_r \cdots \omega_2$. Of course we have $\xi_r = \varepsilon$. Here we have

$$L = \mathbf{Z}\xi_i + \mathbf{Z}\xi_{i+1} \qquad (0 \le i \le r-1), \tag{13.13}$$

because by the relation $\xi_{i-1} = \xi_i \times \frac{1}{\omega_{r+2-i}}$ we have

$$\begin{aligned} \mathbf{Z}\xi_{i-1} + \mathbf{Z}\xi_i &= \xi_i \left(\mathbf{Z}\frac{1}{\omega_{r+2-i}} + \mathbf{Z} \right) \\ &= \xi_i \left(\mathbf{Z}(n_{r+1-i} - \omega_{r+1-i}) + \mathbf{Z} \right) \\ &= \mathbf{Z}\xi_i + \mathbf{Z}\xi_{i+1} \end{aligned}$$

and starting from $i = 1$ inductively we can show the assertion.

Now we define a double zeta function of a lattice L by

$$\Xi_L(s) = \sum_{\theta \in L \bigcap C(1,\varepsilon)} \theta^{-s}. \tag{13.14}$$

Here we denote by $C(1,\varepsilon)$ the simplicial cone given by

$$C(1,\varepsilon) = \{(x + y\varepsilon, x + y\varepsilon') \mid x > 0,\ y \geq 0\}.$$

As we shall see later, $\Xi_L(s)$ is written as a finite sum of double zeta functions explained before, so it converges absolutely for $\mathrm{Re}(s) > 2$ and is holomorphic in that range. Let $\mathbf{R}_+$ be the set of positive numbers. The group of totally positive units $E(L)$ acts on $\mathbf{R}_+ \times \mathbf{R}_+$ by

$$\eta(u,v) = (\eta u, \overline{\eta} v) \qquad (\eta \in E(L),\ (u,v) \in \mathbf{R}_+ \times \mathbf{R}_+).$$

The set $C(1,\varepsilon)$ is a fundamental domain of $\mathbf{R}_+ \times \mathbf{R}_+$ by $E(L)$.

Since $\omega_i > 1$ for each i and since we see $1 < \xi_i < \xi_{i+1} < \epsilon$ and $\overline{\epsilon} < \overline{\xi_{i+1}} < \overline{\xi_i} < 1$, we can write

$$C(1,\varepsilon) = \bigcup_{k=1}^{r} \{x\xi_{k-1} + y\xi_k \mid x > 0,\ y \geq 0\} \quad \text{(disjoint union)},$$

so by (13.13) we get

$$L \cap C(1,\varepsilon) = \bigcup_{k=1}^{r} \{m\xi_{k-1} + n\xi_k \mid m \in \mathbf{Z}_{>0},\ n \in \mathbf{Z}_{\geq 0}\} \quad \text{(disjoint union)}. \tag{13.15}$$

This decomposition was given in Zagier [105]. Here we denote by $\mathbf{Z}_{>0}$ or $\mathbf{Z}_{\geq 0}$ the set of positive integers, or integers not less than 0, respectively. By the decomposition (13.15), we get

$$\Xi_L(s) = \sum_{k=1}^{r} (\xi_{k-1})^{-s} \zeta_2(s, \omega_{r+2-k}, 1). \tag{13.16}$$

By this expression, $\Xi_L(s)$ converges absolutely for $\mathrm{Re}(s) > 2$ and is continued analytically to a meromorphic function which is holomorphic in the whole s-plane except for $s = 1,\ 2$.

Between the functions $\xi(s,\omega)$ and $\Xi_L(s)$, the following functional equation holds.

Theorem 13.5. *Let ω be a reduced number of F. Put $L = \mathbf{Z} + \mathbf{Z}\omega$ and let $\varepsilon > 1$ be a totally positive fundamental unit of the order $\mathcal{O}(L)$ of L. Then we have*

$$\xi(s,\omega) = -\cot(\pi s/2)\zeta(s) - \frac{\gamma(s)}{\varepsilon^{s-1} - 1} \Xi_L(1-s). \tag{13.17}$$

We can also show through this equation that $\xi(s,\omega)$ can be continued analytically to a meromorphic function of s.

Proof. Since we have now $\omega_i = V(n_i)\omega_{i+1}$ and $\omega_i > 0$, by Proposition 13.2, we have an equality

$$(\omega_{r+2-k})^{s-1}\xi(s,\omega_{r+1-k}) - \xi(s,\omega_{r+2-k})$$

$$= \left(1-(\omega_{r+2-k})^{s-1}\right)\cot(\pi s/2)\zeta(s) - \gamma(s)\zeta_2(1-s,\omega_{r+2-k},1)$$

for $1 \leq k \leq r$. Multiplying $(\xi_{k-1})^{s-1}$ by both sides of this equality, then taking the sum from $k=1$ to $k=r$ and using (13.12) and (13.16), we get

$$(\varepsilon^{s-1}-1)\xi(s,\omega) = (1-\varepsilon^{s-1})\cot(\pi s/2)\zeta(s) - \gamma(s)\Xi_L(1-s).$$

□

Remark 13.6. If we cut out the contour integral representation, not only of a double zeta function, but also of a multiple zeta function in the same way as in Sect. 13.2, then a kind of Lerch type zeta function appears. It would be an interesting problem to investigate what kind of modular properties it has.

Chapter 14
Poly-Bernoulli Numbers

In this chapter, we define and study a generalization of Bernoulli numbers referred to as poly-Bernoulli numbers, which is a different generalization than the generalized Bernoulli numbers introduced in Chap. 4.

14.1 Poly-Bernoulli Numbers

We give a definition in a down-to-earth way.

Definition 14.1 (Poly-Bernoulli number [56]). Poly-Bernoulli numbers $\mathbb{B}_n^{(k)} \in \mathbf{Q}$ $(n \geq 0,\ k \in \mathbf{Z})$ are defined by the generating function

$$\frac{Li_k(1-e^{-t})}{1-e^{-t}} = \sum_{n=0}^{\infty} \mathbb{B}_n^{(k)} \frac{t^n}{n!}.$$

Here, for any integer k, $Li_k(t)$ stands for the formal power series $\displaystyle\sum_{n=1}^{\infty} \frac{t^n}{n^k}$.

When $k \geq 1$, the series $Li_k(t)$ is the defining series of a "polylogarithm" function, whereas when $k \leq 0$, it is the Taylor series of the rational function $\left(t\frac{d}{dt}\right)^{-k}\left(\frac{t}{1-t}\right)$ at the origin. If $k = 1$, we have $Li_1(t) = -\log(1-t)$ and $Li_1(1-e^{-t}) = t$, hence by Theorem 1.12 on p. 20, the number $\mathbb{B}_n^{(1)}$ is nothing but the original Bernoulli number B_n. If $k \geq 1$, by making a change of variables $t_i \mapsto 1-e^{-t_i}$ in the following iterated integral expression of $Li_k(t)$,

$$Li_k(t) = \int_0^t \frac{dt_k}{t_k} \int_0^{t_k} \frac{dt_{k-1}}{t_{k-1}} \cdots \int_0^{t_3} \frac{dt_2}{t_2} \int_0^{t_2} \frac{dt_1}{1-t_1},$$

T. Arakawa et al., *Bernoulli Numbers and Zeta Functions*, Springer Monographs in Mathematics, DOI 10.1007/978-4-431-54919-2_14, © Springer Japan 2014

Table 14.1 $\mathbb{B}_n^{(k)}$ $(0 \le k \le 5,\ 0 \le n \le 7)$

$k\backslash n$	0	1	2	3	4	5	6	7
0	1	1	1	1	1	1	1	1
1	1	$\frac{1}{2}$	$\frac{1}{6}$	0	$-\frac{1}{30}$	0	$\frac{1}{42}$	0
2	1	$\frac{1}{4}$	$-\frac{1}{36}$	$-\frac{1}{24}$	$\frac{7}{450}$	$\frac{1}{40}$	$-\frac{38}{2205}$	$-\frac{5}{168}$
3	1	$\frac{1}{8}$	$-\frac{11}{216}$	$-\frac{1}{288}$	$\frac{1243}{54000}$	$-\frac{49}{7200}$	$-\frac{75613}{3704400}$	$\frac{599}{35280}$
4	1	$\frac{1}{16}$	$-\frac{49}{1296}$	$\frac{41}{3456}$	$\frac{26291}{3240000}$	$-\frac{1921}{144000}$	$\frac{845233}{1555848000}$	$\frac{1048349}{59270400}$
5	1	$\frac{1}{32}$	$-\frac{179}{7776}$	$\frac{515}{41472}$	$-\frac{216383}{194400000}$	$-\frac{183781}{25920000}$	$\frac{4644828197}{653456160000}$	$\frac{153375307}{49787136000}$

we have the following expression of the generating function:

$$\frac{Li_k(1-e^{-t})}{1-e^{-t}} = \\ e^t \cdot \frac{1}{e^t-1}\int_0^t \frac{dt_k}{e^{t_k}-1}\int_0^{t_k}\frac{dt_{k-1}}{e^{t_{k-1}}-1}\int_0^{t_{k-1}}\cdots\int_0^{t_3}\frac{dt_2}{e^{t_2}-1}\int_0^{t_2} dt_1. \qquad (14.1)$$

This shows that, in the case of $k \ge 1$, the generating function $\sum_{n=0}^{\infty}\mathbb{B}_n^{(k)}\frac{t^n}{n!}$ of poly-Bernoulli numbers $\mathbb{B}_n^{(k)}$ is obtained, starting with 1, by applying the integration $\int_0^t dt$ and division by $e^t - 1$ successively k times, and finally by multiplying e^t.

Remark 14.2. (1) Our poly-Bernoulli number is different from the "Bernoulli numbers of higher order" defined by Nörlund [75, Chapter 6]. The generating function of the Bernoulli numbers of higher order is $\left(\frac{t}{e^t-1}\right)^k$ and this can be written as a linear combination of derivatives of $\frac{t}{e^t-1}$. Hence the Bernoulli number of higher order is essentially a linear combination of classical Bernoulli numbers. As is shown below, our $\mathbb{B}_n^{(k)}$ can also be expressed in terms of classical Bernoulli numbers, but we need to take products successively.

(2) There is a very interesting object called "multiple zeta values" (cf. [57,98,109]), which generalize the values at positive integer arguments of the Riemann zeta function. The poly-Bernoulli numbers are in some way related to the multiple zeta values [9].

Table 14.1 gives some values of $\mathbb{B}_n^{(k)}$ for $k \ge 0$.

The next two propositions give recurrence formulas for $\mathbb{B}_n^{(k)}$, which can be derived from the definition and the iterated integral expression of the generating function.

Proposition 14.3. *For any $k \in \mathbf{Z}$ and $n \ge 0$, we have*

$$\mathbb{B}_n^{(k)} = \frac{1}{n+1}\left\{\mathbb{B}_n^{(k-1)} - \sum_{m=1}^{n-1}\binom{n}{m-1}\mathbb{B}_m^{(k)}\right\}.$$

We regard the sum on the right as 0 if $n \le 1$.

Proof. We differentiate the relation $Li_k(1-e^{-t}) = (1-e^{-t})\left(\sum_{n=0}^{\infty} \mathbb{B}_n^{(k)} \frac{t^n}{n!}\right)$ which is obtained from the definition by multiplying by $1-e^{-t}$. By the relation $Li_k'(t) = \frac{1}{t} Li_{k-1}(t)$, we have

$$\frac{Li_{k-1}(1-e^{-t})}{1-e^{-t}} e^{-t} = e^{-t} \sum_{n=0}^{\infty} \mathbb{B}_n^{(k)} \frac{t^n}{n!} + (1-e^{-t}) \sum_{n=1}^{\infty} \mathbb{B}_n^{(k)} \frac{t^{n-1}}{(n-1)!}.$$

Multiplying by e^t on both sides and using $\frac{Li_{k-1}(1-e^{-t})}{1-e^{-t}} = \sum_{n=0}^{\infty} \mathbb{B}_n^{(k-1)} \frac{t^n}{n!}$, we have

$$\begin{aligned}
\sum_{n=0}^{\infty} \mathbb{B}_n^{(k-1)} \frac{t^n}{n!} &= \sum_{n=0}^{\infty} \mathbb{B}_n^{(k)} \frac{t^n}{n!} + (e^t - 1) \sum_{n=1}^{\infty} \mathbb{B}_n^{(k)} \frac{t^{n-1}}{(n-1)!} \\
&= \sum_{n=0}^{\infty} \mathbb{B}_n^{(k)} \frac{t^n}{n!} + \sum_{l=1}^{\infty} \frac{t^l}{l!} \sum_{n=1}^{\infty} \mathbb{B}_n^{(k)} \frac{t^{n-1}}{(n-1)!} \\
&= \sum_{n=0}^{\infty} \mathbb{B}_n^{(k)} \frac{t^n}{n!} + \sum_{n=1}^{\infty} \left(\sum_{m=0}^{n-1} \mathbb{B}_{m+1}^{(k)} \frac{1}{(n-m)!m!} \right) t^n \\
&= \mathbb{B}_0^{(k)} + \sum_{n=1}^{\infty} \left(\mathbb{B}_n^{(k)} + \sum_{m=0}^{n-1} \binom{n}{m} \mathbb{B}_{m+1}^{(k)} \right) \frac{t^n}{n!}.
\end{aligned}$$

Hence we have $\mathbb{B}_0^{(k-1)} = \mathbb{B}_0^{(k)}$ if $n = 0$, and if $n \geq 1$ we obtain

$$\mathbb{B}_n^{(k-1)} = \mathbb{B}_n^{(k)} + \sum_{m=0}^{n-1} \binom{n}{m} \mathbb{B}_{m+1}^{(k)} = (n+1)\mathbb{B}_n^{(k)} + \sum_{m=1}^{n-1} \binom{n}{m-1} \mathbb{B}_m^{(k)}.$$

This divided by $n+1$ gives the proposition. □

Proposition 14.4. *For $k \geq 1$ and $n \geq 0$, we have*

$$\mathbb{B}_n^{(k)} = \sum_{m=0}^{n} (-1)^m \binom{n}{m} \mathbb{B}_{n-m}^{(k-1)} \left(\sum_{l=0}^{m} \frac{(-1)^l}{n-l+1} \binom{m}{l} B_l \right).$$

Proof. We use the iterated integral expression of the generating function, which shows

$$\frac{Li_k(1-e^{-t})}{1-e^{-t}} = \frac{e^t}{e^t - 1} \int_0^t e^{-s} \frac{Li_{k-1}(1-e^{-s})}{1-e^{-s}} ds.$$

From this we have

$$\begin{aligned}
\sum_{n=0}^{\infty} \mathbb{B}_n^{(k)} \frac{t^n}{n!} &= \left(\sum_{n=0}^{\infty} B_n \frac{t^{n-1}}{n!}\right) \int_0^t \left(\sum_{n=0}^{\infty} \frac{(-s)^n}{n!}\right) \left(\sum_{n=0}^{\infty} \mathbb{B}_n^{(k-1)} \frac{s^n}{n!}\right) ds \\
&= \left(\sum_{n=0}^{\infty} B_n \frac{t^{n-1}}{n!}\right) \int_0^t \sum_{n=0}^{\infty} \left(\sum_{m=0}^{n} (-1)^{n-m} \binom{n}{m} \mathbb{B}_m^{(k-1)}\right) \frac{s^n}{n!} ds \\
&= \left(\sum_{n=0}^{\infty} B_n \frac{t^{n-1}}{n!}\right) \sum_{n=0}^{\infty} \left(\sum_{m=0}^{n} (-1)^{n-m} \binom{n}{m} \mathbb{B}_m^{(k-1)}\right) \frac{t^{n+1}}{(n+1)!} ds \\
&= \sum_{n=0}^{\infty} \left(\sum_{l=0}^{n} \frac{B_{n-l}}{l+1} \binom{n}{l} \left(\sum_{m=0}^{l} (-1)^{l-m} \binom{l}{m} \mathbb{B}_m^{(k-1)}\right)\right) \frac{t^n}{n!} \\
&= \sum_{n=0}^{\infty} \left(\sum_{m=0}^{n} (-1)^m \mathbb{B}_m^{(k-1)} \left(\sum_{l=m}^{n} \frac{(-1)^l}{l+1} \binom{n}{l} \binom{l}{m} B_{n-l}\right)\right) \frac{t^n}{n!}.
\end{aligned}$$

Use $\binom{n}{l}\binom{l}{m} = \binom{n}{m}\binom{n-m}{n-l}$ and replace $l \to n-l$ and then $m \to n-m$ to obtain

$$\begin{aligned}
&\sum_{m=0}^{n} (-1)^m \mathbb{B}_m^{(k-1)} \left(\sum_{l=m}^{n} \frac{(-1)^l}{l+1} \binom{n}{l} \binom{l}{m} B_{n-l}\right) \\
&\quad = \sum_{m=0}^{n} (-1)^m \binom{n}{m} \mathbb{B}_{n-m}^{(k-1)} \left(\sum_{l=0}^{m} \frac{(-1)^l}{n-l+1} \binom{m}{l} B_l\right),
\end{aligned}$$

which gives the proposition. □

Remark 14.5. In the above two recurrences, we need the numbers with different upper index like $\mathbb{B}_n^{(k-1)}$ in order to express $\mathbb{B}_n^{(k)}$. We do not know if there exists a recurrence formula with the single index k.

The following theorem generalizes Theorem 2.8 (p. 35), which expresses Bernoulli numbers by Stirling numbers.

Theorem 14.6.

$$\mathbb{B}_n^{(k)} = (-1)^n \sum_{m=0}^{n} \frac{(-1)^m m! \left\{ {n \atop m} \right\}}{(m+1)^k}, \qquad (n \geq 0,\ k \in \mathbf{Z}).$$

Proof. The proof goes in the same way as Theorem 2.8. In fact, by using Proposition 2.6 (7) (p. 30) we have

$$\sum_{n=0}^{\infty} \mathbb{B}_n^{(k)} \frac{t^n}{n!} = \frac{Li_k(1-e^{-t})}{1-e^{-t}} = \sum_{m=1}^{\infty} \frac{(1-e^{-t})^{m-1}}{m^k}$$
$$= \sum_{m=0}^{\infty} \frac{(-1)^m (e^{-t}-1)^m}{(m+1)^k} = \sum_{m=0}^{\infty} \frac{(-1)^m m!}{(m+1)^k} \sum_{n=m}^{\infty} \left\{ {n \atop m} \right\} \frac{(-t)^n}{n!}$$
$$= \sum_{n=0}^{\infty} (-1)^n \left(\sum_{m=0}^{n} \frac{(-1)^m m! \left\{ {n \atop m} \right\}}{(m+1)^k} \right) \frac{t^n}{n!}.$$

Comparing coefficients of $\frac{t^n}{n!}$ on both sides, we obtain the theorem. □

14.2 Theorem of Clausen and von Staudt Type

In this section we discuss denominators of poly-Bernoulli numbers. The denominators of the classical Bernoulli numbers are completely determined by the theorem of Clausen and von Staudt (Theorem 3.1, p. 41). The theorem not only determines the denominator but also describes the "fractional part" of B_n. In the case of general $\mathbb{B}_n^{(k)}$ we only have partial results below, but when $k = 2$ ("di-Bernoulli numbers"), the denominators are completely determined. Note that, by Theorem 14.6, no prime greater than $n + 1$ appears in the denominator of $\mathbb{B}_n^{(k)}$.

Theorem 14.7. *Let $k \geq 2$ be an integer and p be a prime number satisfying $k+2 \leq p \leq n+1$.*

(1) *When $p - 1 \mid n$, $p^k \mathbb{B}_n^{(k)}$ is in $\mathbf{Z}_{(p)}$ (the set of rational numbers whose denominators are prime to p) and*

$$p^k \mathbb{B}_n^{(k)} \equiv -1 \mod p.$$

(2) *When $p - 1 \nmid n$, $p^{k-1} \mathbb{B}_n^{(k)}$ is in $\mathbf{Z}_{(p)}$ and*

$$p^{k-1}\mathbb{B}_n^{(k)} \equiv \begin{cases} \dfrac{1}{p}\left\{ {n \atop p-1} \right\} - \dfrac{n}{2^k} \mod p & \textit{if } \; n \equiv 1 \mod (p-1), \\ \dfrac{(-1)^{n-1}}{p}\left\{ {n \atop p-1} \right\} \mod p & \textit{otherwise}. \end{cases}$$

Proof. Denote by $\mathrm{ord}_p(a)$ the p-order of a rational number a. We have $\mathrm{ord}_p(p^t) = t$ and both numerator and denominator of $a \cdot p^{-\mathrm{ord}_p(a)}$ are prime to p. We proceed in the same way as in the proof of Theorem 3.1, using the formula in Theorem 14.6 and calculating the p-order of each term in the formula. Put $b_n^{(k)}(m) = \frac{(-1)^m m! \left\{ {n \atop m} \right\}}{(m+1)^k}$. We prove (1) and (2) simultaneously.

Write $m+1 = ap^e, (a,p) = 1, e \geq 0$. If $e = 0$, then $b_n^{(k)}(m) \in \mathbf{Z}_{(p)}$ and $p^{k-1}b_n^{(k)}(m) \equiv 0 \mod p$ by the assumption $k \geq 2$, hence this term can be discarded. Note that, since the Stirling number $\left\{ {n \atop m} \right\}$ is an integer, we have

$$\mathrm{ord}_p(b_n^{(k)}(m)) \geq \mathrm{ord}_p\left(\frac{m!}{(m+1)^k}\right).$$

First assume $e \geq 2$. We show the congruence $p^k b_n^{(k)}(m) \equiv 0 \mod p$ and further $p^{k-1}b_n^{(k)}(m) \equiv 0 \mod p$ if $p-1 \nmid n$. Using $\mathrm{ord}_p(m!) = \sum_{j=1}^{\infty}\left[\frac{m}{p^j}\right]$, we estimate

$$\begin{aligned}
\mathrm{ord}_p\left(\frac{m!}{(m+1)^k}\right) &= \sum_{j=1}^{\infty}\left[\frac{m}{p^j}\right] - ek \\
&\geq \left[\frac{m}{p}\right] - ek = \left[\frac{ap^e - 1}{p}\right] - ek = ap^{e-1} - 1 - ek \\
&\geq p^{e-1} - 1 - ek = (1 + p - 1)^{e-1} - 1 - ek \\
&\geq 1 + (e-1)(p-1) - 1 - ek = (e-1)(p-1) - ek \\
&\geq (e-1)(k+1) - ek = -k + e - 1 \\
&\geq -k + 1.
\end{aligned}$$

From this we obtain $\mathrm{ord}_p\left(\frac{m!}{(m+1)^k}\right) \geq -k+1$ and thus $p^{k-1}b_n^{(k)}(m) \in \mathbf{Z}_{(p)}$. Therefore $p^k b_n^{(k)}(m) \equiv 0 \mod p$ holds. If at least one inequality in the above chain of inequalities is strict, we get $p^{k-1}b_n^{(k)}(m) \equiv 0 \mod p$. If all the inequalities become equalities, we have $e = 2, m+1 = p^2$, and $p = k+2$. In this case, the lemma below (the case $a = p$) shows $p^{k-1}b_n^{(k)}(m) \equiv 0 \mod p$ if $p - 1 \nmid n$.

Lemma 14.8. *Let n and a be natural numbers. We have the congruence*

$$\left\{ {n \atop ap-1} \right\} \equiv \begin{cases} \dbinom{c-1}{a-1} \mod p & \text{if } n = a - 1 + c(p-1) \text{ for some } c \geq a, \\ 0 \mod p & \text{otherwise.} \end{cases}$$

We use the generating function of the Stirling numbers $\left\{ {n \atop m} \right\}$ (Proposition 2.6 (8) on p. 30)

$$\sum_{n=m}^{\infty}\left\{ {n \atop m} \right\} t^n = \frac{t^m}{(1-t)(1-2t)\cdots(1-mt)}. \tag{14.2}$$

When $m = ap - 1$, the right-hand side reduces modulo p to

$$\frac{t^{ap-1}}{(1-t^{p-1})^a} = t^{ap-1}\sum_{i=0}^{\infty}\binom{a+i-1}{i}t^{i(p-1)} = \sum_{i=0}^{\infty}\binom{a+i-1}{a-1}t^{a-1+(a+i)(p-1)}.$$

Here we used $(1-t)(1-2t)\cdots(1-(p-1)t) \equiv 1-t^{p-1} \mod p$. Putting $a+i=c$, we obtain the lemma. □

Next suppose $e=1$ $(m=ap-1)$. If $a \geq 3$, then $p^2|(ap-1)!$ ensures $\text{ord}_p\left(\frac{m!}{(m+1)^k}\right) > -k+1$ and this gives $p^{k-1}b_n^{(k)}(m) \equiv 0 \mod p$. If $a=2$, we have $\text{ord}_p(m!) = 1$ and $\text{ord}_p\left(\frac{m!}{(m+1)^k}\right) = 1-k$, and hence $\text{ord}_p(b_n^{(k)}(m)) = 1-k+\text{ord}_p\left(\left\{{n \atop m}\right\}\right)$. From this, we have $p^k b_n^{(k)}(m) \equiv 0 \mod p$. If $n \not\equiv 1 \mod (p-1)$, by putting $a=2$ in the lemma above we have $\left\{{n \atop 2p-1}\right\} \equiv 0 \mod p$. This gives us $p^{k-1}b_n^{(k)}(m) \equiv 0 \mod p$ if $n \not\equiv 1 \mod (p-1)$. If $n \equiv 1 \mod (p-1)$, by writing $n = 1+c(p-1)$, we have $c \geq 2$ (note $c=1$ never happens because $n \geq m$) and the lemma again gives $\left\{{n \atop 2p-1}\right\} \equiv c-1 \equiv -n \mod p$. Hence we have

$$\begin{aligned} p^{k-1}b_n^{(k)}(m) &= p^{k-1}\frac{(-1)^{2p-1}(2p-1)!\left\{{n \atop 2p-1}\right\}}{(2p)^k} \\ &\equiv \frac{n}{2^k} \mod p. \end{aligned}$$

(We have used Wilson's[1] theorem $(p-1)! \equiv -1 \mod p$, which implies $(2p-1)!/p \equiv ((p-1)!)^2 \equiv 1 \mod p$.) Lastly, when $a=1$ $(m=p-1)$, we have $b_n^{(k)}(m) = \frac{(p-1)!\left\{{n \atop p-1}\right\}}{p^k}$. By the lemma we have $\left\{{n \atop p-1}\right\} \equiv 0 \mod p$ if $n \not\equiv 0 \mod (p-1)$ and hence $p^{k-1}b_n^{(k)}(m) \equiv -\frac{1}{p}\left\{{n \atop p-1}\right\} \mod p$. If $n \equiv 0 \mod (p-1)$, then we have $\left\{{n \atop p-1}\right\} \equiv 1 \mod p$ and $p^k b_n^{(k)}(m) \equiv -1 \mod p$. Combining all these, we obtain the theorem. □

Remark 14.9. Let k and p be as in the theorem, and assume $n \not\equiv 0, 1 \mod (p-1)$. Denote by n' the unique integer satisfying $n' \equiv n \mod p-1$ and $1 < n' < p$. Then we have

$$p^{k-1}\mathbb{B}_n^{(k)} \equiv (n-n')\frac{B_n}{n} \mod p.$$

If in particular n is odd, we have $p^{k-1}\mathbb{B}_n^{(k)} \equiv 0 \mod p$.

This comes from the congruence

[1] John Wilson (born on August 6, 1741 in Applethwaite, England—died on October 18, 1793 in Kendal, England).

$$\frac{(-1)^{n-1}}{p}\left\{ {n \atop p-1} \right\} \equiv (n-n')\frac{B_n}{n} \mod p,$$

which is a consequence of Eq. (14.3) in the proof of Theorem 14.10.

The denominators of di-Bernoulli numbers are completely determined.

Theorem 14.10. (1) *If n is odd, we have $\mathbb{B}_n^{(2)} = -\frac{(n-2)}{4}B_{n-1}$. (Hence in this case the determination of the denominator is essentially reduced to the theorem of Clausen and von Staudt (Theorem 3.1).)*

(2) *Suppose n is even (≥ 2). For a prime p, denote by $ord(p,n)$ the p*-order *of $\mathbb{B}_n^{(2)}$. Then we have the following:*

(2-1) *If $p > n+1$, then $ord(p,n) \geq 0$ (no p appears in the denominator).*

(2-2) *For p with $5 \leq p \leq n+1$,*

(a) *$ord(p,n) = -2$ if $p-1 \mid n$.*

(b) *Suppose $p-1 \nmid n$.*

(b-1) *We have $ord(p,n) \geq 0$ if either $p \mid \frac{B_n}{n}$ or $n \equiv n' \mod p(p-1)$ for some n' with $1 < n' < p-1$, and*

(b-2) *$ord(p,n) = -1$ otherwise.*

(2-3) *We have* $ord(3,n) \geq 0$ *if $n > 2$ and $n \equiv 2 \mod 3$, and* $ord(3,n) = -2$ *otherwise.*

(2-4) *We have* $ord(2,n) \geq 0$ *if $n > 2$ and $n \equiv 2 \mod 4$,* $ord(2,n) = -1$ *if $n \equiv 0 \mod 4$, and* $ord(2,2) = -2$.

Remark 14.11. The rational number $\frac{B_n}{n}$ in (b-1) is always an element of $\mathbf{Z}_{(p)}$ (Theorem 3.2 (1)).

For the proof, we need a lemma.

Lemma 14.12. *Let $n \geq 2$ be even, $p \geq 5$ a prime, and $2p-1 = m$. We have*

$$(-1)^m m! \left\{ {n \atop m} \right\} \equiv 0 \mod p^2,$$

and hence $\frac{(-1)^m m! \left\{ {n \atop m} \right\}}{(m+1)^2} \in \mathbf{Z}_{(p)}$.

Proof. By Proposition 2.6 (p. 30), we have

$$(-1)^m m! \left\{ {n \atop m} \right\} = \sum_{l=1}^{2p-1} (-1)^l \binom{2p-1}{l} l^n$$

$$= \sum_{l=1}^{p-1} \left\{ (-1)^l \binom{2p-1}{l} l^n + (-1)^{2p-l} \binom{2p-1}{2p-l} (2p-l)^n \right\} + (-1)^p \binom{2p-1}{p} p^n$$

$$\equiv \sum_{l=1}^{p-1} \left\{ (-1)^l \binom{2p-1}{l} l^n + (-1)^l \binom{2p-1}{l-1} (-2npl^{n-1} + l^n) \right\} \quad \bmod p^2.$$

Using $\binom{2p-1}{l} + \binom{2p-1}{l-1} = \frac{2p}{l}\binom{2p-1}{l-1}$, we see the last sum is equal to

$$2p(1-n)\sum_{l=1}^{p-1}(-1)^l \binom{2p-1}{l-1} l^{n-1}.$$

Combining

$$\binom{2p-1}{l-1} \equiv (-1)^{l-1} \qquad \bmod p$$

and $p-1 \nmid n-1$ (since n is even and p is odd), we have

$$\sum_{l=1}^{p-1}(-1)^l \binom{2p-1}{l-1} l^{n-1} \equiv -\sum_{l=1}^{p-1} l^{n-1} \equiv 0 \qquad \bmod p.$$

□

Proof of Theorem 14.10. The iterated integral expression (14.1) of the generating function of $\mathbb{B}_n^{(2)}$ and $\frac{s}{e^s-1} = \sum_{l=0}^{\infty}(-1)^l B_l \frac{s^l}{l!}$ give

$$\begin{aligned}\sum_{n=0}^{\infty} \mathbb{B}_n^{(2)} \frac{t^n}{n!} &= \frac{e^t}{e^t-1}\int_0^t \sum_{l=0}^{\infty}(-1)^l B_l \frac{s^l}{l!} ds \\ &= \sum_{m=0}^{\infty} B_m \frac{t^{m-1}}{m!} \cdot \sum_{l=0}^{\infty}(-1)^l B_l \frac{t^{l+1}}{(l+1)!}.\end{aligned}$$

From this we obtain

$$\mathbb{B}_n^{(2)} = \sum_{l=0}^{n}(-1)^l \binom{n}{l} \frac{B_{n-l}B_l}{l+1}.$$

If n is odd, one of $n-l$ and l is odd. Since $B_l = 0$ for odd $l \geq 3$, we have for $n \geq 3$

$$\mathbb{B}_n^{(2)} = -\frac{n}{2}B_{n-1}B_1 + B_1 B_{n-1} = -\frac{(n-2)}{4}B_{n-1}.$$

This also holds for $n = 1$, and thus (1) is proved.

(2-1) is clear from the formula in Theorem 14.6. By elementary computation, one sees that the only cases where $\frac{m!}{(m+1)^2}$ in that formula is not an integer are $m+1 = 8, 9$, prime, $2 \times$ prime. By Lemma 14.12, one sees that p does not appear in the denominator of $\frac{(-1)^m m! \left\{ {n \atop m} \right\}}{(m+1)^2}$ if $m+1 = 2p$, (p: prime ≥ 5). Next consider the case of $m+1 = p$ (p: prime ≥ 5). In this case, we have

$$(-1)^m m! \left\{ {n \atop m} \right\} = \sum_{l=1}^{p-1} (-1)^l \binom{p-1}{l} l^n \equiv \sum_{l=1}^{p-1} l^n \mod p.$$

This reduces mod p to -1 if $p-1 \mid n$, 0 if $p-1 \nmid n$. Hence the p-order of $\frac{(-1)^m m! \left\{ {n \atop m} \right\}}{(m+1)^2}$ is -2 if $p-1 \mid n$. The other terms being in $\mathbf{Z}_{(p)}$, this establishes (2-2)-(a). Suppose $p-1 \nmid n$. Noting the congruence

$$\binom{p-1}{l} \equiv (-1)^l + (-1)^{l-1} p \sum_{i=1}^{l} \frac{1}{i} \mod p^2$$

(expand $(p-1)(p-2)\cdots(p-l)$), we have

$$\sum_{l=1}^{p-1} (-1)^l \binom{p-1}{l} l^n \equiv \sum_{l=1}^{p-1} l^n - p \sum_{l=1}^{p-1} l^n \sum_{i=1}^{l} \frac{1}{i} \mod p^2.$$

It is known that if n is even and $p-1 \nmid n$, the congruence

$$\sum_{l=1}^{p-1} l^n \equiv pB_n \mod p^2$$

holds (cf. Ireland and Rosen [50, Cor. of Prop. 15.2.2]). On the other hand, if $n \equiv n'$ mod $p-1$, $1 < n' < p-1$, then n' is also even and by Vandiver[2] [93, (63)] (see also Remark 14.24) we have

$$\begin{aligned}\sum_{l=1}^{p-1} l^n \sum_{i=1}^{l} \frac{1}{i} &\equiv \sum_{l=1}^{p-1} l^{n'} \sum_{i=1}^{l} \frac{1}{i} \mod p \\ &\equiv B_{n'} \mod p.\end{aligned}$$

[2] Harry Schultz Vandiver (born on October 21, 1882 in Philadelphia, USA, died on January 9, 1973 in Austin, USA).

This gives

$$(-1)^m m! \left\{ {n \atop m} \right\} \equiv p\,(B_n - B_{n'}) \mod p^2.$$

By the assumption $p-1 \nmid n$, we have $\frac{B_n}{n} \in \mathbf{Z}_{(p)}$ and $B_{n'} \equiv n' \frac{B_n}{n} \mod p$ (Theorem 3.2). We therefore have $(m = p-1)$

$$(-1)^m m! \left\{ {n \atop m} \right\} \equiv p(n-n')\frac{B_n}{n} \mod p^2, \tag{14.3}$$

and (2-2)-(b) follows. As for the 3-order, the terms in the formula of Theorem 14.6 which have 3 in the denominators are

$$\frac{2!\left\{ {n \atop 2} \right\}}{3^2}, \ \frac{-5!\left\{ {n \atop 5} \right\}}{6^2}, \ \frac{8!\left\{ {n \atop 8} \right\}}{9^2}.$$

Computation using Proposition 2.6 (6) (p. 30) gives (2-3). The determination of the 2-order is similar and is omitted. □

By noting the appearance of the condition $p \mid \frac{B_n}{n}$ in (2-2)-(b-1) of Theorem 14.10, we can state the irregularity of a prime p in terms of the denominator of the di-Bernoulli number $\mathbb{B}_n^{(2)}$.

Corollary 14.13. *A prime p (≥ 5) is irregular if and only if there exists an even n with $p+1 \leq n \leq 2p-4$ such that the denominator of $\mathbb{B}_n^{(2)}$ is not divisible by p.*

For classical Bernoulli numbers, odd indexed ones are 0 except for B_1, denominators of even indexed ones are easily determined (Clausen and von Staudt), and the numerators of even indexed ones are mysterious quantities that are closely related to the arithmetic of cyclotomic fields. As for the di-Bernoulli numbers, the odd indexed ones are essentially the classical Bernoulli numbers and the determination of denominators of even indexed ones requires the numerators of the classical Bernoulli numbers. Can one expect any interesting arithmetic concerning the numerators of the di-Bernoulli numbers? We do not know.

14.3 Poly-Bernoulli Numbers with Negative Upper Indices

When k is 0 or negative, the poly-Bernoulli number $\mathbb{B}_n^{(k)}$ is a positive integer (Table 14.2). We give yet another formula for $\mathbb{B}_n^{(k)}$ in this case, and introduce two different combinatorial interpretations of $\mathbb{B}_n^{(k)}$ for non-positive k which were discovered recently.

We have the following formula.

Table 14.2 $\mathbb{B}_n^{(k)}$ $(-5 \le k \le 0,\ 0 \le n \le 7)$

$k\backslash n$	0	1	2	3	4	5	6	7
0	1	1	1	1	1	1	1	1
−1	1	2	4	8	16	32	64	128
−2	1	4	14	46	146	454	1394	4246
−3	1	8	46	230	1066	4718	20266	85310
−4	1	16	146	1066	6902	41506	237686	1315666
−5	1	32	454	4718	41506	329462	2441314	17234438

Theorem 14.14. *For any* $n, k \ge 0$, *we have*

$$\mathbb{B}_n^{(-k)} = \sum_{j=0}^{\min(n,k)} (j!)^2 \left\{ {n+1 \atop j+1} \right\} \left\{ {k+1 \atop j+1} \right\}.$$

In particular, $\mathbb{B}_n^{(-k)}$ is positive and symmetric in n and k.

Corollary 14.15.

$$\mathbb{B}_n^{(-k)} = \mathbb{B}_k^{(-n)}.$$

Proof.[3] We compute the two-variable generating function of $\mathbb{B}_n^{(-k)}$ with the aid of Theorem 14.6 and Proposition 2.6 (7), in the following way.

$$\begin{aligned}
\sum_{n=0}^{\infty}\sum_{k=0}^{\infty} \mathbb{B}_n^{(-k)} \frac{x^n}{n!}\frac{y^k}{k!} &= \sum_{n=0}^{\infty}\sum_{k=0}^{\infty} (-1)^n \sum_{m=0}^{n} (-1)^m m! \left\{ {n \atop m} \right\} (m+1)^k \frac{x^n}{n!}\frac{y^k}{k!} \\
&= \sum_{n=0}^{\infty} (-1)^n \sum_{m=0}^{n} (-1)^m m! \left\{ {n \atop m} \right\} e^{(m+1)y} \frac{x^n}{n!} \\
&= \sum_{m=0}^{\infty} (-1)^m m! e^{(m+1)y} \sum_{n=m}^{\infty} (-1)^n \left\{ {n \atop m} \right\} \frac{x^n}{n!} \\
&= \sum_{m=0}^{\infty} (-1)^m m! e^{(m+1)y} \frac{(e^{-x}-1)^m}{m!} \\
&= e^y \left(\frac{1}{1-(-e^y(e^{-x}-1))} \right) \\
&= \frac{e^{x+y}}{e^x + e^y - e^{x+y}}.
\end{aligned}$$

Furthermore, we have

[3] The following proof, which greatly simplifies the original proof in [10], is due to Hiroyuki Ochiai. The authors would like to thank him for providing this simple proof.

$$\begin{aligned}\frac{e^{x+y}}{e^x+e^y-e^{x+y}} &= \frac{e^{x+y}}{1-(e^x-1)(e^y-1)}\\ &= e^{x+y}\sum_{j=0}^{\infty}(e^x-1)^j(e^y-1)^j\\ &= \sum_{j=0}^{\infty}\frac{1}{(j+1)^2}\frac{d}{dx}(e^x-1)^{j+1}\frac{d}{dy}(e^y-1)^{j+1}.\end{aligned}$$

Applying Proposition 2.6 (7) again, we obtain

$$\begin{aligned}&\sum_{n=0}^{\infty}\sum_{k=0}^{\infty}\mathbb{B}_n^{(-k)}\frac{x^n}{n!}\frac{y^k}{k!}\\ &= \sum_{j=0}^{\infty}(j!)^2\frac{d}{dx}\left(\sum_{n=j}^{\infty}\left\{ {n+1 \atop j+1} \right\}\frac{x^{n+1}}{(n+1)!}\right)\frac{d}{dy}\left(\sum_{k=j}^{\infty}\left\{ {k+1 \atop j+1} \right\}\frac{y^{k+1}}{(k+1)!}\right)\\ &= \sum_{j=0}^{\infty}\sum_{n=j}^{\infty}\sum_{k=j}^{\infty}(j!)^2\left\{ {n+1 \atop j+1} \right\}\left\{ {k+1 \atop j+1} \right\}\frac{x^n}{n!}\frac{y^k}{k!}.\end{aligned}$$

Comparison of the coefficients on both sides gives the theorem (note $\left\{ {n+1 \atop j+1} \right\}\left\{ {k+1 \atop j+1} \right\}=0$ when $j>\min(n,k)$). □

By the theorem, we also have a different generating function for $\mathbb{B}_n^{(-k)}$.

Corollary 14.16. *We have*

$$\sum_{n=0}^{\infty}\sum_{k=0}^{\infty}\mathbb{B}_n^{(-k)}x^n y^k = \sum_{j=0}^{\infty}p_j(x)p_j(y),$$

where

$$p_j(x) = \frac{j!x^j}{(1-x)(1-2x)\cdots(1-(j+1)x)}.$$

Proof. This is a consequence of

$$p_j(x) = j!\sum_{n=j}^{\infty}\left\{ {n+1 \atop j+1} \right\}x^n$$

which follows from Proposition 2.6 (8). □

Computing this generating function in a different manner, we obtain the following.

Proposition 14.17. *For any $n > 0$, we have*

$$\sum_{l=0}^{n}(-1)^l \mathbb{B}_{n-l}^{(-l)} = 0.$$

Proof. Denote the left-hand side of Corollary 14.16 by $B(x,y)$. By Theorem 14.6 and Proposition 2.6 (8), we have

$$\begin{aligned} B(x,y) &= \sum_{n=0}^{\infty}\sum_{k=0}^{\infty}(-1)^n \sum_{m=0}^{n}(-1)^m m! \left\{ {n \atop m} \right\} (m+1)^k x^n y^k \\ &= \sum_{m=0}^{\infty}(-1)^m m! \sum_{n=m}^{\infty}(-1)^n \left\{ {n \atop m} \right\} x^n \frac{1}{1-(m+1)y} \\ &= \sum_{m=0}^{\infty} \frac{m! x^m}{(1+x)(1+2x)\cdots(1+mx)(1-(m+1)y)}. \end{aligned}$$

(The term with $m = 0$ is $\frac{1}{1-y}$.) Here we put $y = -x$ and use Proposition 2.6 (8), $m! = \left[{m+1 \atop 1} \right]$, and Proposition 2.6 (5.1), to obtain $B(x,-x) = 1$ and the proposition follows. □

Remark 14.18. This is trivial when n is odd, because of the symmetry in Corollary 14.15. When n is even, however, this is a non-trivial statement. For example, from Table 14.2, we have $1 - 2 + 1 = 0$, $1 - 8 + 14 - 8 + 1 = 0$, etc.

In recent years, two different combinatorial interpretations of the poly-Bernoulli numbers with negative upper indices have been found. We briefly state these theorems, one by Chad Brewbaker and the other by Stephan Launois. For proofs, see their papers [20, 68].

Definition 14.19. A matrix whose entries are 0 or 1 is called "lonesum" if it is uniquely reconstructed from its row and column sums.

Theorem 14.20 (Brewbaker [20]). *Let n and k be natural numbers. The number of n by k lonesum matrices is equal to $\mathbb{B}_n^{(-k)}$.*

We denote by $\mathfrak{S}_n$ the symmetric group of degree n, viewed as a permutation group on the set $\{1, 2, \cdots, n\}$.

Theorem 14.21 (Launois [68]). *Let n and k be natural numbers. The cardinality of the set*

$$\{\sigma \in \mathfrak{S}_{n+k} \mid -n \leq i - \sigma(i) \leq k,\ 1 \leq \forall i \leq n+k\}$$

is equal to $\mathbb{B}_n^{(-k)}$.

We end this section by proving a special case of a theorem of Vandiver, making use of the formulas in Theorem 14.6 and Corollary 14.15.

Theorem 14.22. *For an odd prime p and an integer i with $1 \le i \le p-2$, we have*

$$B_i \equiv \sum_{m=1}^{p-2} \left(1 + \frac{1}{2} + \cdots + \frac{1}{m}\right) (m+1)^i \mod p.$$

Proof. By Theorem 14.6 and Fermat's little theorem,

$$B_i = \mathbb{B}_i^{(1)} \equiv \mathbb{B}_i^{(2-p)} \mod p.$$

By Corollary 14.15, the right-hand side is equal to $\mathbb{B}_{p-2}^{(-i)}$. Applying Theorem 14.6 again, we obtain

$$B_i \equiv -\sum_{m=0}^{p-2} (-1)^m m! \left\{ {p-2 \atop m} \right\} (m+1)^i \mod p.$$

It therefore suffices to show the following lemma (note $\left\{ {p-2 \atop 0} \right\} = 0$).

Lemma 14.23. *If $1 \le m \le p-2$, then*

$$(-1)^{m-1} m! \left\{ {p-2 \atop m} \right\} \equiv 1 + \frac{1}{2} + \cdots + \frac{1}{m} \mod p.$$

Write $(-1)^{m-1} m! \left\{ {p-2 \atop m} \right\} = b_m$. The recursion of the Stirling numbers (2.1) (p. 26) gives

$$(-1)^{m-1} m! \left\{ {p-1 \atop m} \right\} = m(-b_{m-1} + b_m) \quad (m \ge 2).$$

On the other hand, by Proposition 2.6 (6) (p. 30), we have

$$\begin{aligned} (-1)^{m-1} m! \left\{ {p-1 \atop m} \right\} &= -\sum_{l=1}^{m} (-1)^l \binom{m}{l} l^{p-1} \\ &\equiv -\sum_{l=1}^{m} (-1)^l \binom{m}{l} \mod p \\ &\equiv 1 \mod p \end{aligned}$$

and hence $b_m \equiv b_{m-1} + \frac{1}{m} \mod p$.

This together with $b_1 = \left\{ {p-2 \atop 1} \right\} = 1$ proves the lemma and thus settles the theorem. □

Remark 14.24. For $1 < i \le p-2$, the right-hand side of the theorem is congruent modulo p to

$$\sum_{m=1}^{p-1}\left(1+\frac{1}{2}+\cdots+\frac{1}{m}\right)m^i.$$

Vandiver's original theorem (a special case of [93, (63)]) states that this is congruent modulo p to B_i.

Exercise 14.25. Prove that if $\frac{m!}{(m+1)^2}$ is not an integer, then $m+1 = 8, 9$, prime, or $2 \times$ prime.

Exercise 14.26. Define the numbers $C_n^{(k)}$ by the generating function

$$\frac{Li_k(1-e^{-t})}{e^t-1} = \sum_{n=0}^{\infty} C_n^{(k)}\frac{t^n}{n!}.$$

(1) Prove the formula

$$C_n^{(k)} = (-1)^n \sum_{m=0}^{n} \frac{(-1)^m m! \left\{ {n+1 \atop m+1} \right\}}{(m+1)^k}.$$

(This is regarded as a generalization of Proposition 2.10.)

(2) Prove the duality

$$C_{n-1}^{(-k)} = C_{k-1}^{(-n)} \quad (n, k \ge 1).$$

Exercise 14.27. Prove that the number $\mathbb{B}_n^{(-k)}$ is always even when $n, k \ge 1$. Hint: Use Theorem 14.14.

Exercise 14.28. List all n by k lonesum matrices for small n and k, and check Theorem 14.20.

Exercise 14.29. Prove the congruence stated in Remark 14.24.

Appendix
Curious and Exotic Identities for Bernoulli Numbers

Don Zagier

Bernoulli numbers, which are ubiquitous in mathematics, typically appear either as the Taylor coefficients of $x/\tan x$ or else, very closely related to this, as special values of the Riemann zeta function. But they also sometimes appear in other guises and in other combinations. In this appendix we want to describe some of the less standard properties of these fascinating numbers.

In Sect. A.1, which is the foundation for most of the rest, we show that, as well as the familiar (and convergent) *exponential* generating series[1]

$$\frac{x}{e^x - 1} = \sum_{n=0}^{\infty} \frac{B_n}{n!}\, x^n = 1 - \frac{x}{2} + \frac{x^2}{12} - \frac{x^4}{720} + \frac{x^6}{30240} - \cdots \tag{A.1}$$

defining the Bernoulli numbers, the less familiar (and divergent) *ordinary* generating series

$$\beta(x) = \sum_{n=0}^{\infty} B_n\, x^n = 1 - \frac{x}{2} + \frac{x^2}{6} - \frac{x^4}{30} + \frac{x^6}{42} - \cdots \tag{A.2}$$

also has many virtues and is often just as useful as, or even more useful than, its better-known counterpart (A.1). As a first application, in Sect. A.2 we discuss the "modified Bernoulli numbers"

$$B_n^* = \sum_{r=0}^{n} \binom{n+r}{2r} \frac{B_r}{n+r} \qquad (n \geq 1)\,. \tag{A.3}$$

[1]Here, and throughout this appendix, we use the convention $B_1 = -1/2$ rather than the convention $B_1 = 1/2$ used in the main text of the book.

T. Arakawa et al., *Bernoulli Numbers and Zeta Functions*, Springer Monographs in Mathematics, DOI 10.1007/978-4-431-54919-2, © Springer Japan 2014

These numbers, which arose in connection with the trace formula for the Hecke operators acting on modular forms on $\mathrm{SL}(2,\mathbb{Z})$, have several unexpected properties, including the surprising periodicity

$$B^*_{n+12} = B^*_n \qquad (n \text{ odd}) \tag{A.4}$$

and a modified form of the classical von Staudt–Clausen formula for the value of B_n modulo 1. The following section is devoted to an identity discovered by Miki [A10] (and a generalization due to Gessel [A4]) which has the striking property of involving Bernoulli sums both of type $\sum B_r B_{n-r}$ and $\sum \binom{n}{r} B_r B_{n-r}$, i.e., sums related to both the generating functions (A.1) and (A.2). In Sect. A.4 we look at products of Bernoulli numbers and Bernoulli polynomials in more detail. In particular, we prove the result (discovered by Nielsen) that when a product of two Bernoulli polynomials is expressed as a linear combination of Bernoulli polynomials, then the coefficients are themselves multiples of Bernoulli numbers. This generalizes to a formula for the product of two Bernoulli polynomials in two different arguments, and leads to a further proof, due to I. Artamkin, of the Miki–Gessel identities. Finally, in Sect. A.5 we discuss the continued fraction expansions of various power series related to both (A.1) and (A.2) and, as an extra titbit, describe an unexpected appearance of one of these continued fraction expansions in connection with some recent and amazing discoveries of Yu. Matiyasevich concerning the non-trivial zeros of the Riemann zeta function.

This appendix can be read independently of the main text and we will recall all facts and notations needed. We should also add a warning: if you don't like generating functions, don't read this appendix!

A.1 The "Other" Generating Function(s) for the Bernoulli Numbers

Given a sequence of interesting numbers $\{a_n\}_{n\geq 0}$, one often tries to understand them by using the properties of the corresponding generating functions. The two most popular choices for these generating functions are $\sum_{n=0}^{\infty} a_n x^n$ ("ordinary generating function") and $\sum_{n=0}^{\infty} a_n x^n/n!$ ("exponential generating function"). Usually, of course, at most one of these turns out to have useful properties. For the Bernoulli numbers the standard choice is the exponential generating function (A.1) because it has an expression "in closed form." What is not so well known is that the ordinary generating function of the Bernoulli numbers, i.e., the power series (A.2), even though it is divergent for all non-zero complex values of x, also has extremely attractive properties and many nice applications. The key property that makes it useful, despite its being divergent and not being expressible as an elementary function, is the following functional equation:

Proposition A.1. *The power series* (A.2) *is the unique solution in* $\mathbb{Q}[[x]]$ *of the equation*

$$\frac{1}{1-x}\,\beta\Big(\frac{x}{1-x}\Big) - \beta(x) = x\,. \tag{A.5}$$

Proof. Let $\{B_n\}$ be unspecified numbers and define $\beta(x)$ by the first equality in (A.2). Then comparing the coefficients of x^m in both sides of (A 5) gives

$$\sum_{n=0}^{m-1}\binom{m}{n}B_n = \begin{cases}1 & \text{if } m=1,\\ 0 & \text{if } m>1.\end{cases} \tag{A.6}$$

This is the same as the standard recursion for the Bernoulli numbers obtained by multiplying both sides of (A.1) by $e^x - 1$ and comparing the coefficients of $x^m/m!$ on both sides. □

The functional equation (A.5) can be rewritten in a slightly prettier form by setting

$$\beta_1(x) = x\,\beta(x) = \sum_{n=0}^{\infty} B_n\,x^{n+1}\,,$$

in which case it becomes simply

$$\beta_1\Big(\frac{x}{1-x}\Big) - \beta_1(x) = x^2\,. \tag{A.7}$$

A generalization of this is given by the following proposition.

Proposition A.2. *For each integer* $r \geq 1$, *the power series*

$$\beta_r(x) = \sum_{n=0}^{\infty}\binom{n+r-1}{n}B_n\,x^{n+r} \tag{A.8}$$

satisfies the functional equation

$$\beta_r\Big(\frac{x}{1-x}\Big) - \beta_r(x) = r\,x^{r+1} \tag{A.9}$$

and is the unique power series having this property.

Proof. Equation (A.9) for any fixed value of $r \geq 1$ is equivalent to the recursion (A.6), by the calculation

$$\beta_r\Big(\frac{x}{1-x}\Big) - \beta_r(x) = \sum_{n=0}^{\infty} \binom{n+r-1}{n} B_n \sum_{\ell=n+r}^{\infty} \binom{\ell}{n+r-1} x^{\ell+1}$$
$$= \sum_{\ell=r}^{\infty} \binom{\ell}{r-1} x^{\ell+1} \Big(\sum_{n=0}^{\ell-r} \binom{\ell-r+1}{n} B_n \Big) = r\, x^{r+1}\,.$$

Alternatively, we can deduce (A.9) from (A.7) by induction on r by using the easily checked identity

$$x^2\, \beta_r'(x) = r\, \beta_{r+1}(x) \qquad (r \geq 1) \tag{A.10}$$

and the fact that

$$x^2 \frac{d}{dx} F\Big(\frac{x}{1-x}\Big) = \Big(\frac{x}{1-x}\Big)^2 F'\Big(\frac{x}{1-x}\Big) \tag{A.11}$$

for any power series $F(x)$. □

We observe next that the definition (A.8) makes sense for any r in $\mathbb{Z}$,[2] and that the properties (A.9) and (A.10) still hold. But this extension is not particularly interesting since $\beta_{-k}(x)$ for $k \in \mathbb{Z}_{\geq 0}$ is just a known polynomial in $1/x$:

$$\beta_{-k}(x) = \sum_{n=0}^{\infty} \binom{n-k-1}{n} B_n\, x^{n-k} = \sum_{n=0}^{k} (-1)^n \binom{k}{n} \frac{B_n}{x^{k-n}}$$
$$= B_k\Big(\frac{1}{x}\Big) + \frac{k}{x^{k-1}} = B_k\Big(\frac{1}{x}+1\Big) = (-1)^k\, B_k\Big(-\frac{1}{x}\Big),$$

where $B_k(X)$ is the kth Bernoulli polynomial. (One can also prove these identities by induction on k, using either (A.10) or else (A.9) together with the uniqueness statement in Proposition A.2 and the corresponding well-known functional equation for the Bernoulli polynomials.) However, there is a different and more interesting way to extend the definition of β_r to non-positive integral values of r. For $k \in \mathbb{Z}$, define

$$\gamma_k(x) = \sum_{n \geq \max(1,-k)} \frac{(n-1)!}{(n+k)!} B_{n+k}\, x^n \quad \in x\,\mathbb{Q}[[x]]\,.$$

Then one easily checks that $\gamma_{-r}(x) = (r-1)!\,\beta_r(x)$ for $r > 0$, so that the negative-index power series γ_k are just renormalized versions of the positive-index power series β_r. But now we do get interesting power series (rather than merely polynomials) when $k \geq 0$, e.g.

[2] Or even in $\mathbb{C}$ if we work formally in $x^r\,\mathbb{Q}[[x]]$.

$$\gamma_0(x) = \sum_{n=1}^{\infty} \frac{B_n\, x^n}{n}\,, \quad \gamma_1(x) = \sum_{n=1}^{\infty} \frac{B_{n+1}\, x^n}{n(n+1)}\,, \quad \gamma_2(x) = \sum_{n=1}^{\infty} \frac{B_{n+2}\, x^n}{n(n+1)(n+2)}\,. \tag{A.12}$$

The properties of these new functions corresponding to (A.10) and (A.9) are given by:

Proposition A.3. *The power series $\gamma_k(x)$ satisfy the differential recursion*

$$x^2\, \gamma_k'(x) = \gamma_{k-1}(x) - \frac{B_k}{k!}\, x \qquad (k \geq 0) \tag{A.13}$$

(with $\gamma_{-1}(x) = \beta_1(x)$) as well as the functional equations

$$\gamma_0\Big(\frac{x}{1-x}\Big) - \gamma_0(x) = \quad \log(1-x) + x\,, \tag{A.14}$$

$$\gamma_1\Big(\frac{x}{1-x}\Big) - \gamma_1(x) = -\Big(\frac{1}{x} - \frac{1}{2}\Big)\, \log(1-x) - 1\,,$$

and more generally for $k \geq 1$

$$\gamma_k\Big(\frac{x}{1-x}\Big) - \gamma_k(x) = \frac{(-1)^k}{k!}\,\Big[B_k\Big(\frac{1}{x}\Big)\, \log(1-x) + P_{k-1}\Big(\frac{1}{x}\Big)\Big]\,, \tag{A.15}$$

where $P_{k-1}(X)$ is a polynomial of degree $k-1$, the first few values of which are $P_0(X) = 1$, $P_1(X) = X - \frac{1}{2}$, $P_2(X) = X^2 - X + \frac{1}{12}$, $P_3(X) = X^3 - \frac{3}{2}X^2 + \frac{1}{3}X + \frac{1}{12}$ and $P_4(X) = X^4 - 2X^3 + \frac{3}{4}X^2 + \frac{1}{4}X - \frac{13}{360}$.

Proof. Equation (A.13) follows directly from the definitions, and then Eqs. (A.14) and (A.15) (by induction over k) follow successively from (A.7) using the general identity (A.11). □

We end this section with the observation that, although $\beta(x)$ and the related power series $\beta_r(x)$ and $\gamma_k(x)$ that we have discussed are divergent and do not give the Taylor or Laurent expansion of any elementary functions, they are related to the *asymptotic* expansions of very familiar, "nearly elementary" functions. Indeed, Stirling's formula in its logarithmic form says that the logarithm of Euler's Gamma function has the asymptotic expansion

$$\log\Gamma(X) \sim \Big(X - \frac{1}{2}\Big)\log X - X + \frac{1}{2}\log(2\pi) + \sum_{n=2}^{\infty} \frac{B_n}{n(n-1)}\, X^{-n+1}$$

as $X \to \infty$, and hence that its derivative $\psi(X)$ ("digamma function") has the expansion

$$\psi(X) := \frac{\Gamma'(X)}{\Gamma(X)} \sim \log X - \frac{1}{2X} - \sum_{n=2}^{\infty} \frac{B_n}{n}\, X^{-n} = \log X - \gamma_0\Big(-\frac{1}{X}\Big)$$

as $X \to \infty$, with $\gamma_0(x)$ defined as in Eq. (A.12), and the functions $\beta_r(x)$ correspond similarly to the derivatives of $\psi(x)$ ("polygamma functions"). The transformation $x \mapsto x/(1-x)$ occurring in the functional equations (A.5), (A.9), (A.14) and (A.15) corresponds under the substitution $X = -1/x$ to the translation $X \mapsto X+1$, and the compatibility equation (A.11) simply to the fact that this translation commutes with the differential operator d/dX, while the functional equations themselves reflect the defining functional equation $\Gamma(X+1) = X\Gamma(X)$ of the Gamma function.

A.2 An Application: Periodicity of Modified Bernoulli Numbers

The "modified Bernoulli numbers" defined by (A.3) were introduced in [A14]. These numbers, as already mentioned in the introduction, occurred naturally in a certain elementary derivation of the formula for the traces of Hecke operators acting on modular forms for the full modular group [A15]. They have two surprising properties which are parallel to the two following well-known properties of the ordinary Bernoulli numbers:

$$n > 1 \text{ odd} \qquad \Rightarrow \qquad B_n = 0\,, \tag{A.16}$$

$$n > 0 \text{ even} \qquad \Rightarrow \qquad B_n \equiv -\sum_{\substack{p \text{ prime} \\ (p-1)|n}} \frac{1}{p} \pmod{1} \tag{A.17}$$

(von Staudt–Clausen theorem). These properties are given by:

Proposition A.4. *Let B_n^* $(n > 0)$ be the numbers defined by* (A.3). *Then for n odd we have*

$$B_n^* = \begin{cases} \pm 3/4 & \textit{if } n \equiv \pm 1 \pmod{12}, \\ \mp 1/4 & \textit{if } n \equiv \pm 3 \textit{ or } \pm 5 \pmod{12}, \end{cases} \tag{A.18}$$

and for n even we have the modified von Staudt–Clausen formula

$$2nB_n^* - B_n \equiv \sum_{\substack{p \text{ prime} \\ (p+1)|n}} \frac{1}{p} \pmod{1}\,. \tag{A.19}$$

Remark. The modulo 12 periodicity in (A.18) is related, via the above-mentioned connection with modular forms on the full modular group $\mathrm{SL}(2,\mathbb{Z})$, with the well-known fact that the space of these modular forms of even weight $k > 2$ is the sum of $k/12$ and a number that depends only on $k \pmod{12}$.

Proof. The second assertion is an easy consequence of the corresponding property (A.17) of the ordinary Bernoulli numbers and we omit the proof. (It is given in [A15].) To prove the first, we use the generating functions for Bernoulli numbers introduced in Sect. A.1. Specifically, for $\lambda \in \mathbb{Q}$ we define a power series $g_\lambda(t) \in \mathbb{Q}[[t]]$ by the formula

$$g_\lambda(t) = \gamma_0\Big(\frac{t}{1-\lambda t+t^2}\Big) - \log(1-\lambda t+t^2),$$

where $\gamma_0(x) = \sum_{n>0} B_n x^n/n$ is the power series defined in (A.12). For $\lambda = 2$ this specializes to

$$g_2(t) = \sum_{r=1}^{\infty} \frac{B_r}{r}\,\frac{t^r}{(1-t)^{2r}} - 2\log(1-t) = 2\sum_{n=1}^{\infty} B_n^* t^n. \tag{A.20}$$

with B_n^* as in (A.3). On the other hand, the functional equation (A.14) applied to $x = t/(1-\lambda t+t^2)$, together with the parity property $\gamma_0(x)+x = \gamma_0(-x)$, which is a restatement of (A.16), implies the two functional equations

$$g_{\lambda+1}(t) = g_\lambda(t) + \frac{t}{1-\lambda t+t^2} = g_{-\lambda}(-t)$$

for the power series g_λ. From this we deduce

$$\begin{aligned} g_2(t)-g_2(-t) &= \big(g_2(t)-g_1(t)\big) + \big(g_1(t)-g_0(t)\big) + \big(g_0(t)-g_{-1}(t)\big) \\ &= \frac{t}{1-t+t^2} + \frac{t}{1+t^2} + \frac{t}{1+t+t^2} = \frac{3t-t^3-t^5+t^7+t^9-3t^{11}}{1-t^{12}}, \end{aligned}$$

and comparing this with (A.20) immediately gives the desired formula (A.18) for B_n^*, n odd. □

We mention one further result about the modified Bernoulli numbers from [A15]. The ordinary Bernoulli numbers satisfy the asymptotic formula

$$B_n \sim (-1)^{(n-2)/2}\,\frac{2\,n!}{(2\pi)^n} \qquad (n\to\infty,\ n \text{ even}). \tag{A.21}$$

As one might expect, the modified ones have asymptotics given by a very similar formula:

$$B_n^* \sim (-1)^{(n-2)/2}\,\frac{(n-1)!}{(2\pi)^n} \qquad (n\to\infty,\ n \text{ even}). \tag{A.22}$$

The (small) surprise is that, while the asymptotic formula (A.21) holds to all orders in $1/n$ (because the ratio of the two sides equals $\zeta(n) = 1+O(2^{-n})$), this is not true of the new formula (A.22), which only acquires this property if the right-hand side is replaced by $(-1)^{n/2}\pi\, Y_n(4\pi)$, where $Y_n(x)$ is the nth Bessel function of the second kind.

Here is a small table of the numbers B_n^* and $\tilde{B}_n = 2nB_n^* - B_n$ for n even:

n	2	4	6	8	10	12	14	16	18	20	22
B_n^*	$\frac{1}{24}$	$-\frac{27}{80}$	$-\frac{29}{1260}$	$\frac{451}{1120}$	$-\frac{65}{264}$	$-\frac{6571}{12012}$	$\frac{571}{312}$	$-\frac{181613}{38080}$	$\frac{23663513}{1220940}$	$-\frac{10188203}{83600}$	$\frac{564133}{552}$
$\tilde{B}_n$	0	$-\frac{8}{3}$	$-\frac{3}{10}$	$\frac{136}{21}$	-5	$-\frac{4249}{330}$	$\frac{651}{13}$	$-\frac{3056}{21}$	$\frac{109269}{170}$	$-\frac{247700}{57}$	38775

A.3 Miki's Identity

The surprising identity described in this section was found and proved by Miki [A10] in an indirect and non-elementary way, using p-adic methods. In this section we describe two direct proofs of it, or rather, of it and of a very similar identity discovered by Faber and Pandharipande in connection with Chern numbers of moduli spaces of curves. The first, which is short but not very enlightening, is a variant of a proof I gave of the latter identity [A2] (but which with a slight modification works for Miki's original identity as well). The second one, which is more natural, is a slight reworking of the proof given by Gessel [A4] based on properties of Stirling numbers of the second kind. In fact, Gessel gives a more general one-parameter family of identities, provable by the same methods, of which both the Miki and the Faber–Pandharipande identities are special cases. In Sect. A.4 we will give yet a third proof of these identities, following I. Artamkin [A1].

Proposition A.5 (Miki). *Write $\mathcal{B}_n = (-1)^n B_n/n$ for $n > 0$. Then for all $n > 2$ we have*

$$\sum_{i=2}^{n-2} \mathcal{B}_i \mathcal{B}_{n-i} = \sum_{i=2}^{n-2} \binom{n}{i} \mathcal{B}_i \mathcal{B}_{n-i} + 2H_n \mathcal{B}_n\,, \tag{A.23}$$

where $H_n = 1 + \frac{1}{2} + \cdots + \frac{1}{n}$ denotes the nth harmonic number.

(Faber–Pandharipande). *Write $b_g = (2 - 2^{2g})\,\dfrac{B_{2g}}{(2g)!}$ for $g \geq 0$. Then for all $g > 0$ we have*

$$\sum_{\substack{g_1+g_2=g \\ g_1,\, g_2>0}} \frac{(2g_1-1)!\,(2g_2-1)!}{2\,(2g-1)!}\, b_{g_1} b_{g_2} = \sum_{n=1}^{g} \frac{2^{2n} B_{2n}}{2n\,(2n)!}\, b_{g-n} + H_{2g-1}\, b_g\,. \tag{A.24}$$

First proof. We prove (A.24), following [A2]. Write the identity as $a(g) = b(g) + c(g)$ in the obvious way, and let $A(x) = \sum_{g=1}^{\infty} a(g)\,x^{2g-1}$, $B(x) = \sum_{g=1}^{\infty} b(g)\,x^{2g-1}$ and $C(x) = \sum_{g=1}^{\infty} c(g)\,x^{2g-1}$ be the corresponding odd generating functions. Using the identity $\sum_{g=0}^{\infty} b_g\,x^{2g-1} = \dfrac{1}{\sinh x}$, we obtain

$$\begin{aligned}
A(x) &= \frac{1}{2}\sum_{g_1,g_2>0} b_{g_1} b_{g_2} \int_0^x t^{2g_1-1}\,(x-t)^{2g_2-1}\,dt \quad \text{(by Euler's beta integral)}\\
&= \frac{1}{2}\int_0^x \Big(\frac{1}{t} - \frac{1}{\sinh t}\Big)\Big(\frac{1}{x-t} - \frac{1}{\sinh(x-t)}\Big)\,dt\,,\\
B(x) &= \frac{1}{\sinh x}\sum_{n=1}^{\infty} \frac{2^{2n} B_{2n}}{2n\,(2n)!}\,x^{2n} = \frac{1}{\sinh x}\,\log\Big(\frac{\sinh x}{x}\Big)\,,\\
C(x) &= \sum_{g=1}^{\infty} b_g \int_0^x \frac{x^{2g-1} - t^{2g-1}}{x-t}\,dt = \int_0^x \Big[\frac{1}{x-t}\Big(\frac{1}{\sinh x} - \frac{1}{\sinh t}\Big) + \frac{1}{xt}\Big]\,dt\,,
\end{aligned}$$

and hence, symmetrizing the integral giving $C(x)$ with respect to $t \to x - t$,

$$\begin{aligned}
2\,A(x) - 2\,C(x) &= \int_0^x \Big\{\Big(\frac{1}{t} - \frac{1}{\sinh t}\Big)\Big(\frac{1}{x-t} - \frac{1}{\sinh(x-t)}\Big)\\
&\qquad - \Big(\frac{1}{x-t} + \frac{1}{t}\Big)\Big(\frac{1}{\sinh x} + \frac{1}{x}\Big)\\
&\qquad + \frac{1}{x-t}\,\frac{1}{\sinh t} + \frac{1}{t}\,\frac{1}{\sinh(x-t)}\Big\}\,dt\\
&= \int_0^x \Big(\frac{1}{\sinh(t)\sinh(x-t)} - \frac{x}{\sinh x}\,\frac{1}{t\,(x-t)}\Big)\,dt\\
&= \frac{1}{\sinh x}\,\log\Big(\frac{\sinh t}{t}\cdot\frac{x-t}{\sinh(x-t)}\Big)\Big|_{t=0}^{t=x} = 2\,B(x)\,.
\end{aligned}$$

A similar proof can be given for Miki's original identity (A.23), with "sinh" replaced by "tanh". □

Second proof. Now we prove (A.23), following the method in [A4]. Recall that the *Stirling number of the second kind* $S(k,m)$ is defined as the number of partitions of a set of k elements into m non-empty subsets or, equivalently, as $1/m!$ times the number of surjective maps from the set $\{1,2,\ldots,k\}$ to the set $\{1,2,\ldots,m\}$. It can be given either by the closed formula

$$S(k,m) = \frac{1}{m!}\sum_{\ell=0}^{m}(-1)^{m-\ell}\binom{m}{\ell}\ell^k \tag{A.25}$$

(this follows immediately from the second definition and the inclusion-exclusion principle, since ℓ^k is the number of maps from $\{1,2,\ldots,k\}$ to a given set of ℓ elements) or else by either of the two generating functions

$$\begin{aligned}\sum_{k=0}^{\infty} S(k,m)\,x^k &= \frac{x^m}{(1-x)(1-2x)\cdots(1-mx)}\,,\\ \sum_{k=0}^{\infty} S(k,m)\,\frac{x^k}{k!} &= \frac{(e^x-1)^m}{m!}\,,\end{aligned} \tag{A.26}$$

both of which can be deduced easily from (A.25). (Of course all of these formulas are standard and can be found in many books, including Chap. 2 of this one, where $S(k,m)$ is denoted using Knuth's notation $\left\{ {k \atop m} \right\}$.) From either generating function one finds easily that $S(k,m)$ vanishes for $k < m$, $S(m,m) = 1$, $S(m+1,m) = \frac{m^2+m}{2}$, and more generally that $S(m+n,m)$ for a fixed value of n is a polynomial in m (of degree $2n$, and without constant term if $n > 0$). Gessel's beautiful and very natural idea was to compute the first few coefficients of this polynomial using each of the generating functions in (A.26) and to equate the two expressions obtained. It turned out that this gives nothing for the coefficients of m^0 and m^1 (which are found from either point of view to be 0 and $\mathcal{B}_n$, respectively), but that the equality of the coefficients of m^2 obtained from the two generating functions coincides precisely with the identity that Miki had discovered!

More precisely, from the first formula in (A.26) we obtain

$$\begin{aligned}\log\Bigl(\sum_{n=0}^{\infty} S(m+n,m)\,x^n\Bigr) &= \sum_{j=1}^{m}\log\Bigl(\frac{1}{1-jx}\Bigr) = \sum_{r=1}^{\infty}\frac{1^r+2^r+\cdots m^r}{r}\,x^r\\ &= \sum_{r=1}^{\infty}\Bigl(\frac{B_r}{r}\,m + \frac{(-1)^{r-1}B_{r-1}}{2}\,m^2+\cdots\Bigr)x^r\end{aligned}$$

(the last line by the Bernoulli–Seki formula) and hence, exponentiating,

$$S(m+n,m) = \mathcal{B}_n\,m + \Bigl(n\mathcal{B}_{n-1} + \sum_{i=2}^{n-2}\mathcal{B}_i\mathcal{B}_{n-i}\Bigr)\frac{m^2}{2}+\cdots \quad (n\ge 3)\,, \tag{A.27}$$

while from the second formula in (A.26) and the expansion $\log\bigl((e^x-1)/x\bigr) = \sum\limits_{n>0}\mathcal{B}_n x^n/n!$ we get

$$\begin{aligned} S(m+n,m) &\\ &= \Big(1+\frac{m}{1}\Big)\Big(1+\frac{m}{2}\Big)\cdots\Big(1+\frac{m}{n}\Big) \times \text{Coefficient of } \frac{x^n}{n!} \text{ in } \Big(\frac{e^x-1}{x}\Big)^m \\ &= \Big(1+H_n m+\cdots\Big)\Big(\mathcal{B}_n m + \Big(\sum_{i=1}^{n-1}\binom{n}{i}\mathcal{B}_i\mathcal{B}_{n-i}\Big)\frac{m^2}{2}+\cdots\Big) \\ &= \mathcal{B}_n\, m + \Big(2H_n\mathcal{B}_n + \sum_{i=1}^{n-1}\binom{n}{i}\mathcal{B}_i\mathcal{B}_{n-i}\Big)\frac{m^2}{2}+\cdots \quad (n\ge 1)\,. \end{aligned} \tag{A.28}$$

Comparing the coefficients of $m^2/2$ in (A.27) and (A.28) gives Eq. (A.23). □

Finally, we state the one-parameter generalization of (A.23) and (A.24) given in [A4]. For $n > 0$ denote by $\mathcal{B}_n(x)$ the polynomial $B_n(x)/n$.

Proposition A.6 (Gessel). *For all $n > 0$ one has*

$$\frac{n}{2}\Big(B_{n-1}(x) + \sum_{i=1}^{n-1}\mathcal{B}_i(x)\mathcal{B}_{n-i}(x)\Big) = \sum_{i=1}^{n}\binom{n}{i}\mathcal{B}_i\,\mathcal{B}_{n-i}(x) + H_{n-1}\,\mathcal{B}_n(x)\,. \tag{A.29}$$

Gessel does not actually write out the proof of this identity, saying only that it can be obtained in the same way as his proof of (A.23) and pointing out that, because $\mathcal{B}_n(1) = \mathcal{B}_n$ and $2^{2g}\mathcal{B}_{2g}(1/2) = (2g-1)!\,b_g$, it implies (A.23) and (A.24) by specializing to $x = 1$ and $x = 1/2$, respectively.

A.4 Products and Scalar Products of Bernoulli Polynomials

If A is any algebra over $\mathbb{Q}$ and $e_0, e_1, \dots$ is an additive basis of A, then each product $e_i e_j$ can be written uniquely as a (finite) linear combination $\sum_k c_{ij}^k e_k$ for certain numbers $c_{ij}^k \in \mathbb{Q}$ and the algebra structure on A is completely determined by specifying the "structure constants" c_{ij}^k. If we apply this to the algebra $A = \mathbb{Q}[x]$ and the standard basis $e_i = x^i$, then the structure constants are completely trivial, being simply 1 if $i + j = k$ and 0 otherwise. But the Bernoulli polynomials also form a basis of $\mathbb{Q}[x]$, since there is one of every degree, and we can ask what the structure constants defined by $B_i(x)B_j(x) = \sum_k c_{ij}^k B_k(x)$ are. It is easy to see that c_{ij}^k can only be non-zero if the difference $r{:=}i + j - k$ is non-negative (because $B_i(x)B_j(x)$ is a polynomial of degree $i + j$) and even (because the nth Bernoulli polynomial is $(-1)^n$-symmetric with respect to $x \mapsto 1 - x$). The surprise is that, up to an elementary factor, c_{ij}^k is equal simply to the kth Bernoulli number, except when $k = 0$. This fact, which was discovered long ago by Nielsen [A11, p. 75] (although I was not aware of this reference at the time when Igor Artamkin and I had the discussions that led to the formulas and proofs described below), is stated in a precise form in the following proposition. The formula turns out to be somewhat simpler if we use the renormalized Bernoulli polynomials $\mathcal{B}_n(x) = \frac{B_n(x)}{n}$ rather than the $B_n(x)$ themselves when $n > 0$. (For $n = 0$ there is nothing to be calculated since the product of any $B_i(x)$ with $B_0(x) = 1$ is just $B_i(x)$.)

Proposition A.7. *Let i and j be strictly positive integers. Then*

$$\mathcal{B}_i(x)\,\mathcal{B}_j(x) = \sum_{0\le \ell < \frac{i+j}{2}} \left[\frac{1}{i}\binom{i}{2\ell} + \frac{1}{j}\binom{j}{2\ell}\right] B_{2\ell}\,\mathcal{B}_{i+j-2\ell}(x) + \frac{(-1)^{i-1}(i-1)!\,(j-1)!}{(i+j)!}\,B_{i+j}\,. \tag{A.30}$$

Note that, despite appearances, the (constant) second term in this formula is symmetric in i and j, because if $B_{i+j} \neq 0$ then i and j have the same parity.

Proof. Write $\mathcal{B}_{i,j}(x)$ for the right-hand side of (A.30). We first show that the difference between $\mathcal{B}_{i,j}(x)$ and $\mathcal{B}_i(x)\mathcal{B}_j(x)$ is constant. This can be done in two different ways. First of all, using $\mathcal{B}_n(x+1) - \mathcal{B}_n(x) = x^{n-1}$ we find

$$\begin{aligned}\mathcal{B}_{i,j}(x+1) - \mathcal{B}_{i,j}(x) &= \sum_{0\le \ell < \frac{i+j}{2}} \left[\frac{1}{i}\binom{i}{2\ell} + \frac{1}{j}\binom{j}{2\ell}\right] B_{2\ell}\,x^{i+j-2\ell-1}\\ &= x^{j-1}\left(\mathcal{B}_i(x) + \frac{1}{2}x^{i-1}\right) + x^{i-1}\left(\mathcal{B}_j(x) + \frac{1}{2}x^{j-1}\right)\\ &= \mathcal{B}_i(x+1)\mathcal{B}_j(x+1) - \mathcal{B}_i(x)\mathcal{B}_j(x)\,.\end{aligned}$$

It follows that the $\mathcal{B}_{i,j}(x) - \mathcal{B}_i(x)\mathcal{B}_j(x)$ is periodic and hence, since it is also polynomial, constant. Alternatively, we can use that $\mathcal{B}_n'(x)$ equals 1 for $n = 1$ and $(n-1)\mathcal{B}_{n-1}(x)$ for $n > 1$ to show by induction on $i+j$ that $\mathcal{B}_{i,j}(x)$ and $\mathcal{B}_i(x)\mathcal{B}_j(x)$ have the same derivative (we omit the easy computation) and hence again that their difference is constant. To show that this constant vanishes, it suffices to show that the integrals of the two sides of (A.30) over the interval [0,1] agree. Since the integral of $\mathcal{B}_n(x)$ over this interval vanishes for any $n > 0$, this reduces to the following statement, in which to avoid confusion with $i = \sqrt{-1}$ we have changed i and j to r and s. □

Proposition A.8. *Let r and s be positive integers. Then*

$$\int_0^1 B_r(x)\,B_s(x)\,dx = (-1)^{r-1}\frac{r!\,s!}{(r+s)!}\,B_{r+s}\,. \tag{A.31}$$

Proof. Here again we give two proofs. The first uses the Fourier development

$$B_k(x) = -\frac{k!}{(2\pi i)^k}\sum_{\substack{n\in\mathbb{Z}\\ n\neq 0}}\frac{e^{2\pi i n x}}{n^k} \qquad (0 < x < 1,\ k \ge 1) \tag{A.32}$$

discussed in Chap. 4, Theorem 4.11 of this book. (For $k = 1$ the sum converges only conditionally and one has to be a little careful.) Since the integral $\int_0^1 e^{2\pi i k x}\,dx$ equals $\delta_{k,0}$, this gives

$$\int_0^1 B_r(x)\,B_s(x)\,dx = (-1)^r \frac{r!\,s!}{(2\pi i)^{r+s}} \sum_{\substack{n\in\mathbb{Z}\\ n\neq 0}} \frac{1}{n^{r+s}} = (-1)^{r-1} \frac{r!\,s!}{(r+s)!} B_{r+s}$$

as desired. (The second equality, giving the well-known connection between Bernoulli numbers and the values at positive even integers of the Riemann zeta function, is just the case $k = r + s$, $x \to 0$ of (A.32).) The second proof, using generating functions, is just as short. Denote the left-hand side of (A.31), also for r or s equal to 0, by $I_{r,s}$. Then we have

$$\begin{aligned}\sum_{r,s\geq 0} I_{r,s} \frac{t^{r-1}}{r!}\frac{u^{s-1}}{s!} &= \int_0^1 \frac{e^{xt}}{e^t-1}\frac{e^{xu}}{e^u-1}\,dx = \frac{1}{e^t-1}\frac{1}{e^u-1}\frac{e^{t+u}-1}{t+u}\\ &= \frac{1}{t+u}\left[\frac{1}{e^t-1} - \frac{1}{e^{-u}-1}\right] = \sum_{k=0}^{\infty} \frac{B_k}{k!}\frac{t^{k-1}-(-u)^{k-1}}{t+u}\\ &= \frac{1}{tu} + \sum_{k\geq 2}\frac{B_k}{k!}\sum_{\substack{r,s\geq 1\\ r+s=k}} t^{r-1}(-u)^{s-1},\end{aligned}$$

and Eq. (A.31) follows by equating the coefficients of $t^{r-1}u^{s-1}$.

Before continuing, we show that Proposition A.7 immediately yields another proof of the identities of Miki and Gessel discussed in the preceding section. This method is due to I. Artamkin [A1] (whose proof, up to a few small modifications, we have followed here). Indeed, summing (A.30) over all $i,\ j \geq 1$ with $i + j = n$, and using the easy identities

$$\sum_{i=1}^{n-1} \frac{1}{i}\binom{i}{r} = \frac{1}{r}\binom{n-1}{r} \qquad (r > 0)$$

and

$$\begin{aligned}\sum_{\substack{i,j\geq 1\\ i+j=n}} (-1)^{i-1}\frac{(i-1)!\,(j-1)!}{(n-1)!} &= \sum_{i=1}^{n-1}\int_0^1 (-x)^{i-1}(1-x)^{n-i-1}dx\\ &= \int_0^1 \left[(1-x)^{n-1} - (-x)^{n-1}\right]dx = \frac{1+(-1)^n}{n}\end{aligned}$$

(where the first equation is the beta integral again), we obtain

$$\frac{1}{2}\sum_{\substack{i,j\geq 1\\ i+j=n}} \mathcal{B}_i(x)\mathcal{B}_j(x) = H_{n-1}\,\mathcal{B}_n(x) + \sum_{r=2}^{n-1}\binom{n-1}{r}\mathcal{B}_r(0)\,\mathcal{B}_{n-r}(x) + \frac{\mathcal{B}_n(0)}{n}, \tag{A.33}$$

which is equivalent to Gessel's identity (A.29).

Proposition A.8 describes the scalar products among the Bernoulli polynomials with respect to the scalar product $(f, g) = \int_0^1 f(x)g(x)dx$. It is more natural to replace the Bernoulli polynomials $B_k(x)$ by their periodic versions $\overline{B}_k(x)$ (defined for $x \notin \mathbb{Z}$ as $B_k(x - [x])$ or by the right-hand side of (A.32), and for $x \in \mathbb{Z}$ by continuity if $k \neq 1$ and as zero if $k = 1$), since then the scalar product is simply the integral of $\overline{B}_r(x)\overline{B}_s(x)$ over the whole domain of definition $\mathbb{R}/\mathbb{Z}$. The first proof just given then carries over almost unchanged to give the following more general result:

Proposition A.9. *Let r and s be integers ≥ 1 and α, β two real numbers. Then*

$$\int_0^1 \overline{B}_r(x+\alpha)\,\overline{B}_s(x+\beta)\,dx = (-1)^{r-1}\,\frac{r!\,s!}{(r+s)!}\,\overline{B}_{r+s}(\alpha-\beta)\,. \tag{A.34}$$

Using this, one finds, with almost the same proof as before, the following generalization of Proposition A.7:

Proposition A.10. *Let i and j be positive integers. Then for any two variables x and y we have*

$$\mathcal{B}_i(x)\,\mathcal{B}_j(y) = \sum_{m=0}^{\max(i,j)}\left[\frac{1}{i}\binom{i}{m}\mathcal{B}_{i+j-m}(y) + \frac{(-1)^m}{j}\binom{j}{m}\mathcal{B}_{i+j-m}(x)\right]B_m^+(x-y)$$
$$+\,(-1)^{j-1}\,\frac{(i-1)!\,(j-1)!}{(i+j)!}\,B_{i+j}^+(x-y)\,, \tag{A.35}$$

where $B_m^+(x)$ denotes the symmetrized Bernoulli polynomial

$$B_m^+(x) = \frac{B_m(x) + (-1)^m B_m(-x)}{2} = \frac{B_m(x+1) + B_m(x)}{2} = B_m(x) + \frac{m}{2}x^{m-1}.$$

The same calculation as was used above to deduce (A.33) from (A.30), but now applied to (A.35) instead of (A.30), gives the following generalization of Gessel's identity (A.29):

$$\sum_{\substack{i,j\ge 1\\ i+j=n}} \mathcal{B}_i(x)\mathcal{B}_j(y) - H_{n-1}\Big(\mathcal{B}_n(x)+\mathcal{B}_n(y)\Big)$$
$$= \sum_{m=1}^{n-1} \binom{n-1}{m} \Big(\mathcal{B}_{n-m}(y) + (-1)^m\, \mathcal{B}_{n-m}(x)\Big)\, \frac{B_m^+(x-y)}{m}$$
$$+ \frac{1+(-1)^n}{n^2}\, B_n^+(x-y)\,. \qquad \text{(A.36)}$$

We observe that Eq. (A.36) was also found by Hao Pan and Zhi-Wei Sun [A12] in a slightly different form, the right-hand side in their formula being

$$\sum_{m=1}^{n} \binom{n-1}{m-1} \Big(B_{n-m}(y)\, \frac{B_m(x-y)}{m^2} + B_{n-m}(x)\, \frac{B_m(y-x)}{m^2}\Big)$$
$$+ \frac{1}{n}\, \frac{B_n(x)-B_n(y)}{x-y}\,, \qquad \text{(A.37)}$$

which is easily checked to be equal to the right-hand side of (A.36); their formula has the advantage of being more visibly symmetric in x and y and of using only the Bernoulli polynomials $B_m(x)$ rather than the symmetrized Bernoulli polynomials $B_m^+(x)$, but the disadvantage of having a denominator $x-y$ (which of course disappears after division into the numerator $B_n(x)-B_n(y)$) rather than being written in an explicitly polynomial form.

We end this section by giving a beautifully symmetric version of the multiplication law for Bernoulli polynomials given by the same authors in [A13].

Proposition A.11 (Sun–Pan). *For each integer $n \ge 0$ define a polynomial $\begin{bmatrix} r & s \\ x & y \end{bmatrix}_n$ in four variables r, s, x and y by*

$$\begin{bmatrix} r & s \\ x & y \end{bmatrix}_n = \sum_{\substack{i,j\ge 0\\ i+j=n}} (-1)^i \binom{r}{i} \binom{s}{j} B_j(x)\, B_i(y)\,. \qquad \text{(A.38)}$$

Then for any six variables r, s, t, x, y and z satisfying $r+s+t=n$ and $x+y+z=1$ we have

$$t \begin{bmatrix} r & s \\ x & y \end{bmatrix}_n + r \begin{bmatrix} s & t \\ y & z \end{bmatrix}_n + s \begin{bmatrix} t & r \\ z & x \end{bmatrix}_n = 0\,. \qquad \text{(A.39)}$$

First proof (sketch). We can prove (A.39) in the same way as (A.36) was proved above, replacing the product $B_j(x)B_i(y)$ in (A.38) for i and j positive using formula (A.35) (with x and y replaced by $1-y$ and x) and then using elementary binomial coefficient identities to simplify the result. We do not give the full calculation, which is straightforward but tedious. □

Second proof. An alternative, and easier, approach is to notice that, since the left-hand side of (A.39) is a polynomial in the variables x, y and $z = 1 - x - y$, it is enough to prove the identity for $x,\ y,\ z > 0$ with $x + y + z = 1$. But for x and y between 0 and 1 we have from (A.32)

$$(2\pi i)^n \begin{bmatrix} r & s \\ x & y \end{bmatrix}_n = \sum_{a,\,b \in \mathbb{Z}} C_n(r, s; a, b)\, e^{2\pi i(bx - ay)}$$

with

$$C_n(r, s; a, b) = \begin{cases} \sum_{i,\,j \geq 1,\ i+j=n} (r)_i (s)_j a^{-i} b^{-j} & \text{if } a \neq 0, b \neq 0 \\ -(r)_n a^{-n} & \text{if } a \neq 0, b = 0 \\ -(s)_n b^{-n} & \text{if } a = 0, b \neq 0 \\ 0 & \text{if } a = 0, b = 0 \end{cases}$$

where $(x)_m = x(x-1)\cdots(x-m+1)$ is the descending Pochhammer symbol. Equation (A.39) then follows from the identity

$$t\, C_n(r, s; a, b) + r\, C_n(s, t; b, c) + s\, C_n(t, r; c, a) = 0 \quad (a+b+c = 0,\ r+s+t = n)\,.$$

whose elementary proof (using partial fractions if $abc \neq 0$) we omit. □

We end by remarking on a certain formal similarity between the cyclic identity (A.39) and a reciprocity law for generalized Dedekind sums proved in [A5]. The classical Dedekind sums, introduced by Dedekind while posthumously editing some unpublished calculations of Riemann's, are defined by

$$s(b, c) = \sum_{h \ (\mathrm{mod}\ c)} \overline{B}_1\Big(\frac{h}{c}\Big)\, \overline{B}_1\Big(\frac{bh}{c}\Big) \qquad (b,\ c \in \mathbb{N} \text{ coprime}),$$

where $\overline{B}_1(x)$ as usual is the periodic version of the first Bernoulli polynomial (equal to $x - \frac{1}{2}$ if $0 < x < 1$, to 0 if $x = 0$, and periodic with period 1), and satisfy the famous Dedekind reciprocity relation

$$s(b, c) + s(c, b) = \frac{b^2 + c^2 + 1}{12bc} - \frac{1}{4}\,.$$

This was generalized by Rademacher, who discovered that if a, b and c are pairwise coprime integers then the sum

$$s(a, b; c) = \sum_{h \ (\mathrm{mod}\ c)} \overline{B}_1\Big(\frac{ah}{c}\Big)\, \overline{B}_1\Big(\frac{bh}{c}\Big) \qquad \text{(A.40)}$$

which equals $s(a', c)$ for any a' with $aa' \equiv b \pmod{c}$ or $ba' \equiv a \pmod{c}$, satisfies the identity

$$s(a,b;c) + s(b,c;a) + s(c,a;b) = \frac{a^2+b^2+c^2}{12abc} - \frac{1}{4}. \qquad \text{(A.41)}$$

A number of further generalizations, in which the functions $\overline{B}_1$ in (A.40) are replaced by periodic Bernoulli polynomials with other indices and/or the arguments of these polynomials are shifted by suitable rational numbers, were discovered later. The one given in [A5] concerns the sums

$$S_{m,n}\begin{pmatrix} a & b & c \\ x & y & z \end{pmatrix} = \sum_{h \pmod{c}} \overline{B}_m\Big(a\,\frac{h+z}{c} - x\Big)\,\overline{B}_n\Big(b\,\frac{h+z}{c} - y\Big), \qquad \text{(A.42)}$$

where m and n are non-negative integers, a, b and c natural numbers with no common factor, and x, y and z elements of $\mathbb{T}:=\mathbb{R}/\mathbb{Z}$. (The hth summand in (A.42) depends on z modulo c, not just modulo 1, but the whole sum has period 1 in z.) For fixed m and n these sums do not satisfy any relation similar to the 3-term relation (A.41) for the case $m = n = 1$, but if we assemble all of the functions $S_{m,n}$ $(m, n \geq 0)$ into a single generating function

$$\mathfrak{S}\begin{pmatrix} a & b & c \\ x & y & z \\ X & Y & Z \end{pmatrix} = \sum_{m,n\geq 0} \frac{1}{m!\,n!} S_{m,n}\begin{pmatrix} a & b & c \\ x & y & z \end{pmatrix} \Big(\frac{X}{a}\Big)^{m-1} \Big(\frac{Y}{b}\Big)^{n-1}, \qquad \text{(A.43)}$$

in which X, Y and Z (which does not appear explicitly on the right) are formal variables satisfying $X + Y + Z = 0$, then we have the following relation:

Proposition A.12 ([A5]). *Let a, b, c be three natural numbers with no common factor, x, y, z three elements of $\mathbb{T}$, and X, Y, Z three formal variables satisfying $X + Y + Z = 0$. Then*

$$\mathfrak{S}\begin{pmatrix} a & b & c \\ x & y & z \\ X & Y & Z \end{pmatrix} + \mathfrak{S}\begin{pmatrix} b & c & a \\ y & z & x \\ Y & Z & X \end{pmatrix} + \mathfrak{S}\begin{pmatrix} c & a & b \\ z & x & y \\ Z & X & Y \end{pmatrix} = \begin{cases} 1/4 & \text{if } (x,y,z) \in (a,b,c)\mathbb{T}, \\ 0 & \text{otherwise.} \end{cases}$$

We do not give the proof of this relation, since three different proofs (all similar in spirit to various of the proofs that have been given in this appendix) are given in [A5], but we wanted to at least mention this generalized Dedekind–Rademacher reciprocity law because of its formal resemblance, and perhaps actual relationship, to the Sun–Pan reciprocity law (A.39).

A.5 Continued Fraction Expansions for Generating Functions of Bernoulli Numbers

There are several classical formulas expressing various versions of the standard (exponential) generating functions of the Bernoulli numbers as continued fractions. A simple example is

$$\tanh x \quad \left(= \sum_{n\geq 2} \frac{2^n(2^n-1)B_n}{n!} x^{n-1} \right) \quad = \cfrac{x}{1+\cfrac{x^2}{3+\cfrac{x^2}{5+\cfrac{x^2}{\ddots}}}}\,, \tag{A.44}$$

whose proof is recalled below, and a somewhat more complicated one, whose proof we omit, is

$$\frac{x/2}{\tanh x/2} \quad \left(= \sum_{n\geq 0} \frac{B_{2n}}{(2n)!} x^{2n} \right) \quad = \cfrac{1}{1+\cfrac{a_1x^2}{1+\cfrac{a_2x^2}{\ddots}}} \tag{A.45}$$

with a_n defined by

$$a_n = \begin{cases} -\dfrac{1}{12} & \text{if } n = 1\,, \\[2mm] \dfrac{(n+1)(n+2)}{(2n-2)(2n-1)(2n)(2n+1)} & \text{if } n \text{ is even}\,, \\[2mm] \dfrac{(n-2)(n-1)}{(2n-1)(2n)(2n+1)(2n+2)} & \text{if } n > 1 \text{ is odd}\,. \end{cases}$$

It was discovered by M. Kaneko that the convergents $P_n(x)/Q_n(x)$ of the continued fraction (A.45) could be given in a simple closed form, namely

$$P_n(x) = \sum_{i=0}^{n/2} \binom{n}{2i} \binom{2n+1}{2i}^{-1} \frac{x^i}{(2i+1)!}$$

$$Q_n(x) = \sum_{i=0}^{n/2} \binom{n+1}{2i} \binom{2n+2}{2i}^{-1} \frac{x^i}{(2i)!}$$

if n is even and a similar but slightly more complicated expression if n is odd. (It was in connection with this discovery that he found the short recursion formula for Bernoulli numbers discussed in Sect. 1.2 of the book.) Again we omit the proof, which is given in [A6].

What is perhaps more surprising is that there are also nice continued fraction expansions for certain non-standard (ordinary) generating functions of Bernoulli numbers of the type considered in Sect. A.1, and these are in some sense of even more interest because the continued fractions, unlike the power series themselves, converge for positive real values of the argument (and give the appropriate derivatives of $\psi(X)$ as discussed in the last paragraph of Sect. A.1). For instance, on the cover of the Russian original of Lando's beautiful book on generating functions [A7] one finds the pair of formulas[3]

$$1\cdot x + 2\cdot\frac{x^3}{3!} + 16\cdot\frac{x^5}{5!} + 272\cdot\frac{x^7}{7!} + \cdots = \tan x$$

$$1\cdot x + 2\cdot x^3 + 16\cdot x^5 + 272\cdot x^7 + \cdots = \cfrac{x}{1 - \cfrac{1\cdot 2x^2}{1 - \cfrac{2\cdot 3x^2}{1 - \cfrac{3\cdot 4x^2}{1 - \cdots}}}}$$

The numbers 1, 2, 16, 272, ... defined by the first of these two formulas are just the numbers $(4^n - 2^n)|B_n|/n$, so the second formula gives a continued fraction expansion for the non-exponential generating function for essentially the Bernoulli numbers. Again we omit the proof, referring for this to the book cited, mentioning only the following alternative and in some ways prettier form of the formula:

$$\frac{1}{X} - \frac{2}{X^3} + \frac{16}{X^5} - \frac{272}{X^7} + \cdots = \cfrac{1}{X + \cfrac{1}{\cfrac{X}{2} + \cfrac{1}{\cfrac{X}{3} + \cdots}}} \qquad \text{(A.46)}$$

in which the continued fraction is convergent and equal to $1-\frac{X}{2}\left(\psi\left(\frac{X+4}{4}\right)-\psi\left(\frac{X+2}{4}\right)\right)$ for all $X > 0$.

Other continued fraction expansions for non-exponential Bernoulli number generating functions that can be found in the literature include the formulas

[3] In the English translation [A8] (which we highly recommend to the reader) this formula has been relegated to the exercises: Chapter 5, Problem 5.6, page 85.

$$\sum_{n=1}^{\infty} B_{2n}(4x)^n = \cfrac{x}{1+\frac{1}{2}+\cfrac{x}{\frac{1}{2}+\frac{1}{3}+\cfrac{x}{\frac{1}{3}+\frac{1}{4}+\cfrac{x}{\ddots}}}},$$

or the equivalent but less appealing identity

$$\sum_{n=0}^{\infty} B_n x^n = \cfrac{1}{1+\cfrac{x}{\frac{2}{1}-\cfrac{x}{3+\cfrac{2x}{\frac{2}{2}-\cfrac{2x}{5+\cfrac{3x}{\frac{2}{3}-\cfrac{3x}{7+\cfrac{4x}{\frac{2}{4}-\cfrac{4x}{9+\cfrac{5x}{\ddots}}}}}}}}},$$

and

$$\sum_{n=1}^{\infty} (2n+1)\, B_{2n}\, x^n = \cfrac{x}{1+1+\cfrac{x}{1+\frac{1}{2}+\cfrac{x}{\frac{1}{2}+\frac{1}{2}+\cfrac{x}{\frac{1}{2}+\frac{1}{3}+\cfrac{x}{\frac{1}{3}+\frac{1}{3}+\cfrac{x}{\ddots}}}}}}$$

all given by J. Frame [A3] in connection with a statistical problem on curve fitting.

For good conscience's sake we give the proofs of one continued fraction of each of the two above types, choosing for this purpose the two simplest ones (A.44) and (A.46). We look at (A.44) first. Define functions $I_0, I_1, \ldots$ on $(0, \infty)$ by

$$I_n(a) = \int_0^a \frac{t^n(1-t/a)^n}{n!}\, e^t\, dt \quad \bigl(n \in \mathbb{Z}_{\geq 0}\,,\ a \in \mathbb{R}_{>0}\bigr)\,.$$

Integrating by parts twice, we find that

$$\begin{aligned} I_{n+1}(a) &= \int_0^a e^t \frac{d^2}{dt^2}\left[\frac{t^{n+1}(1-t/a)^{n+1}}{(n+1)!}\right] dt \\ &= \int_0^a e^t \left[\frac{t^{n-1}(1-t/a)^{n-1}}{(n-1)!} - \frac{4n+2}{a}\frac{t^n(1-t/a)^n}{n!}\right] dt \\ &= I_{n-1}(a) - \frac{4n+2}{a} I_n(a) \end{aligned}$$

for $n > 0$. Rewriting this as $\dfrac{I_{n-1}(a)}{I_n(a)} = \dfrac{4n+2}{a} + \dfrac{I_{n+1}(a)}{I_n(a)}$ and noting that

$$I_0(a) = e^a - 1\,, \quad I_1(a) = e^a\left(1 - \frac{2}{a}\right) + \left(1 + \frac{2}{a}\right)$$

by direct calculation, we obtain

$$\frac{1}{\tanh x} = \frac{e^{2x}+1}{e^{2x}-1} = \frac{1}{x} + \frac{I_1(2x)}{I_0(2x)} = \frac{1}{x} + \cfrac{1}{\cfrac{3}{x} + \cfrac{1}{\cfrac{5}{x} + \cfrac{1}{\ddots}}}\,,$$

which is equivalent to (A.44). Similarly, for (A.46), we define functions $J_0,\ J_1,\ \ldots$ on $(0,\infty)$ by

$$J_n(X) = \int_0^\infty \big(\tanh(t/X)\big)^n e^{-t}\, dt \quad \big(n \in \mathbb{Z}_{\geq 0}\,,\ X \in \mathbb{R}_{>0}\big).$$

This time $J_0(X)$ is simply the constant function 1, while $J_1(X)$ has the exact evaluation

$$J_1(X) = 1 - \frac{X}{2}\,\psi\Big(\frac{X}{4} + 1\Big) + \frac{X}{2}\,\psi\Big(\frac{X}{4} + \frac{1}{2}\Big), \qquad \text{(A.47)}$$

as is easily deduced from Euler's integral representation

$$\psi(x) = -\gamma + \int_0^1 \frac{1-t^{x-1}}{1-t}\, dt\,,$$

as well as the asymptotic expansion

$$\begin{aligned} J_1(X) &\sim \int_0^\infty \left(\frac{1}{X}t - \frac{2}{X^3}\frac{t^3}{3!} + \frac{16}{X^5}\frac{t^5}{5!} - \frac{272}{X^7}\frac{t^7}{7!} + \cdots\right) e^{-t}\, dt \\ &\sim \frac{1}{X} - \frac{2}{X^3} + \frac{16}{X^5} - \frac{272}{X^7} + \cdots \end{aligned}$$

as $X \to \infty$. (This last expression can be written as $1 - X\gamma_0(2/X) + X\gamma_0(4/X)$ with γ_0 as in (A.12), in accordance with (A.47) and the relationship between $\gamma_0(X)$ and $\psi(X)$ given at the end of Sect. A.1.) On the other hand, integrating by parts and using $\tanh(x)' = 1 - \tanh(x)^2$, we find

$$\begin{aligned} J_n(X) &= \int_0^\infty e^{-t} \frac{d}{dt}\Big(\big(\tanh(t/X)\big)^n\Big)\,dt \\ &= \frac{n}{X}\int_0^\infty e^{-t}\,\big(\tanh(t/X)\big)^{n-1}\Big(1 - \big(\tanh(t/X)\big)^2\Big)\,dt \\ &= \frac{n}{X}\big(J_{n-1}(X) - J_{n+1}(X)\big) \end{aligned}$$

for $n > 0$, and rewriting this as $\dfrac{J_{n-1}(X)}{J_n(X)} = \dfrac{X}{n} + \dfrac{J_{n+1}(X)}{J_n(X)}$ we obtain that $J_1(X) = \dfrac{J_1(X)}{J_0(X)}$ has the continued fraction expansion given by the right-hand side of (A.46), as claimed. □

We end this appendix by describing an appearance of the continued fraction (A.46) in connection with the fantastic discovery of Yuri Matiyasevich that "the zeros of the Riemann zeta function know about each other." Denote the zeros of $\zeta(s)$ on the critical line $\Re(s) = \frac{1}{2}$ by ρ_n and $\overline{\rho_n}$ with $0 < \Im(\rho_1) \le \Im(\rho_2) \le \cdots$ and for $M \ge 1$ consider the finite Dirichlet series $\Delta_M(s)$ defined as the $N \times N$ determinant[4]

$$\Delta_M(s) = \begin{vmatrix} 1 & 1 & \cdots & 1 & 1 & 1 \\ \vdots & \vdots & \ddots & \vdots & \vdots & \vdots \\ n^{-\rho_1} & n^{-\overline{\rho_1}} & \cdots & n^{-\rho_M} & n^{-\overline{\rho_M}} & n^{-s} \\ \vdots & \vdots & \ddots & \vdots & \vdots & \vdots \\ N^{-\rho_1} & N^{-\overline{\rho_1}} & \cdots & N^{-\rho_M} & N^{-\overline{\rho_M}} & N^{-s} \end{vmatrix},$$

where $N = 2M + 1$. This function clearly vanishes when $s = \rho_n$ or $\overline{\rho_n}$ for $1 \le n \le M$, but Matiyasevich's discovery (for which we refer to [A9] and the other papers and talks listed on his website) was that its subsequent zeros are incredibly close to the following zeros of the Riemann zeta function, e.g., the first zero of Δ_{50} on $\frac{1}{2} + \mathbb{R}_{>0}$ following ρ_{50} differs in absolute value from ρ_{51} by less than 4×10^{-15}, the first zero of Δ_{1500} after ρ_{1500} differs in absolute value from ρ_{1501} by less than 5×10^{-1113}, and even the 300th zero of Δ_{1500} after ρ_{1500} differs in absolute value from ρ_{1801} by less than 5×10^{-766}! Moreover, if we write the Dirichlet series $\Delta_M(s)$ as $c_M \sum_{n=1}^N a_{M,n} n^{-s}$ with the normalizing constant c_M chosen to make $a_{M,1} = 1$,

[4] We have changed Matiyasevich's notations slightly for convenience of exposition.

then it turns out that the function $c_M^{-1}\Delta_M(s)$ not only has almost the same zeros, but is itself a very close approximation to $(1-2^{1-s})\zeta(s)$ over a long interval of the critical line.

In studying this latter function, Matiyasevich was led to consider the real numbers ν_M defined by $\nu_M = 4M\sum_{n=1}^{2M}\mu_{M,n}/n$, where $\mu_{M,n}$ denotes the coefficient of n^{-s} in the Dirichlet series $c_M^{-1}\Delta_M(s)/\zeta(s)$. Since by the nature of his investigation he was working to very high precision, he obtained very precise decimal expansions of these numbers, and in an attempt to recognize them, he computed the beginning of their continued fraction expansions. (Recall that rational numbers and real quadratic irrationalities can be recognized numerically by the fact that they have terminating or periodic continued fraction expansions.) To his surprise, when M was highly composite these numbers had very exceptional continued fraction expansions. For instance, for $2M = \text{l.c.m.}\{1, 2, \ldots, 10\} = 2520$, the number ν_M has a decimal expansion beginning $0.9998015873172093\cdots$ and a continued fraction expansion beginning [0, 1, 5039, 2520, 1680, 1260, 1008, 840, 720, 630, 560, 504]. In view of the fact that nearly all real numbers (in a very precise metrical sense) have continued fraction expansions with almost all partial quotients very small, this is certainly not a coincidence, and it is even more obviously not one when we notice that the numbers 5040, 2520, ... 504 are $5040/n$ for $n = 1, 2, \ldots, 10$. This leads one immediately to the continued fraction (A.46) with $X = 4M$ and hence, in view of the evaluation of that continued fraction given above, to the (conjectural) approximation $\nu_M \approx \frac{1}{2}\psi\big(M+1\big) - \frac{1}{2}\psi\big(M+\frac{1}{2}\big)$, which turns out indeed to be a very good one for M large, the two numbers differing only by one part in 10^{108} in the above-named case $2M = 2520$. We take this somewhat unusual story as a fitting place to end our survey of curious and exotic identities connected with Bernoulli numbers.

References

[A1] И.В. Артамкин : Элементарное доказательство тождества Мики–Загира–Гесселя, *Успехи мат. наук*, **62** (2007), 165–166. [I. Artamkin : An elementary proof of the Miki–Zagier–Gessel identity, *Russian Mathematical Surveys*, **62** (2007), 1195–1196.]

[A2] C. Faber and R. Pandharipande : Logarithmic series and Hodge integrals in the tautological ring (with an appendix by D. Zagier, Polynomials arising from the tautological ring), *Mich. Math. J.*, **48** (2000), 215–252 (appendix: pp. 240–252).

[A3] J. Frame : The Hankel power sum matrix inverse and the Bernoulli continued fraction, *Math. Comp.*, **33** (1979), 15–826.

[A4] I. Gessel : On Miki's identity for Bernoulli numbers, *J. Number Theory*, **110** (2005), 75–82.

[A5] R. Hall, J. Wilson and D. Zagier : Reciprocity formulae for general Dedekind–Rademacher sums, *Acta Arithmetica*, **73** (1995), 389–396.

[A6] M. Kaneko : A recurrence formula for the Bernoulli numbers, *Proc. Japan Acad.*, **71** (1995), 192–193.

[A7] С.К. Ландо : Лекции о производящих функциях, *Москва, МЦНМО* (2002), 144 pages.

[A8] S. Lando : Lectures on Generating Functions, *Student Mathematical Library, Amer. Math. Soc.*, **27** (2003), 150 pages. [=translation of [A7]]

[A9] Yu. Matiyasevich : Calculation of Riemann's zeta function via interpolating determinants (talk given at the MPI in March 2013). On the author's website at http://logic.pdmi.ras.ru/~yumat/personaljournal/artlessmethod.

[A10] H. Miki : A relation between Bernoulli numbers, *J. Number Theory*, **10** (1978), 297–302.

[A11] N. Nielsen : Traité élémentaire des nombres de Bernoulli, *Gauthiers-Villars, Paris* (1923), 398 pages.

[A12] H. Pan and Z.-W. Sun : New identities involving Bernoulli and Euler polynomials, *J. Comb. Theory*, **113** (2006), 156–175.

[A13] Z.-W. Sun and H. Pan : Identities concerning Bernoulli and Euler polynomials, *Acta Arithmetica*, **125** (2006), 21–39.

[A14] D. Zagier : Hecke operators and periods of modular forms, *Israel Math. Conf. Proc.*, **3** (1990), 321–336.

[A15] D. Zagier : A modified Bernoulli number, *Nieuw Archief voor Wiskunde*, **16** (1998), 63–72.

References

1. Ahlfors , L.: Complex Analysis, 3rdedn. McGraw Hill (1979)
2. Akiyama, S., Tanigawa, Y.: Multiple zeta values at non-positive integers. Ramanujan J. **5**(4), 327–351 (2001)
3. Andrianov, A.N.: Quadratic Forms and Hecke Operators. Grundlehren der mathematischen Wissenshaften, vol. 286. Springer (1987)
4. Ankeny, N.C., Artin, E., Chowla, S.: The class-number of real quadratic number fields. Ann. Math. **56**(2), 479–493 (1952)
5. Apéry, R.: Irrationalité de $\zeta(2)$ et $\zeta(3)$. Astérisque **61**, 11–13 (1979)
6. Arakawa, T.: Generalized eta-functions and certain ray class invariants of real quadratic fields. Math. Ann. **260**, 475–494 (1982)
7. Arakawa, T.: Dirichlet series $\sum_{n=1}^{\infty}(\cot \pi n\alpha)/n^s$, Dedekind sums, and Hecke L-functions for real quadratic fields. Comment. Math. Univ. St. Pauli **37**, 209–235 (1988)
8. Arakawa, T.: A noteon the Hirzebruch sum. Comment. Math. Univ. St. Pauli **42**, 81–92 (1993)
9. Arakawa, T., Kaneko, M.: Multiple zeta values, poly-Bernoulli numbers, and related zeta functions. Nagoya Math. J. **153**, 189–209 (1999)
10. Arakawa, T., Kaneko, M.: On poly-Bernoulli numbers. Comment. Math. Univ. St. Pauli **48**, 159–167 (1999)
11. Barnes, E.: The genesis of the double gamma functions. Proc. Lond. Math. Soc. **31**, 358–381 (1899)
12. Barnes, E.: The theory of the double gamma functions. Philos. Trans. Roy. Soc. (A) **196**, 265–388 (1901)
13. Berggren, L., Borwein, J., Borwein, P.: Pi, A Source Book. Springer, New York (1997)
14. Berndt, B.C.: Dedekind sums and a paper of G. H. Hardy. J. Lond. Math. Soc. **13**, 129–137 (1976)
15. Berndt, B.C., Evans, R.J., Williams, K.S.: Gauss and Jacobi Sums. Canadian Math. Soc. Series of Monographs and Advanced Texts, vol. 21. Wiley-Interscience (1998)
16. Bernoulli, J.: Ars Conjectandi, in Werke, vol. 3, pp. 107–286. Birkhäuser (1975)
17. Biermann, K.-R.: Thomas Clausen, Mathematiker und Astronom. J. Reine Angew. Math. **216**, 159–198 (1964)
18. Biermann, K.-R.: Kummer, Ernst Eduard, in Dictionary of Scientific Biography, vols. 7 & 8. Charles Scribner's Sons, New York (1981)
19. Bjerknes, C.A.: Niels-Henrik Abel, Tableau de sa vie et de son Action Scientifique. Cambridge University Press, Cambridge (2012)
20. Brewbaker, C.: Lonesum (0, 1)-matrices and poly-Bernoulli numbers of negative index. Master's thesis, Iowa State University (2005)
21. Buchmann, J., Vollmer, U.: Binary quadratic forms. An algorithmic approach. Algorithms and Computation in Mathematics, vol. 20. Springer, Berlin (2007)

T. Arakawa et al., *Bernoulli Numbers and Zeta Functions*, Springer Monographs in Mathematics, DOI 10.1007/978-4-431-54919-2, © Springer Japan 2014

22. Carlitz, L.: Arithmetic properties of generalized Bernoulli numbers. J. Reine Angew. Math. **202**, 174–182 (1959)
23. Clausen, T.: Ueber die Fälle, wenn die Reihe von der Form $y = 1+\frac{\alpha}{1}\cdot\frac{\beta}{\gamma}x+\frac{\alpha.\alpha+1}{1.2}\cdot\frac{\beta.\beta+1}{\gamma.\gamma+1}x^2+$ etc. ein Quadrat von der Form $z = 1 + \frac{\alpha'}{1} \cdot \frac{\beta'}{\gamma'}\frac{\delta'}{\varepsilon'}x + \frac{\alpha'.\alpha'+1}{1.2} \cdot \frac{\beta'.\beta'+1}{\gamma'.\gamma'+1} \cdot \frac{\delta'.\delta'+1}{\varepsilon'.\varepsilon'+1}x^2 +$ etc. hat. J. Reine Angew. Math. **3**, 89–91 (1828)
24. Clausen, T.: Über die Function $\sin\varphi + \frac{1}{2^2}\sin 2\varphi + \frac{1}{3^2}\sin 3\varphi +$ etc.. J. Reine Angew. Math. **8**, 298–300 (1832)
25. Clausen, T.: Beweis, daß die algebraischen Gleichungen Wurzeln von der Form $a+bi$ haben. Astron. Nachr. **17**, 325–330 (1840)
26. Clausen, T.: Lehrsatz aus einer Abhandlung über die Bernoullischen Zahlen. Astron. Nachr. **17**, 351–352 (1840)
27. Coates, J., Wiles, A.: Kummer's criterion for Hurwitz numbers, in Algebraic number theory (Kyoto Internat. Sympos., RIMS, Kyoto, 1976), pp. 9–23. Japan Soc. Promotion Sci., Tokyo (1977)
28. Dilcher, K.: A Bibliography of Bernoulli Numbers, available online, http://www.mscs.dal.ca/~dilcher/bernoulli.html
29. Dirichlet, P.G.L.: Lectures on Number Theory, with Supplements by R. Dedekind. History of Mathematics Sources, vol. 16. Amer. Math. Soc. (1999)
30. Edwards, A.W.F.: A quick route to sums of powers. Am. Math. Monthly **93**, 451–455 (1986)
31. Edwards, H.W.: Fermat's Last Theorem. A Genetic Introduction to Algebraic Number Theory. Springer, New York-Berlin (1977)
32. Euler, L.: Remarques sur un beau rapport entre les séries des puissances tant directes que réciproques. Opera Omnia, series prima **XV**, 70–90 (1749)
33. Euler, L.: De numero memorabili in summatione progressionis harmonicae naturalis occurrente. Opera Omnia, series prima **XV**, 567–603 (1785)
34. Freudenthal, H.: Hurwitz, Adolf, in Dictionary of Scientific Biography, vols. 5 & 6. Charles Scribner's Sons, New York (1981)
35. Gauss, C.F.: DisquisitionesArithmeticae (1801)
36. Gould, H.W.: Explicit formulas for Bernoulli numbers. Am. Math. Monthly **79**, 44–51 (1972)
37. Gouvéa, F.Q.: p-adic Numbers, an Introduction. Springer
38. Graham, R., Knuth, D., Patashnik, O.: Concrete Mathematics. Addison-Wesley (1989)
39. Hashimoto, K.: Representation of the finite symplectic group $Sp(2, \mathbf{F}_p)$ in the space of Siegel modular forms. Contemporary Math. **53**, 253–276 (1986)
40. Hensel, K.: Festschrift zur Feier des 100. Geburtstages Eduard Kummers, B.G. Teubner, 1910. (Kummer's collected papers, 33–69)
41. Hilbert, D.: Adolf Hurwitz, Gedächtnisrede, Nachrichten von der k. Gesellshaft der Wissenschaften zu Göttingen (1920), pp. 75–83 (Hurwitz's Mathematische Werke, I xiii–xx)
42. Hungerford, T.W.: Algebra, Graduate Texts in Mathematics, vol. 73. Springer (1974)
43. Hurwitz, A.: Einige Eigenschaften der Dirichlet'schen Funktionen $F(s) = \sum \left(\frac{D}{n}\right) \cdot \frac{1}{n^s}$, die bei der Bestimmung der Klassenanzahlen binärer quadratischer Formen auftreten. Zeitschrift für Math. und Physik **27**, 86–101 (1882). (Mathematische Werke I, 72–88)
44. Hurwitz, A.: Über die Entwicklungskoeffizienten der lemniskatischen Funktionen. Math. Ann. **51**, 196–226 (1899). (Mathematische Werke II, 342–373)
45. Ibukiyama, T.: On some elementary character sums. Comment. Math. Univ. St. Pauli **47**, 7–13 (1998)
46. Ibukiyama, T., Saito, H.: On zeta functions associated to symmetric matrices and an explicit conjecture on dimensions of Siegel modular forms of general degree. Int. Math. Res. Notices **8**, 161–169 (1992)
47. Ibukiyama, T., Saito, H.: On a L-functions of ternary zero forms and exponential sums of Lee and Weintraub. J. Number Theory **48-2**, 252–257 (1994)
48. Ibukiyama, T., Saito, H.: On zeta functions associated to symmetric matrices I. Am. J. Math. **117-5**, 1097–1155 (1995); II: Nagoya Math. J. **208**, 265–316 (2012); III: Nagoya Math. J. **146**, 149–183 (1997)

49. Ibukiyama, T., Saito, H.: On "easy" zeta functions (trans. by Don Zagier). Sugaku Exposition, **14**(2), 191–204 (2001). Originally in Sugaku, **50–1**, 1–11 (1998)
50. Ireland, K., Rosen, M.: A Classical Introduction to Modern Number Theory, 2nd edn. Graduate Texts in Mathematics, vol. 84. Springer (1990)
51. Iwasawa, K.: Lectures on p-adic L-functions. Annals of Math. Studies, vol. 74. Princeton University Press, Princeton (1972)
52. Jacobi, C.G.J.: De usu legitimo formulae summatoriae Maclaurinianae J. Reine Angew. Math. **12**, 263–272 (1834). (Mathematische Werke VI, 64–75)
53. Jordan, Ch.: Calculus of Finite Differences, Chelsea Publication, New York (1965). (First edition, Budapest, 1939)
54. Kaneko, M.: A generalization of the Chowla-Selberg formula and the zeta functions of quadratic orders. Proc. Jpn. Acad. **66(A)-7**, 201–203 (1990)
55. Kaneko, M.: A recurrence formula for the Bernoulli numbers. Proc. Jpn. Acad. **71(A)-8**, 192–193 (1995)
56. Kaneko, M.: Poly-Bernoulli numbers. J. Th. Nombre Bordeaux **9**, 199–206 (1997)
57. Kaneko, M.: Multiple zeta values. Sugaku Expositions **18**(2), 221–232 (2005)
58. Katz, N.: The congruences of Clausen-von Staudt and Kummer for Bernoulli-Hurwitz numbers. Math. Ann. **216**, 1–4 (1975)
59. Knuth, D.: Two notes on notation. Am. Math. Monthly **99**, 403–422 (1992)
60. Knuth, D.: Johann Faulhaber and sum of powers. Math. Comp. **61**(203), 277–294 (1993)
61. Knuth, D., Buckholtz, T.J.: Computation of tangent Euler and Bernoulli numbers. Math. Comp. **21**, 663–688 (1967)
62. Kronecker, L.: Bemerkungen zur Abhandlung des Herrn Worpitzky. J. Reine Angew. Math. **94**, 268–270 (1883). (Mathematische Werke II, 405–407)
63. Kummer, E.E.: Über die hypergeometrische Reihe $1 + \frac{\alpha\cdot\beta}{1\cdot\gamma}x + \frac{\alpha(\alpha+1)\beta(\beta+1)}{1\cdot 2\cdot\gamma(\gamma+1)}x^2 + \frac{\alpha(\alpha+1)(\alpha+2)\beta(\beta+1)(\beta+2)}{1\cdot 2\cdot 3\cdot\gamma(\gamma+1)(\gamma+2)}x^3 + \ldots$. J. Reine Angew. Math. **15**, 39–83, 127–172 (1836). (Collected papers II, 75–166)
64. Kummer, E.E.: Über eine allgemeine Eigenschaft der rationalen Entwicklungscoefficienten einer bestimmten Gattung analytischer Functionen. J. Reine Angew. Math. **41**, 368–372 (1851). (Collected papers I, 358–362)
65. Lampe, E.: Nachruf für Ernst Eduard Kummer. Jahresbericht der Deutschen Mathematiker-vereinigung **3**, 13–21 (1894). (Kummer's collected papers I, 15–30)
66. Lang, S.: Cyclotomic Fields, Graduate Texts in Mathematics, vol. 59. Springer (1980)
67. Lang, S.: Elliptic Functions, 2nd edn. Graduate Texts in Mathematics, vol. 112. Springer (1987). (Original edition, Addison-Wesley Publishing, 1973)
68. Launois, S.: Combinatorics of $\mathcal{H}$-primes in quantum matrices. J. Algebra **309**(1), 139–167 (2007)
69. Lee, R., Weintraub, S.H.: On a generalization of a theorem of Erich Hecke. Proc. Natl. Acad. Sci. USA **79**, 7955–7957 (1982)
70. Leopoldt, H.-W.: Eine Verallgemeinerung der Bernoullischen Zahlen. Abh. Math. Sem. Univ. Hamburg **22**, 131–140 (1958)
71. Lindemann, F.: Ueber die Zahl π. Math. Ann. **20**, 213–225 (1882)
72. Meissner, E.: Gedächtnisrede auf Adolf Hurwitz. Vierteljahrsschrift der Naturforschenden Gesellschaft in Zürich. **64**, 855–857 (1919)
73. Nielsen, N.: Handbuch der Theorie der Gammafunktion. Chelsea Publication, New York (1965). (First edition, Leipzig, 1906)
74. Noether, M.: Zur Erinnerung an Karl Georg Christian von Staudt. Jahresbericht d. Deutschen Mathem.-Vereinigung **32**, 97–119 (1923)
75. Nörlund, N.E.: Vorlesungen über Differenzenrechnung. Chelsea Publication, New York (1954). (First edition, Springer, Berlin, 1924)
76. Pólya, G.: Some mathematicians I have known. Am. Math. Monthly **76**, 746–753 (1969)
77. Rademacher, H.: Lectures on Elementary Number Theory, Die Grundlehren der mathematischen Wissenschaften, vol. 169. Springer (1773)

78. Reid, C.: Hilbert. Springer, Berlin-Heidelberg-New York (1970)
79. Ribenboim, P.: 13 Lectures on Fermat's Last Theorem. Springer, New York-Heidelberg-Berlin (1979)
80. Rivoal, T.: La fonction zêta de Riemann prend une infinité de valeurs irrationnelles aux entiers impairs. C. R. Acad. Sci. Paris, Sér. l. Math. **331**, 267–270 (2000)
81. Scherk, H.F.: Über einen allgemeinen, die Bernoullischen Zahlen und die Coëfficienten der Secantenreihe zugleich darstellenden Ausdruck. J. Reine Angew. Math. **4**, 299–304 (1829)
82. Serre, J.-P.: Formes modulaires et fonctions zêta p-adiques. Lect. Notes in Math., vol. 350, pp. 191–268. Springer (1973). (Collected papers III, 95–172)
83. Serre, J.-P.: Cours d'arithmétique, Presses Universitaires de France, 1970. English translation: A course in arithmetic, Graduate Text in Mathematics, vol. 7. Springer (1973)
84. Seki, T.: CollectedWorks of Takakazu Seki. In: Hirayama, A., Shimodaira, K., Hirose, H. (eds.) Osaka Kyouiku Tosho (1974)
85. Shintani, T.: On a Kronecker limit formula for real quadratic fields. J. Fac. Sci. Univ. Tokyo, Sec. IA **24**, 167–199 (1977)
86. Shintani, T.: On special values of L-functions of number fields. Sugaku **29–3**, 204–216 (1977)
87. Shintani, T.: On certain ray class invariants of real quadratic fields. J. Math. Soc. Jpn. **30**, 139–167 (1978)
88. Slavutskii, I.Sh.: Staudt and arithmetical properties of Bernoulli numbers. Historia Scientiarum **5–1**, 69–74 (1995)
89. Stark, H.M.: L-functions at $s = 1$, III. Totally real fields and Hilbert's twelfth problem. Advances in Math. **22**, 64–84 (1976)
90. Titchmarsh, E.C.: TheTheory of the Riemann Zeta-function, 2nd edn. (revised by D.R. Heath-Brown). Oxford, (1986)
91. Tsushima, R.: The spaces of Siegel cusp forms of degree two and the representation of $Sp(2, \mathbf{F}_p)$. Proc. Jpn. Acad. **60**, 209–211 (1984)
92. Tsushima, R.: Dimension formula for the spaces of Siegel cusp forms and a certain exponential sum. Mem. Inst. Sci. Tech. Meiji Univ. **36**, 1–56 (1997)
93. Vandiver, H.S.: On developments in an arithmetic theory of the Bernoulli and allied numbers. Scripta Math. **25**, 273–303 (1961)
94. Volkenborn, A.: Ein p-adisches Integral und seine Anwendungen. I. Manuscripta Math. **7**, 341–373 (1972)
95. Volkenborn, A.: Ein p-adisches Integral und seine Anwendungen. II. Manuscripta Math. **12**, 17–46 (1974)
96. von Staudt, K.G.C.: Beweis eines Lehrsatzes, die Bernoullischen Zahlen betreffend. J. Für Reine u. Angew. Math. **21**, 372–374 (1840)
97. von Staudt, K.G.C.: Beweis des Satzes, daß jede algebraische rationale ganze Function von einer Veränderlichen in Faktoren vom ersten Grade aufgelöst werden kann. J. Für Reine u. Angew. Math. **29**, 97–102 (1845)
98. Waldschmidt, M.: Valeurs zêta multiples. Une introduction. Colloque International de Théorie des Nombres (Talence, 1999). J. Théor. Nombres Bordeaux **12**(2), 581–595 (2000)
99. Waldshcmidt, M.: Open diophantine problems. Moscow Math. J. **4–1**, 245–300 (2004)
100. Washington, L.C.: Introduction to Cyclotomic Fields. Graduate Text in Mathematics, vol. 83. Springer (1982)
101. Weber, H.: Lehrbuchder Algebra, vol. III, Chelsea Publication, New York. (First edition, Friedrich Vieweg und Sohn, Braunschweig, 1908)
102. Weil, A.: Basic Number Theory. Springer, New York (1973)
103. Weil, A.: Number Theory: An Approach Through History; From Hammurapi to Legendre. Birkhäuser, Boston (1983)
104. Whittaker, E.T., Watson, G.N.: A Course of Modern Analysis. Cambridge University Press, Cambridge (1927)
105. Zagier, D.: A Kronecker limit formula for real quadratic fields. Math. Ann. **213**, 153–184 (1975)

106. Zagier, D.: Modular forms whose Fourier coefficients involve zeta functions of quadratic fields, in Modular functions of one variable VI. Lect. Notes in Math., vol. 627, pp. 105–169. Springer (1977)
107. Zagier, D.: Zetafunktionen und Quadratische Körper. Springer (1981)
108. Zagier, D.: A one-sentence proof that every prime $p \equiv 1 \pmod 4$ is a sum of two squares. Am. Math. Monthly **97–2**, 144 (1990)
109. Zagier, D.: Values of zeta functions and their applications, in ECM volume. Progress Math. **120**, 497–512 (1994)

Index

Page numbers shown in bold indicate biographical note.

T. Arakawa et al., *Bernoulli Numbers and Zeta Functions*, Springer Monographs in Mathematics, DOI 10.1007/978-4-431-54919-2, © Springer Japan 2014

Printed by Printforce, the Netherlands